Privacy Enhancing Techniques

Xun Yi • Xuechao Yang • Xiaoning Liu •
Andrei Kelarev • Kwok-Yan Lam •
Mengmeng Yang • Xiangning Wang • Elisa Bertino

Privacy Enhancing Techniques

Practices and Applications

Xun Yi
RMIT University
Melbourne, VIC, Australia

Xiaoning Liu
School of Computing Technologies
RMIT University
Melbourne, VIC, Australia

Kwok-Yan Lam
Nanyang Technological University
Singapore, Singapore

Xiangning Wang
NTU Singapore
Singapore, Singapore

Xuechao Yang
RMIT University
Melbourne, VIC, Australia

Andrei Kelarev
Victoria, Australia

Mengmeng Yang
CSIRO Australia
Clayton, VIC, Australia

Elisa Bertino
Purdue University
West Lafayette, IN, USA

ISBN 978-3-031-95139-8   ISBN 978-3-031-95140-4   (eBook)
https://doi.org/10.1007/978-3-031-95140-4

© The Editor(s) (if applicable) and The Author(s), under exclusive license to Springer Nature Switzerland AG 2025

This work is subject to copyright. All rights are solely and exclusively licensed by the Publisher, whether the whole or part of the material is concerned, specifically the rights of translation, reprinting, reuse of illustrations, recitation, broadcasting, reproduction on microfilms or in any other physical way, and transmission or information storage and retrieval, electronic adaptation, computer software, or by similar or dissimilar methodology now known or hereafter developed.
The use of general descriptive names, registered names, trademarks, service marks, etc. in this publication does not imply, even in the absence of a specific statement, that such names are exempt from the relevant protective laws and regulations and therefore free for general use.
The publisher, the authors and the editors are safe to assume that the advice and information in this book are believed to be true and accurate at the date of publication. Neither the publisher nor the authors or the editors give a warranty, expressed or implied, with respect to the material contained herein or for any errors or omissions that may have been made. The publisher remains neutral with regard to jurisdictional claims in published maps and institutional affiliations.

This Springer imprint is published by the registered company Springer Nature Switzerland AG
The registered company address is: Gewerbestrasse 11, 6330 Cham, Switzerland

If disposing of this product, please recycle the paper.

To our families

Preface

Data privacy generally means the ability of a person to determine for themselves when, how, and to what extent personal information about them is shared with or communicated to others. This personal information can be one's name, location, contact information, or online or real-world behaviour.

Data privacy is a crucial concern for individuals, businesses, and organisations of all sizes. With the increasing reliance on technology for storing and transmitting sensitive information, the threat of data breaches and theft has become a major issue. In 2022, there were more than 2000 publicly disclosed data breaches, with 60% being a result of hacking. Affected companies and individuals were at risk of financial and reputation losses, compromised data, and sometimes legal liability.

To mitigate these risks, some techniques have been developed to protect data from unauthorised access and manipulation. For example, encryption, backup and recovery, access control, network security, and physical security have been used to protect data privacy. In addition, compliance regulations can help ensure that user's privacy requests are carried out by companies, and companies are responsible to take measures to protect private user data. However, these are not enough.

This book presents privacy enhancing techniques and their practices and applications. It is designed to serve as a reference book for undergraduate- or graduate-level courses in computer science or mathematics departments, as a general introduction suitable for self-study (especially for beginning graduate students), and as a reference for students, researchers, and practitioners.

Melbourne, VIC, Australia	Xun Yi
Melbourne, VIC, Australia	Xuechao Yang
Melbourne, VIC, Australia	Xiaoning Liu
Melbourne, VIC, Australia	Andrei Kelarev
Singapore, Singapore	Kwok-Yan Lam
Clayton, VIC, Australia	Mengmeng Yang
Singapore, Singapore	Xiangning Wang
West Lafayette, IN, USA	Elisa Bertino
February 2025	

Contents

1 **Introduction** 1
   1.1   Data Privacy 1
   1.2   Previous Data Privacy Protection Techniques 3
   1.3   Privacy Enhancing Techniques 5
   1.4   Book Organisation 6
   References 7

2 **Homomorphic Encryption** 9
   2.1   Public Key Encryption Definition 9
   2.2   Homomorphic Encryption Definition 11
   2.3   Partial Homomorphic Encryption Schemes 12
   2.4   Fully Homomorphic Encryption 19
   2.5   Implementations of Homomorphic Encryption 35
   References 49

3 **Secure Multiparty Computation** 51
   3.1   Secure Multiparty Computation Definition 51
   3.2   Security Requirements for MPC Protocols 52
   3.3   Adversaries Against MPC Protocols 54
   3.4   Assumptions on MPC Protocols 55
   3.5   Oblivious Transfer 56
   3.6   Yao's Garbled Circuits 59
   3.7   Secure Multiparty Computation with Secret Sharing 67
   3.8   Secure Multiparty Computation Implementations 78
   References 83

4 **Differential Privacy** 87
   4.1   Introduction to Differential Privacy 87
   4.2   Relaxed Differential Privacy 96
   4.3   Local Differential Privacy 99
   4.4   Differential Privacy Implementations 106
   References 111

5 Privacy-Preserving Association Rule Mining in Cloud Computing ... 113
- 5.1 Introduction ... 113
- 5.2 Related Work ... 115
- 5.3 Preliminaries ... 117
- 5.4 Model for Association Rule Mining with Data Privacy ... 120
- 5.5 Association Rule Mining with Item Privacy ... 123
- 5.6 Association Rule Mining with Transaction Privacy ... 125
- 5.7 Association Rule Mining with Database Privacy ... 130
- 5.8 Conclusion ... 132
- References ... 132

6 Single-Database Private Information Retrieval from FHE ... 135
- 6.1 Introduction ... 135
- 6.2 Prelimaries ... 139
- 6.3 Generic Single-Database PIR from FHE ... 142
- 6.4 Practical Single-Database PIR from FHE ... 146
- 6.5 Conclusion ... 149
- References ... 149

7 Privacy-Preserving Lightweight Deep Learning in Medical Diagnostic Service ... 153
- 7.1 Introduction ... 153
- 7.2 Related Work ... 155
- 7.3 Preliminaries ... 155
- 7.4 Model for Privacy-Preserving Machine Learning ... 156
- 7.5 Our Proposed Design ... 157
- 7.6 Conclusion ... 164
- References ... 165

8 Privacy-Preserving User Profile Matching in Social Networks ... 167
- 8.1 Introduction ... 167
- 8.2 Related Work ... 169
- 8.3 Model for Privacy-Preserving User Profile Matching ... 172
- 8.4 User Profile Matching Protocol ... 175
- 8.5 Extended Profile Matching Protocol ... 182
- 8.6 Conclusion ... 185
- References ... 186

9 Private k Nearest Neighbor Queries with Location Privacy ... 189
- 9.1 Introduction ... 189
- 9.2 Related Works ... 191
- 9.3 Backgrounds ... 193
- 9.4 Model for Private k Nearest Neighbor Queries ... 194
- 9.5 Basic Private k Nearest Neighbor Queries ... 199
- 9.6 Genetic Private k Nearest Neighbor Queries ... 203
- 9.7 Conclusion ... 208
- References ... 208

10 Differentially Private Distributed Frequency Estimation 211
10.1 Introduction 211
10.2 Related Work 214
10.3 Preliminaries 215
10.4 Proposed Solution 218
10.5 Conclusion 233
References 233

11 Privacy and Digital Trust in VANETs 237
11.1 Introduction 237
11.2 Privacy and Trust Requirements 239
11.3 Privacy and Trust Challenges: A Review of Current Research 241
11.4 Survey of Authentication Schemes for Privacy and Trust 243
11.5 Conclusion 248
References 249

12 Conclusions and Acknowledgements 253
12.1 Summary 253
12.2 Acknowledgements 253

Chapter 1
Introduction

Abstract This chapter introduces the concept of data privacy, provides a brief overview of existing data privacy protection techniques, and surveys privacy-enhancing technologies.

1.1 Data Privacy

Data privacy refers to an individual's ability to control when, how, and to what extent their personal information is shared or disclosed. Personal information, also known as personal data, encompasses any data that identifies a specific individual. Examples include names, mailing addresses, email addresses, phone numbers, and medical records.

With the widespread adoption of the Internet, the importance of data privacy has grown significantly. Websites, applications, and social media platforms often require users to provide personal data to access services. However, some platforms may collect more information than users expect or fail to safeguard collected data adequately, leading to potential breaches and privacy violations.

In many jurisdictions, privacy is considered a fundamental human right, supported by laws designed to protect personal data. Data privacy is also essential for building trust; users are more likely to engage with organizations that handle their data responsibly.

Personal data can be exploited in several ways if privacy is not maintained or individuals cannot control how their data is used:

- Criminals may exploit personal data for fraud or harassment.
- Organizations might sell personal data to advertisers or third parties without consent, leading to intrusive marketing.
- Monitoring or tracking of individuals' activities can suppress freedom of expression, especially under oppressive regimes.

For individuals, these outcomes can be harmful. For businesses, privacy violations can result in reputational damage, legal penalties, and financial losses. Beyond these tangible consequences, privacy holds intrinsic value as a human right essential to freedom in society.

In addition to the real-world implications of privacy infringements, many people and countries hold that privacy has intrinsic value: that privacy is a human right fundamental to a free society, like the right to free speech.

What are some of the challenges users face when protecting their online privacy?

- Online tracking: Cookies and similar technologies often track users' activities, sometimes beyond their knowledge or consent.
- Loss of control over data: Users may be unaware of how their data is shared beyond the services they interact with and may lack a say in its usage.
- Opaque privacy policies: Many privacy policies are dense and difficult to understand, leaving users unaware of how their data is handled.
- Social media risks: Social media platforms can inadvertently expose more personal information than intended, and many collect significant user data.
- Cybercrime: Attackers may steal user data for fraudulent activities or sell it on illicit markets. Techniques such as phishing and system compromises are commonly used.

What are some of the challenges businesses face when protecting user privacy?

- Effective communication: Clearly explaining data usage policies to users can be challenging.
- Cybercrime: Organizations that collect personal data are prime targets for attackers.
- Data breaches: Breaches can lead to massive privacy violations, with attackers constantly evolving their methods.
- Insider threats: Employees or contractors may misuse access to sensitive data if protections are insufficient.

Governments worldwide have implemented laws to regulate data collection, usage, and protection. Key frameworks include:

- General Data Protection Regulation (GDPR) [1]: Governs how personal data of individuals in the EU is collected, stored, and processed, granting data subjects rights such as the "right to be forgotten."
- National data protection laws: Countries like Canada, Japan, Australia, and Brazil have enacted comprehensive laws similar to the GDPR, such as Brazil's General Law for the Protection of Personal Data (LGPD) [2].
- California Consumer Privacy Act (CCPA)[3]: Requires transparency about data collection practices and gives consumers control over their personal data, including the right to opt out of data sales.

Certain industries are also subject to specific regulations. For instance, in the U.S., the Health Insurance Portability and Accountability Act (HIPAA) [4] governs the handling of healthcare data.

Many existing data protection laws draw on foundational principles such as the Fair Information Practices [5], first proposed in 1973 and later adopted by the international Organization for Economic Cooperation and Development (OECD). These principles include:

- Collection limitation: Restricting the volume of collected data.
- Data quality: Ensuring collected data is accurate and relevant.
- Purpose specification: Clearly defining the purpose for data usage.
- Use limitation: Using data solely for specified purposes.
- Security safeguards: Protecting data from unauthorized access or breaches.
- Openness: Maintaining transparency about data collection and usage practices.
- Individual participation: Allowing individuals to access, correct, and request deletion of their personal data.
- Accountability: Ensuring data collectors adhere to these principles.

Despite these measures, many privacy advocates argue that individuals still lack sufficient control over their personal data.

1.2 Previous Data Privacy Protection Techniques

The protection of data is a critical concern for individuals, businesses, and organizations of all sizes. As reliance on technology for storing and transmitting sensitive information increases, the risk of data breaches and theft becomes a pressing issue. In 2022 alone, over 2000 publicly disclosed data breaches occurred, with 60% attributed to hacking, leading to financial losses, reputational damage, and potential legal liabilities for affected individuals and organizations.

To mitigate these risks, several methods have been developed to safeguard data from unauthorized access and manipulation. Historically, four primary techniques have been employed for data protection:

1. **Encryption**: Encryption serves as a foundational method for protecting sensitive data. By converting data into an encoded format, encryption ensures that only authorized users with the correct decryption key can access the original information. Common encryption algorithms, such as AES [6] and RSA [7], are widely used to secure data during transmission over the internet and on physical devices like laptops and smartphones.

    Encryption's primary advantage lies in its robustness: even if encrypted data is compromised, it remains unintelligible to attackers. Moreover, encryption enables organizations to comply with regulatory standards such as the General Data Protection Regulation (GDPR) [1] and the Payment Card Industry Data Security Standard (PCI DSS) [8].

    However, encryption is not without limitations. Proper implementation is critical; for example, losing or mismanaging encryption keys can render data inaccessible, even to rightful owners.

2. **Access Control [9]**: Access control restricts access to sensitive data, allowing only authorized users to retrieve or manipulate information. This can be

achieved using techniques like passwords, multi-factor authentication, and role-based access control. By ensuring that only those with appropriate credentials can access data, access control significantly reduces the risk of breaches and unauthorized access.

A major benefit of access control is its ability to foster accountability within organizations, enabling monitoring of who accesses data and performs specific actions. Additionally, access control streamlines the management of permissions, enhancing operational efficiency.

To be effective, access control systems must be properly implemented and maintained. For instance, passwords should be strong and unique, multi-factor authentication should be employed, and systems must undergo regular updates and testing to remain secure.

3. **Network Security [10]**: Network security encompasses measures designed to protect information and assets on computer networks from unauthorized access, theft, or damage. This includes deploying firewalls, implementing intrusion detection systems, encrypting transmitted data, and conducting regular software updates. Employee training also plays a crucial role in mitigating cyber threats.

    Network security enhances confidentiality by safeguarding sensitive information from unauthorized access and breaches. It also helps organizations comply with standards like HIPAA and PCI DSS and serves as an integral part of risk management by identifying and mitigating potential security vulnerabilities.

    Effective network security requires skilled IT personnel, as the systems can be complex to configure and maintain. Furthermore, the dynamic nature of cyber threats necessitates constant updates and improvements to keep pace with evolving attack methods.

4. **Physical Security [11]**: Physical security focuses on securing the physical devices and facilities that store sensitive data. Measures include locking devices in secure storage, implementing biometric or key card access controls, and installing security cameras and alarms. For portable devices like laptops and smartphones, additional protections such as encryption, strong passwords, and remote wipe capabilities are essential.

    Physical security provides confidence in the integrity and availability of data backups, reducing risks associated with data loss or corruption. Additionally, effective physical security minimizes costs associated with data recovery efforts.

    However, human error remains a significant challenge; for example, misplaced or unsecured backup media can undermine physical security measures. Organizations must ensure proper tracking and secure storage of backup materials to mitigate these risks.

Each of these methods plays a crucial role in safeguarding sensitive data. However, as cyber threats continue to evolve, a multi-layered approach combining these techniques is essential to ensure robust protection.

Data privacy protection has become a paramount concern for both individuals and organizations in today's interconnected digital era. To address this challenge effectively, it is essential to adopt a comprehensive, multi-layered approach that safeguards sensitive information throughout its lifecycle. This approach should encompass key measures such as encryption, access control, network security, and physical security, ensuring the confidentiality and integrity of data at every stage.

Moreover, it is vital to conduct regular assessments and updates of security protocols to align with technological advancements and the ever-evolving threat landscape. By staying proactive, organizations can mitigate potential vulnerabilities and maintain robust defenses against emerging cyber risks.

While these technologies represent significant advancements in protecting user privacy and securing data, they alone are insufficient. True data privacy protection requires a combination of technology, sound policy frameworks, user education, and a culture of accountability to comprehensively address the complex challenges in safeguarding sensitive information.

1.3 Privacy Enhancing Techniques

Traditional data privacy protection methods are effective at safeguarding data in storage from unauthorized access. However, these techniques fall short when data needs to be processed by third parties, such as cloud service providers. For instance, when a data owner outsources their data to a cloud platform for tasks like data mining or machine learning, how can they ensure that their sensitive information remains private? Similarly, when multiple data owners collaborate to analyze joint datasets, how can they prevent one another from accessing each other's private data?

To address these challenges, advanced privacy-enhancing techniques such as homomorphic encryption, secure multiparty computation, and differential privacy have been developed. These approaches enable third-party processing while preserving data privacy.

1. **Homomorphic Encryption [12]**: Homomorphic encryption allows computations to be performed directly on encrypted data without the need for decryption. The results of these computations remain encrypted, and when decrypted, yield the same outputs as if the operations were conducted on plaintext data. This makes homomorphic encryption particularly useful for privacy-preserving outsourced storage and computation.

    By enabling data to remain encrypted during processing, homomorphic encryption mitigates risks of exposure due to privilege escalation attacks. For instance, in healthcare, this technique can facilitate privacy-preserving predictive analytics by allowing third-party service providers to process encrypted medical data without accessing the decryption keys. Even if the provider's system is compromised, the data remains secure, addressing critical privacy concerns in data sharing.

2. **Secure Multiparty Computation (MPC) [13]**: Secure multiparty computation is a cryptographic method that allows multiple parties to jointly compute a function over their inputs while keeping those inputs private. Unlike traditional

cryptography that protects against external adversaries, MPC ensures privacy within the group, preventing participants from learning each other's private inputs.

The field of MPC originated in the late 1970s with protocols designed for specific tasks, formalized by Andrew Yao in 1982 with his secure two-party computation model. Over time, its applications have expanded to include voting systems, digital signatures, auctions, and more. Recently, MPC has been employed in securing digital assets, with initiatives such as the MPC Alliance promoting its adoption.

3. **Differential Privacy (DP) [14]**: Differential privacy provides a rigorous mathematical framework to share statistical insights about datasets while preserving individual privacy. By introducing carefully calibrated noise to statistical computations, DP ensures that the output remains useful while limiting the risk of re-identification of individuals within the dataset.

Differential privacy is widely used in contexts like government statistical reporting and corporate user behavior analysis, where maintaining confidentiality is paramount. For instance, government agencies use DP algorithms to publish demographic statistics while ensuring survey response privacy. Similarly, organizations employ DP techniques to balance the need for aggregate insights with robust privacy protections.

A differentially private algorithm ensures that its outputs are indistinguishable regardless of whether a specific individual's data is included, offering strong resistance to re-identification attacks and guaranteeing data confidentiality.

1.4 Book Organisation

This book is structured into three parts. The first part introduces fundamental privacy-enhancing techniques. The second part explores practical applications of these techniques. The final part focuses on the interplay between privacy and digital trust in Vehicular Ad Hoc Networks (VANETs).

In the first part, Chaps. 2 to 4 cover three core privacy-enhancing techniques: homomorphic encryption, secure multiparty computation, and differential privacy.

The second part, spanning Chaps. 5 to 10, delves into the practical applications of privacy-enhancing techniques. These include privacy-preserving data mining, private information retrieval, machine learning, social networks, location-based services with location privacy protection, and differentially private frequency estimation.

The last part uses VANETs as a case study in Chap. 11 to examine the critical relationship between privacy and digital trust, highlighting real-world challenges and potential solutions.

References

1. https://en.wikipedia.org/wiki/General_Data_Protection_Regulation
2. https://en.wikipedia.org/wiki/General_Personal_Data_Protection_Law
3. https://en.wikipedia.org/wiki/California_Consumer_Privacy_Act
4. https://en.wikipedia.org/wiki/Health_Insurance_Portability_and_Accountability_Act
5. https://en.wikipedia.org/wiki/FTC_fair_information_practice
6. https://en.wikipedia.org/wiki/Advanced_Encryption_Standard
7. https://en.wikipedia.org/wiki/RSA_(cryptosystem)
8. https://en.wikipedia.org/wiki/Payment_Card_Industry_Data_Security_Standard
9. https://en.wikipedia.org/wiki/Access_control
10. https://en.wikipedia.org/wiki/Network_security
11. https://en.wikipedia.org/wiki/Physical_security
12. https://en.wikipedia.org/wiki/Homomorphic_encryption
13. https://en.wikipedia.org/wiki/Secure_multi\discretionary-party_computation
14. https://en.wikipedia.org/wiki/Differential_privacy

Chapter 2
Homomorphic Encryption

Abstract This chapter provides an overview of homomorphic encryption, beginning with an introduction to public key encryption and the foundational concepts of homomorphic encryption. It then explores partial homomorphic encryption schemes, highlighting their key features and applications. The chapter further delves into the concept of fully homomorphic encryption, presenting its definition and examining notable schemes in this domain. Through this structured progression, the chapter offers a comprehensive understanding of homomorphic encryption techniques and their practical significance in privacy-enhancing technologies.

2.1 Public Key Encryption Definition

In the early stages of encryption, secure communication between two parties relied on a shared secret key, exchanged through a secure method, such as face-to-face meetings or via a trusted courier. While effective, this approach presented several practical challenges, particularly in securely distributing keys.

Public key encryption was introduced to address these limitations, enabling users to communicate securely over public channels without the need to share a secret key beforehand. This approach revolutionized secure communication by providing a mechanism that allows for secure exchanges without relying on a pre-shared key.

The concept of public key encryption was introduced in 1976 by Whitfield Diffie and Martin Hellman [3], who, drawing on Ralph Merkle's work on public-key distribution, disclosed a method for public-key agreement. Their groundbreaking work paved the way for modern cryptographic systems and the development of the public key encryption model, as depicted in Fig. 2.1.

Public key encryption, also known as asymmetric key encryption, involves a pair of keys: one public and one private. These keys are mathematically linked, yet distinct. The public key is used to encrypt plaintext, while the private key is used to decrypt ciphertext. The term "asymmetric" highlights the use of different keys for encryption and decryption, in contrast to symmetric encryption, where the same key is used for both operations.

Fig. 2.1 Public key encryption model

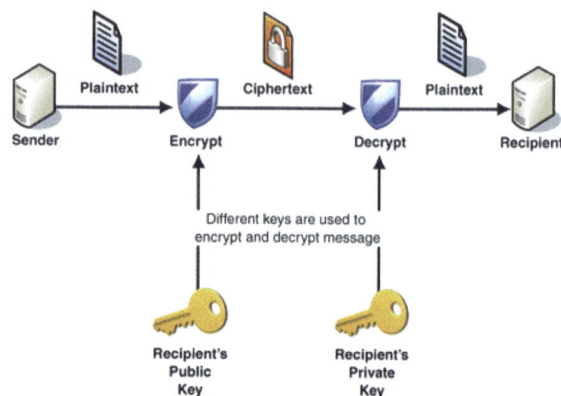

A public key cryptosystem, typically associated with a key space (**K**), a plaintext space **M**, and a ciphertext space **C**, consists of three key algorithms:

- **Key Generation Algorithm (KG)**—Given a security parameter k, a pair of public and private keys (pk, sk) is generated, where $sk \in \mathbf{K}$. The public key pk is made publicly available, while the private key sk is kept secret by the owner.
- **Encryption Algorithm (E)**—Given a plaintext message $m \in \mathbf{M}$ and a public key pk, a ciphertext c is produced, denoted as $c = E(m, pk)$, where $c \in \mathbf{C}$.
- **Decryption Algorithm (D)**—Given a ciphertext $c = E(m, pk)$ and the private key sk, the plaintext message m is recovered, denoted as $m = D(c, sk)$.

The encryption algorithm E, which maps plaintext to ciphertext, must be a trap-door one-way function. For virtually all ciphertexts $c = E(m, pk)$, it must be computationally infeasible to recover the plaintext m from pk and c. However, to ensure that the legitimate recipient can recover the message, the function must be invertible. Each encryption function E must have a corresponding inverse D, which can be easily computed when the additional secret information sk (known as the trap-door) is available. The ability to efficiently compute $m = D(c, sk)$ for all ciphertexts c is a key requirement for the security of the system.

Trap-door functions rely on difficult mathematical problems, such as integer factorization, discrete logarithms, and elliptic curve relationships, which are currently not solvable efficiently. This computational complexity ensures the security of the system. Generating a public and private key pair is computationally simple, and using these keys for encryption and decryption is also straightforward. The strength of public key encryption lies in the fact that it is computationally infeasible to derive the private key from its corresponding public key. As a result, the public key can be freely shared without compromising security, while the private key must remain secret to prevent unauthorized decryption of messages.

Unlike symmetric key encryption, public key encryption does not require a secure initial exchange of keys. This absence of a shared secret key exchange

2.2 Homomorphic Encryption Definition

In abstract algebra, a homomorphism is a structure-preserving map between two algebraic structures, such as groups.

A group is defined as a set, G, together with an operation $\circ$ (referred to as the group law of G), which combines any two elements a and b to form another element, denoted $a \circ b$. For a set and operation, $(G, \circ)$, to qualify as a group, they must satisfy four fundamental properties known as the group axioms:

- **Closure:** For all $a, b \in G$, the result of the operation, $a \circ b$, must also belong to G.
- **Associativity:** For all $a, b, c \in G$, it must hold that $(a \circ b) \circ c = a \circ (b \circ c)$.
- **Identity Element:** There exists an element e in G such that for every element $a \in G$, the equation $e \circ a = a \circ e = a$ holds. This element is unique and is referred to as the identity element.
- **Inverse Element:** For each $a \in G$, there exists an element $b \in G$ such that $a \circ b = b \circ a = e$, where e is the identity element.

The identity element of a group G is often denoted as 1.

The operation within a group may depend on the order of the operands. That is, the result of combining element a with element b need not yield the same result as combining element b with element a. The equation $a \circ b = b \circ a$ does not always hold. However, in the case of the group of integers under addition, this equation holds true because $a + b = b + a$ for any two integers (commutativity of addition). Groups for which the commutative equation $a \circ b = b \circ a$ always holds are called abelian groups.

Given two groups $(G, \diamond)$ and $(H, \circ)$, a group homomorphism from $(G, \diamond)$ to $(H, \circ)$ is defined as a function $F : G \rightarrow H$ such that for all $a, b \in G$, the following condition holds:

$$F(a \diamond b) = F(a) \circ F(b) \tag{2.1}$$

This concept of group homomorphism can be visually represented in Fig. 2.2.

Now, consider a public key encryption scheme (P, C, KG, E, D), where P and C represent the plaintext and ciphertext spaces, respectively, KG is the key generation algorithm, and E and D are the encryption and decryption algorithms. If the plaintexts form a group $(P, \diamond)$ and the ciphertexts form a group $(C, \circ)$, then the encryption algorithm can be viewed as a map from the group P to the group C, i.e., $E : P \rightarrow C$.

Fig. 2.2 Group homomorphism

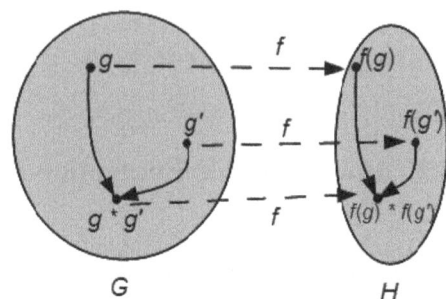

For all $a, b \in P$, if the following equation holds:

$$E(a, pk) \circ E(b, pk) = E(a \diamond b, pk) \tag{2.2}$$

where pk is the public key generated by KG, the public key encryption scheme is said to be **homomorphic**.

2.3 Partial Homomorphic Encryption Schemes

2.3.1 Goldwasser-Micali Encryption Scheme

The Goldwasser-Micali (GM) cryptosystem [5] is a public key encryption algorithm introduced in 1982 by Shafi Goldwasser and Silvio Micali. It is recognized as the first probabilistic public-key encryption scheme that is provably secure under standard cryptographic assumptions. However, the GM cryptosystem is not efficient, as its ciphertexts are significantly larger than the plaintext. To formalize its security properties, Goldwasser and Micali introduced the widely adopted concept of semantic security.

The GM encryption scheme consists of three key algorithms: a probabilistic key generation algorithm that produces a public-private key pair, a probabilistic encryption algorithm, and a deterministic decryption algorithm.

The scheme is based on determining whether a given value x is a quadratic residue modulo N, provided the factorization (p, q) of N is known. This determination uses the following procedure: compute $x_p = x \mod p$ and $x_q = x \mod q$. If $x_p^{(p-1)/2} = 1 \mod p$ and $x_q^{(q-1)/2} = 1 \mod q$, then x is a quadratic residue modulo N.

- **Key Generation.** The modulus N used in GM encryption is generated similarly to RSA (refer to the RSA key generation process).

  Alice generates two distinct large prime numbers p and q independently and computes $N = p \times q$. She then identifies a non-residue x such that $x_p^{(p-1)/2} = -1 \mod p$ and $x_q^{(q-1)/2} = -1 \mod q$. If $p, q \equiv 3 \mod 4$ (i.e., N is a Blum

2.3 Partial Homomorphic Encryption Schemes

integer), then $N - 1$ is guaranteed to satisfy this property. The public key consists of (x, N), while the private key is the factorization (p, q).

- **Message Encryption.** Suppose Bob wishes to send a message m to Alice:
  1. Bob encodes m as a string of bits $(m_1, \ldots, m_n)$.
  2. For each bit m_i, Bob generates a random value y_i such that $\gcd(y_i, N) = 1$. He computes:
  $$c_i = y_i^2 x^{m_i} \mod N$$
  3. Bob sends the ciphertext $(c_1, \ldots, c_n)$ to Alice.

- **Message Decryption.** Alice receives the ciphertext $(c_1, \ldots, c_n)$ and decrypts it as follows:
  1. For each c_i, using her private key (p, q), Alice determines whether c_i is a quadratic residue modulo N. If it is, $m_i = 0$; otherwise, $m_i = 1$.
  2. Alice reconstructs the message $m = (m_1, \ldots, m_n)$.

The GM encryption scheme is a probabilistic encryption algorithm [6]. Probabilistic encryption uses randomness during encryption, ensuring that the same plaintext produces different ciphertexts when encrypted multiple times. This randomness is essential for semantic security, which prevents adversaries from inferring partial information about plaintexts.

Probabilistic encryption is particularly critical in public key systems. For instance, without randomness, an adversary could encrypt guessed plaintexts using the public key and compare the results to the observed ciphertext. To mitigate this, public key schemes incorporate randomness, ensuring that each plaintext maps to multiple possible ciphertexts.

2.3.1.1 Semantic Security and Quadratic Residuosity Assumption

The GM cryptosystem achieves semantic security under the assumption that the quadratic residuosity problem is computationally hard. This problem involves determining whether a value x is a quadratic residue modulo $N = p \times q$, where p and q are large primes, without knowledge of the factorization. The GM scheme encrypts bits as either quadratic residues or non-residues modulo N, ensuring security against adversaries without access to the factorization.

2.3.1.2 Homomorphic Properties

The GM cryptosystem possesses homomorphic properties. Specifically, given encryptions c_0 and c_1 of bits m_0 and m_1, the product $c_0 \cdot c_1 \mod N$ is an encryption of $m_0 \oplus m_1$ (addition modulo 2). This property is demonstrated as follows:

$$c_0 \cdot c_1 = (y_0^2 x^{m_0}) \cdot (y_1^2 x^{m_1}) \mod N = (y_0 y_1)^2 x^{m_0 + m_1} \mod N$$

When $m_0 + m_1 \equiv 2 \mod 2$, the ciphertext corresponds to an encryption of 0. Thus, the GM scheme supports exclusive-or operations directly on ciphertexts, a characteristic crucial for many applications.

2.3.2 ElGamal Encryption Scheme

The ElGamal encryption scheme [4] is a public key encryption algorithm developed by Taher ElGamal in 1985, based on the Diffie-Hellman key exchange. This scheme has found application in various systems, including GNU Privacy Guard, recent versions of PGP, and other cryptographic frameworks. The ElGamal encryption scheme can be applied to any cyclic group G, with its security relying on the computational difficulty of discrete logarithms in G.

ElGamal encryption consists of three components: a key generation process, an encryption algorithm, and a decryption algorithm.

- **Key Generation:** Alice follows these steps to generate her keys:
    1. Generate a cyclic group G of order q, with generator g.
    2. Choose a random $x \in \{1, \ldots, q-1\}$ as her private key.
    3. Compute:
    $$y = g^x$$
    4. Publish the public key (G, q, g, y) and keep x as the private key.

- **Encryption:** Bob encrypts a message m under Alice's public key as follows:
    1. Select a random $r \in \{1, \ldots, q-1\}$.
    2. Compute:
    $$c_1 = g^r \quad \text{and} \quad c_2 = m \cdot y^r$$
    3. The ciphertext is:
    $$(c_1, c_2) = (g^r, m \cdot y^r)$$

    Each encryption uses a new random value r, often referred to as an ephemeral key, to ensure security.

- **Decryption:** Alice decrypts the ciphertext (c_1, c_2) using her private key x:
    1. Compute the shared secret:
    $$s = c_1^x$$
    2. Recover the plaintext message m as:
    $$m = c_2 \cdot s^{-1}$$

2.3 Partial Homomorphic Encryption Schemes

Here, s^{-1} is the modular multiplicative inverse of s in G.

The correctness of decryption is ensured because:

$$c_2 \cdot s^{-1} = m \cdot y^r \cdot (g^{xr})^{-1} = m \cdot g^{xr} \cdot g^{-xr} = m$$

2.3.2.1 Properties and Security

ElGamal encryption is probabilistic, allowing a plaintext to map to multiple possible ciphertexts. This property is achieved through the use of the random value r in encryption, ensuring semantic security when paired with the decisional Diffie-Hellman (DDH) assumption in the cyclic group G.

While ElGamal encryption provides semantic security, it is malleable, making it vulnerable to chosen ciphertext attacks. For example, an adversary with access to a ciphertext (c_1, c_2) can construct $(c_1, 2c_2)$, a valid encryption of $2m$. To address this, secure padding schemes or modifications, such as the Cramer-Shoup cryptosystem [2], are often employed.

2.3.2.2 Homomorphic Properties

ElGamal encryption exhibits homomorphic properties. Given two ciphertexts:

$$(c_{11}, c_{12}) = (g^{r_1}, m_1 \cdot y^{r_1}), \quad (c_{21}, c_{22}) = (g^{r_2}, m_2 \cdot y^{r_2}),$$

their product:

$$(c_{11} \cdot c_{21}, c_{12} \cdot c_{22}) = (g^{r_1+r_2}, (m_1 \cdot m_2) \cdot y^{r_1+r_2}),$$

is a valid encryption of $m_1 \cdot m_2$, demonstrating its homomorphic capability.

2.3.2.3 Example

Consider the following example:

- Alice selects $p = 2879$, $g = 2585$, and $x = 47$ (her private key).
- She computes:

$$y = g^x \mod p = 2585^{47} \mod 2879 = 2826$$

Her public key is (p, g, y).
- Bob wants to encrypt $m = 77$. He selects $r = 65$ and computes:

$$c_1 = g^r \mod p = 2585^{65} \mod 2879 = 319,$$

$$c_2 = m \cdot y^r \mod p = 77 \cdot 2826^{65} \mod 2879 = 472$$

The ciphertext is $(c_1, c_2) = (319, 472)$.
- Alice decrypts the ciphertext:

$$s = c_1^x \mod p = 319^{47} \mod 2879,$$

$$m = c_2 \cdot s^{-1} \mod p = 472 \cdot s^{-1} \mod 2879 = 77$$

ElGamal encryption, while computationally intensive, provides a foundation for hybrid cryptosystems and advanced cryptographic applications.

2.3.3 Paillier Encryption Scheme

The Paillier encryption scheme [8], named after and invented by Pascal Paillier in 1999, is a probabilistic public key algorithm. Its security is based on the decisional composite residuosity assumption, which postulates the computational difficulty of deciding whether an integer is an n-th residue modulo n^2. This cryptosystem is notable for its additive homomorphic properties, allowing the encryption of $m_1 + m_2$ to be computed directly from the ciphertexts of m_1 and m_2 without decrypting them.

- **Key Generation**
    1. Choose two large prime numbers p and q such that $\gcd(pq, (p-1)(q-1)) = 1$.
    2. Compute $n = p \cdot q$ and $\lambda = \text{lcm}(p-1, q-1)$, where lcm denotes the least common multiple.
    3. Select a random integer $g \in \mathbb{Z}_{n^2}^*$ and ensure n divides the order of g by checking the modular multiplicative inverse:

    $$\mu = (L(g^\lambda \mod n^2))^{-1} \mod n,$$

    where the function L is defined as:

    $$L(u) = \frac{u-1}{n}.$$

    4. The public key is (n, g), and the private key is (λ, μ).

2.3 Partial Homomorphic Encryption Schemes

- **Encryption**
  Given a message $m \in \mathbb{Z}_n$ and a random $r \in \mathbb{Z}_n^*$:

$$c = g^m \cdot r^n \bmod n^2,$$

  where c is the ciphertext.
- **Decryption**
  Given the ciphertext $c \in \mathbb{Z}_{n^2}^*$:

$$m = L(c^\lambda \bmod n^2) \cdot \mu \bmod n.$$

2.3.3.1 Homomorphic Properties

The Paillier cryptosystem has remarkable homomorphic properties:

- **Addition of Plaintexts:** The product of two ciphertexts decrypts to the sum of their corresponding plaintexts:

$$D(E(m_1, r_1) \cdot E(m_2, r_2) \bmod n^2) = m_1 + m_2 \bmod n.$$

- **Scalar Multiplication of Plaintexts:** Raising a ciphertext to the power of a constant decrypts to the product of the plaintext and the constant:

$$D(E(m, r)^k \bmod n^2) = k \cdot m \bmod n.$$

2.3.3.2 Example

For simplicity, consider small primes for key generation:
- Let $p = 7, q = 11$, then $n = p \cdot q = 77$.
- Choose $g = 5652$ and compute:

$$\lambda = \text{lcm}(p - 1, q - 1) = \text{lcm}(6, 10) = 30,$$

$$\mu = (L(g^\lambda \bmod n^2))^{-1} \bmod n = 74.$$

Encryption:

- Let $m = 42$ and choose a random $r = 23$.
- Compute:

$$c = g^m \cdot r^n \bmod n^2 = 5652^{42} \cdot 23^{77} \bmod 5929 = 4624.$$

Decryption:

- Compute:

$$m = L(c^\lambda \bmod n^2) \cdot \mu \bmod n,$$

$$L(c^\lambda \bmod n^2) = L(4624^{30} \bmod 5929) = 4852,$$

$$m = 4852 \cdot 74 \bmod 77 = 42.$$

This example demonstrates the correctness of the Paillier cryptosystem. Despite its computational overhead, Paillier encryption is widely used in applications requiring additive homomorphic properties, such as secure voting and private data aggregation.

2.3.4 Boneh-Goh-Nissim Encryption Scheme

The Boneh-Goh-Nissim (BGN) encryption scheme [1] shares similarities with the Paillier [8] and Okamoto-Uchiyama [7] encryption schemes. Notably, the BGN scheme was the first to support both additions and a single multiplication with a constant-size ciphertext. This capability comes with a limitation: while multiple additions are permitted, only one multiplication is allowed.

A key idea of the BGN system is leveraging elliptic curve groups whose order is a composite number N, difficult to factor. The scheme comprises three primary algorithms: key generation, encryption, and decryption.

- **Key Generation** $KeyGen(k)$
  Given a security parameter $k \in \mathbb{Z}^+$, execute $KeyGen(k)$ to generate a tuple $(q_1, q_2, \mathbb{G}, \mathbb{G}_1, e)$. Set $N = q_1 q_2$. Choose two random generators g and u from $\mathbb{G}$, and compute $h = u^{q_2}$. The value h serves as a random generator of the subgroup of $\mathbb{G}$ with order q_1. The public key is $PK = \{N, \mathbb{G}, \mathbb{G}_1, e, g, h\}$. The private key is $SK = q_1$.
- **Encryption** $Encrypt(m, PK)$
  Assume the message space consists of integers in the set $\{0, 1, \cdots, T\}$, with $T < q_2$. To encrypt a message m using the public key PK, select a random r from $\{1, 2, \cdots, N\}$ and compute:

$$C = g^m h^r \in \mathbb{G}. \tag{2.3}$$

Output C as the ciphertext.
- **Decryption** $Decrypt(C, SK)$
  To decrypt a ciphertext C using the private key $SK = q_1$, observe that:

$$C^{q_1} = (g^m h^r)^{q_1} = (g^{q_1})^m. \tag{2.4}$$

2.4 Fully Homomorphic Encryption

Recovering the message m requires computing the discrete logarithm of C^{q_1} to the base g^{q_1}. Given $0 \leq m \leq T$, this can be achieved in expected time $O(\sqrt{T})$ using Pollard's lambda method [12].

Homomorphic Properties

The BGN scheme exhibits additive homomorphism. Given two ciphertexts $C_1, C_2 \in \mathbb{G}$, representing messages $m_1, m_2 \in \{0, 1, \cdots, T\}$ respectively, one can compute a uniformly distributed encryption of $m_1 + m_2 \pmod{N}$ as:

$$C = C_1 C_2 h^r, \tag{2.5}$$

where r is a random value in $\{1, 2, \cdots, N-1\}$.

Moreover, the BGN scheme allows a single multiplication using a bilinear map. Let $g_1 = e(g, g)$ and $h_1 = e(g, h)$. Then g_1 is of order N, and h_1 is of order q_1. Suppose the ciphertexts $C_1 = g^{m_1} h^{r_1} \in \mathbb{G}$ and $C_2 = g^{m_2} h^{r_2} \in \mathbb{G}$ encrypt messages m_1 and m_2, respectively. To compute an encryption of $m_1 m_2 \pmod{N}$:

1. Choose a random $r \in \mathbb{Z}_N$.
2. Compute $C = e(C_1, C_2) h_1^r \in \mathbb{G}_1$.

Then:

$$\begin{aligned} C &= e(C_1, C_2) h_1^r \\ &= e(g^{m_1} h^{r_1}, g^{m_2} h^{r_2}) h_1^r \\ &= e(g, g)^{m_1 m_2} h_1^{r + m_1 r_2 + m_2 r_1 + \alpha q_2 r_1 r_2}. \end{aligned} \tag{2.6}$$

Here, the term $r + m_1 r_2 + m_2 r_1 + \alpha q_2 r_1 r_2$ is uniformly distributed in $\mathbb{Z}_N$. Consequently, C is a valid encryption of $m_1 m_2 \pmod{N}$ in $\mathbb{G}_1$. Additionally, the BGN scheme retains its additive homomorphic property in $\mathbb{G}_1$.

2.4 Fully Homomorphic Encryption

As we discussed in the previous chapter, partial homomorphic encryption is a very useful tool with a number of attractive applications. However, the main drawback of such homomorphic encryption schemes lies in their limited operation support. Only a specific set of operations, often limited to either addition or multiplication but not both, is supported, which makes it unsuitable for applications that require a combination of both.

Many practical applications such as data analysis, machine learning, and AI model inference include both addition and multiplication. Therefore, it would be more powerful if the homomorphic encryption scheme could evaluate both addition and multiplication over encrypted data without decrypting it. In fact, this would enable someone to evaluate any efficiently computable function over the encrypted data without knowing the secret key.

In this chapter, we introduce fully homomorphic encryption (FHE), which supports both addition and multiplication operations and can perform arbitrary computation on the ciphertext. The concept of FHE was introduced by Rivest [13] under the name privacy homomorphisms. The problem of constructing a scheme with these properties remained unsolved until 2009, when Gentry [11] presented his breakthrough result. His scheme supports both addition and multiplication operations on ciphertexts, from which it is possible to construct circuits to perform arbitrary computations. The chapter begins with an introduction of FHE model and definitions, following by the most well-known constructions of FHE schemes.

We begin our examples of constructing a fully homomorphic encryption scheme with the original construction by Gentry [11] and the subsequent simplification by using two elements to represent the basis matrix by Smart-Vercauteren [14]. This is followed by the simplification using the integer ring by van Dijk et al. [10]. Next we introduce a scheme proposed by Brakerski-Vaikuntanathan [9]. This scheme is based on relatively simple problem called learning with errors (LWE). We conclude with three FHE schemes based on ring-LWE: the BGV scheme, the BFV scheme and the CKKS scheme.

2.4.1 Fully Homomorphic Encryption Model

A fully homomorphic encryption (FHE) scheme is formulated as follows. It consists of a tuple of four algorithms ($KeyGen, Encrypt, Decrypt, Evaluate$).

- $KeyGen(\lambda)$ algorithm takes λ (the security parameter) as input and outputs a pair of public key and secret key (pk, sk).
- $Encrypt(m, pk)$ function takes a message m and the public key pk as input and outputs a ciphertext c.
- $Decrypt(c, sk)$ takes a ciphertext c and the secret key sk as input, and computes a corresponding message. For correctness, FHE requires that for all messages m we have $Decrypt(Encrypt(m)) = m$.
- $Evaluate(pk, f, \langle c_1, c_2, c_3, \cdots, c_n \rangle)$ It takes the public key pk, a function f, and a tuple of ciphertexts $\langle c_1, c_2, c_3, \cdots, c_n \rangle$ as input and outputs a new ciphertext c'. For correctness, FHE requires that for all messages m and functions f we have $Decrypt(Evaluate(pk, f, \langle c_1, c_2, \cdots, c_n \rangle)) = f(m_1, \ldots, m_n)$, where $Decrypt(c_i)) = m_i, i \in 1, \ldots, n$.

As shown in Fig. 2.3, the correctness of FHE guarantees that the decrypted result of the evaluation output should be the same as the result of evaluating the corresponding functions directly on the plaintext.

2.4 Fully Homomorphic Encryption

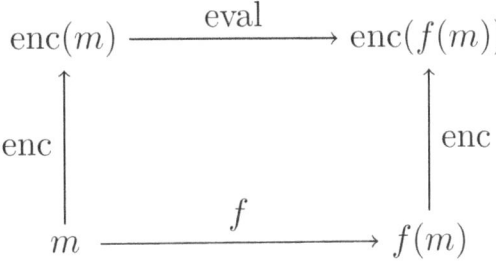

Fig. 2.3 Evaluating a function on a ciphertext

Both homomorphic addition and multiplication are supported by FHE. This is defined by the following equations:

$$Decrypt[Encrypt(m_1) \oplus Encrypt(m_2)] = Decrypt[Encrypt(m_1 + m_2)] \tag{2.7}$$

$$Decrypt[Encrypt(m_1) \otimes Encrypt(m_2)] = Decrypt[Encrypt(m_1 \times m_2)] \tag{2.8}$$

where m_1 and m_2 are two plaintexts and the public and secret keys have been omitted for aesthetics.

The compactness of FHE refers to the size of the ciphertexts that the scheme generates during the evaluations, which means that the size of ciphertexts that $Evaluate(\cdot, f, \cdot)$ outputs does not depend on the size or the depth of the function f. It ensures that the ciphertext remains small after multiple homomorphic operations. An FHE scheme is compact if the size of the ciphertext c' outputted by $Evaluate(pk, f, \langle c_1, c_2, c_3, \cdots, c_n \rangle)$ is not more than $b(\lambda)$ bits, where $b(\lambda)$ is a fixed polynomial bound and independent of the depth of f.

The security notion of FHE scheme we consider is semantic security, namely security w.r.t. passive adversaries. Informally speaking, semantic security guarantees that given two ciphertexts, then they are indistinguishable by a computationally bounded adversary with respect to a reasonable security level.

We are now ready to introduce the definition of FHE and its related relaxed definition for leveled fully homomorphic encryption [15]. The definition of the fully homomorphic encryption scheme is defined over $GF(2)$ because it simplifies the construction. Extending the definition to higher finite fields should be trivial.

Definition 2.1 A cryptosystem is *fully homomorphic* if it satisfies two requirements: (1) it is compact according to the definition above; and (2) it is homomorphic over $GF(2)$ (the Boolean finite field).

Definition 2.2 A cryptosystem is called a *leveled fully homomorphic* if it can evaluate any L-depth circuit C (over the $GF(2)$ field as above). The size of the ciphertext must remain independent of the level parameter L.

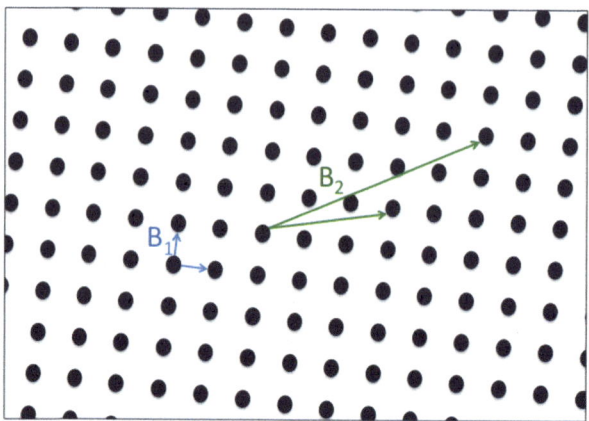

Fig. 2.4 2-dimension lattice

2.4.2 Gentry's Fully Homomorphic Encryption

Craig Gentry [11] using lattice-based cryptography showed the first fully homomorphic encryption scheme as announced by IBM on June 25, 2009.

A lattice is a n-dimensional periodic grid of points. It is the set of all integer linear combinations of vectors from a basis $\{\mathbf{b_1}, \ldots, \mathbf{b_n}\}$, which is typically arranged as a matrix $n \times n$, called the lattice basis $B = (\mathbf{b_1}, \ldots, \mathbf{b_n})$, so we denote the lattice as $\mathcal{L}(B)$. One of the most important lattice-based computational problem is the shortest vector problem (SVP), whose task is to find the shortest vectors in the lattice. This problem is thought to be hard to solve efficiently, even with quantum computer, making lattice-based cryptography a promising candidate for post-quantum security. Figure 2.4 shows an illustration of a 2-dimensional lattice. B_1 and B_2 demonstrate that this lattice can be defined by two different set of bases, where one (B_1) is short and orthogonal while the other (B_2) is long and less orthogonal. Therefore, there is an infinite number of bases of lattice dimension $n \geq 2$.

As the basis B for a lattice $\mathcal{L}$ is not unique, this property can be used to create a trapdoor, and further, allows us to construct public-key cryptosystems that are based on the difficulty of solving a difficult mathematical problem. Examples of these types of hard problems include SVP and GapSVP.[1]

A point on the n-dimensional lattice is a n-dimensional vector $\mathbf{x} \in \mathbb{Z}^n$. When two vectors are added component-wise, it forms a (closed) group under addition, which is a direct consequence of the base ring. In order to define multiplication, we need to embed an ideal into the lattice. An ideal $\mathcal{I}$ is a sub-ring inside the lattice. Given a definition of an ideal, we can separate (or remove) an ideal component within some ring. This provides a way to mask the encoding of a message as $m + x\mathcal{I}$. Since $\mathcal{I}$

---

[1] SVP stands for Shortest Vector Problem.

2.4 Fully Homomorphic Encryption

is defined, this can be done without corrupting m. Thus if we have two encryptions c_1, c_2 that are encryptions of m_1, m_2, respectively, we have [9]:

$$\begin{aligned} c_1 c_2 &= (m_1 + x_1 I)(m_2 + x_2 I) \\ &= m_1 m_2 + (m_1 x_2 + m_2 x_1 + x_1 x_2 I) I \\ &= m_1 m_2 + z I \end{aligned}$$

Now we show the first plausible construction of a fully homomorphic encryption scheme proposed by Gentry [11]. This scheme consists of four high-level algorithms: KeyGen, Encrypt, Decrypt, and Eval.

$KeyGen(R, B_I)$ Takes as input a ring R and basis B_I of I. It sets $(B_J^{sk}, B_J^{pk}) \xleftarrow{R} IdealGen(R, B_I)$. Here $IdelGen(R, B_I)$ generates public and secret bases (B_J^{sk}, B_J^{pk}) of some ideal $J \subset R$ such that I and J are relatively prime. The plaintext space $\mathcal{P}$ is (a subset of) R mod B_I. The public key pk includes R, B_I, B_J^{pk}, and the function $Samp(x, B_I, R, B_J^{pk})$. The function $Samp$ samples from coset $x + I$. The secret key sk includes B_J^{sk}.

$Encrypt(pk, \pi)$ Takes as input the public key pk and plaintext $\pi \in \mathcal{P}$. It sets $\psi' \leftarrow Samp(\pi, B_I, R, B_J^{pk})$ and outputs $\psi \leftarrow \psi' \bmod B_J^{pk}$.

$Decrypt(sk, \psi)$ Takes as input the secret key sk and a ciphertext ψ. It outputs

$$\pi \leftarrow (\psi \bmod B_J^{sk}) \bmod B_I$$

$Eval(pk, C, \Psi)$ Takes as input the public key pk, a circuit C in some permitted set $C_\mathcal{E}$ of circuits composed of Add_{B_I} and $Mult_{B_I}$ gates and a set of input ciphertexts Ψ. It invokes Add and $Mult$, given below, in the proper sequence to compute the output ciphertext Ψ. (We will describe C_E when we consider correctness below. If desired, one could use different arithmetic gates.)

$$Add(pk, \psi_1, \psi_2). \text{Outputs } \psi_1 + \psi_2 \bmod B_J^{pk}$$

$$Mult(pk, \psi_1, \psi_2). \text{Outputs } \psi_1 \times \psi_2 \bmod B_J^{pk}$$

Correctness of the homomorphic encryption properties follows from our discussion of ideals in rings. The security of this scheme is based on a lattice problem called the Ideal Coset problem given in [11]. Basically, the security definition says that nothing about the plaintext can be learned from the ciphertext.

Definition 2.3 (Ideal Coset Problem (ICP) [11]) Fix R, B_I, algorithm $IdealGen$, and an algorithm $Samp_1$ that efficiently samples R. The challenger sets $b \xleftarrow{R} \{0, 1\}$ and $(B_J^{sk}, B_J^{pk}) \xleftarrow{R} IdealGen(R, B_I)$. If $b = 0$, it sets $r \xleftarrow{R} Samp_1(R)$ and $t \leftarrow r \bmod B_J^{pk}$. If $b = 1$, it samples t uniformly from R mod B_J^{pk}. The problem: guess b given (t, B_J^{pk}).

This scheme exhibits both additive and multiplicative properties. However, as previously discussed, the noise grows with each homomorphic operation until the point at which decryption fails. Gentry then shows how to modify the scheme to make it bootstrappable, which means that the scheme can evaluate its own decryption circuit homomorphically to refreshes the ciphertext. By bootstrapping the ciphertext periodically before the noise grows too large, Gentry's scheme can compute an arbitrary number of additions and multiplications while maintaining the correctness.

The key insight here is that since the scheme can evaluate polynomial bounded circuits, it could evaluate its 'own' decryption circuit without direct access to the secret key. The solution is to add a 'hint' about the secret key to the public. More specifically, we augment the public key with *an encryption of the secret key* (under the same or different public key). With the encrypted secret key, Gentry's scheme can evaluate the decryption circuit homomorphically and efficiently, thus results in a fully homomorphic scheme.

Since the (encrypted) secret key becomes a part of the public key, another assumption, called the Sparse Subset Problem (SSSP), is necessary when proving the security via a reduction. Informally, this means that within a set of objects, a small subset sums up to an integer, and the problem is to determine the which elements contribute to the sum. The definition of SSSP is given by Definition 2.4. The function $\gamma_{set}(n)$ defines the size of a set of vectors, such that there exists a subset that adds up to the secret key. The size of the subset is given by the function $\gamma_{subset}(n)$. Obviously, we require $\gamma_{subset}(n) < \gamma_{set}(n)$.

Definition 2.4 (Sparse Subset Sum Problem (SSSP) [11]) Let $\gamma_{set}(n)$ and $\gamma_{subset}(n)$ be functions as above. Let q be a positive integer. The challenger sets $b \xleftarrow{R} \{0, 1\}$. If $b = 0$ it generates τ as a set of $\gamma_{set}(n)$ integers $\{a_1, \ldots, a_{\gamma_{set}(n)}\}$ in $[-q/2, q/2]$ that are uniformly random, except that there exists a subset $S \subseteq \{1, \ldots, \gamma_{set}(n)\}$ of cardinality $\gamma_{subset}(n)$ such that $\sum_{i \in S} a_i \mod q$. If $b = 1$, it sets the elements without the constraint. The problem: guess b given τ.

2.4.3 Smart-Vercauteren Fully Homomorphic Encryption

SV-FHE is introduced by Nigel Smart and Frederik Vercauteren [14], which focuses on reducing the complexity and making the bootstrapping more efficient. The way they used is principle ideal. Basically this reduces the entire matrix to a two-element representation.

Before providing the description of the scheme, we define some notation. First we define the length of a given polynomial $g(x) = \sum_{i=0}^{t} g_i x^i$, which can also be viewed as a vector $(g_0, g_1, \cdots, g_t)$, as the 2-norm $(\|g(x)\|_2)$ and ∞-norm $(\|g(x)\|_\infty)$. Respectively these norms are as follows.

2.4 Fully Homomorphic Encryption

$$||g(x)||_2 = \sqrt{\sum_{i=0}^{t} g_i^2} \qquad (2.9)$$

$$||g(x)||_\infty = \max_{i=0,\cdots,t} |g_i| \qquad (2.10)$$

With respect to these norms, which can be considered the length of a polynomial, we define some boundaries for the purposes of sampling polynomials. These are centered at the origin. The input parameter r is known as the radius.

$$\mathcal{B}_{2,N}(r) = \left\{ \sum_{i=0}^{N-1} a_i x^i : \sum_{i=0}^{N-1} a_i^2 \le r^2 \right\} \qquad (2.11)$$

$$\mathcal{B}_{\infty,N}(r) = \left\{ \sum_{i=0}^{N-1} a_i x^i : -r \le a_i \le r \right\} \qquad (2.12)$$

With these notations in mind we can now give the definition of the scheme. Note that the scheme is parametrised by three values (N, η, μ), where N defines the degree of the polynomials, η defines large randomness used in key generation, and μ defines the small randomness used in encryption. The authors of the scheme suggest that $(N, \eta, \mu) = (N, 2^{\sqrt{N}}, \sqrt{N})$.

$KeyGen(1^k)$: Let the plaintext space be $\mathcal{P} = \{0, 1\}$. Fix a random irreducible polynomial $F(x) \in \mathbb{Z}[x]$ of degree N. Repeat the following assignments:

- $S(x) \leftarrow_R \mathcal{B}_{\infty,N}(\eta/2)$
- $G(x) \leftarrow 1 + 2 \cdot S(x)$
- $p \leftarrow \text{resultant}(G(x), F(x))$

until p is prime.
Compute $D(x) \leftarrow \gcd(G(x), F(x))$ over $\mathbb{F}_p$. Set α to be the unique root of $D(x)$. Use the extended euclidean algorithm over $\mathbb{Q}$ to find the polynomial $Z(x) = \sum_{i=0}^{N-1} z_i x^i \in \mathbb{Z}[x]$ subject to

$$Z(x) \cdot G(x) = p \mod F(x).$$

Set $B \leftarrow z_0 \pmod{2p}$. Return the public key as $PK = (p, \alpha)$, and the secret key as $SK = (p, B)$.

$Encrypt(M, PK)$ If $PK = (p, \alpha)$ and $M \in \{0, 1\}$, then continue, otherwise abort. Set $R(x) \leftarrow_R \mathcal{B}_{\inf,N}(\mu/2)$ and $C(x) \leftarrow M + 2 \cdot R(x)$. Output c as $c \leftarrow C(\alpha) \pmod{p}$.

$Decrypt(c, SK)$ If $SK = (p, B)$ then continue, otherwise abort. Output message M as $M \leftarrow (c - \lfloor c \cdot B/p \rceil) \pmod 2$

$Add(c_1, c_2)$ If $PK = (p, \alpha)$ then continue, otherwise abort. Output c_3 as $c_3 \leftarrow (c_1 + c_2) \pmod{p}$.
$Mult(c_1, c_2)$ If $PK = (p, \alpha)$ then continue, otherwise abort. Output c_3 as $c_3 \leftarrow (c_1 \times c_2) \pmod{p}$.

We must reiterate that this scheme is essentially the same as the scheme using a basis to represent the lattice. The $n \times n$ matrix, required by Gentry's scheme, has been replaced by a two element representation. Therefore, SV-FHE has the same homomorphic properties with Gentry's scheme. The main problem with this scheme is that the key generation is slow as it is searching for when resultant is prime.

2.4.4 Brakerski-Vaikuntanathan Fully Homomorphic Encryption

We now present a construction that is based on the Learning With Errors (LWE) assumption, and this formulation leads to a radically different cryptosystem when compared with the other lattice-based FHE schemes. The LWE problem was introduced by Oded Regev in 2005 [26]. Regev showed that the LWE problem is as hard to solve as several worst-case lattice problems.

This hardness assumption of LWE problem is built upon the solution to a system of equations. More precisely, when given n equations with n unknowns, solve for the n variables. As long as these unknowns are independent, this problem is trivial. When we introduce error ϵ, this becomes significantly harder. More formally, given $s \in \mathbb{Z}_2^n$ and a list of 'equations with errors':

$$\langle s, a_1 \rangle \approx_\epsilon b_1 \pmod{2}$$

$$\langle s, a_2 \rangle \approx_\epsilon b_2 \pmod{2}$$

output s. As the value of ϵ increases, this problem becomes more difficult. This initial version is in terms of vectors.

The specific scheme we will introduce is based on a special kind of LWE, called Ring-LWE (RLWE) [9]. RLWE involves a polynomial ring, typically $R_q = \mathbb{Z}_q[x]/<f(x)>$. In RLWE, the goal is to solve an equation similar to LWE, but in the polynomial ring environment:

$$a \cdot s + e = b \pmod{q},$$

where a is a polynomial sampled from a random distribution over the polynomial ring, s is the secret polynomial we want to solve, e is the error polynomial sampled from an error distribution.

The ring structure allows for the use of optimization algorithms like number-theoretic transforms (NTT), to perform efficient polynomial multiplications, making

2.4 Fully Homomorphic Encryption

it more practical for use in cryptographic protocols such as FHE and other lattice-based schemes.

Now we introduce the BV-FHE scheme. The scheme requires the following parameters: a prime number q as ciphertext modulus, and a integer as plaintext modulus $t \in \mathbb{Z}_q^*$, a degree n polynomial $f(x) \in \mathbb{Z}[x]$, and an error distribution χ that defines sampling from the ring $R_q = \mathbb{Z}_q[x]/\langle f(x) \rangle$. A derived parameter D defines the maximal degree of homomorphism permitted. Let the message space be defined as $R_t = \mathbb{Z}_t[x]/\langle f(x) \rangle$. For the sake of understandability, consider $t = 2$ first. We show the symmetric key scheme as follows and remark that transforming this scheme from a symmetric somewhat homomorphic encryption scheme to a public-key somewhat homomorphic encryption scheme is straight forward, where the public key can be generated by encrypting '0' under the secret key.

$KeyGen(1^k)$: Sample a ring element $s \xleftarrow{R} \chi$ and set the secret key $sk := s$. Define the secret key vector as $\mathbf{s} := (1, s, s^2, \cdots, s^D) \in R_q^{D+1}$, which is efficiently computable given s and will be used in the decryption process.

$Encrypt(sk, m)$: Recall that our message space is R_t. Namely, we encode our message as a degree n polynomial with coefficients in $\mathbb{Z}_t$. To encrypt, sample $(a, b = as + te) \in R_q^2$, where $a \xleftarrow{R} R_q$ and $e \xleftarrow{R} \chi$. Compute

$$c_0 := b + m \in R_q \text{ and } c_1 := -a$$

and output the ciphertext $\mathbf{c} := (c_0, c_1) \in R_q^2$.

$Decrypt(sk, \mathbf{c})$: We set that the general form of a decryptable ciphertext is w.l.o.g $\mathbf{c} = (c_0, c_1, \cdots, c_D) \in R_q^{D+1}$, e.g. by padding. To decrypt, we first compute $\langle \mathbf{c}, \mathbf{s} \rangle = \sum_{i=0}^{D} c_i s^i \in R_q$, which can be interpreted as inner product over R_q^{D+1}, and output $m = \langle \mathbf{c}, \mathbf{s} \rangle \pmod{t}$ as the message.

$Eval(p(\xi_1, \cdots, \xi_\ell), (\mathbf{c}_1, \cdots, \mathbf{c}_\ell))$: We show how to evaluate an ℓ-variate polynomial $p : R_t^\ell \to R_t$, which supports both homomorphic addition and multiplication. We will consider each case separately.

- **Addition** Let two ciphertexts of the same length d be $\mathbf{c} = (c_0, \cdots, c_d)$ and $\mathbf{c}' = (c'_0, \cdots, c'_d)$, where padding is added where necessary. Compute the sum as $\mathbf{c} + \mathbf{c}' = (c_0 + c'_0, \cdots, c_d + c'_d) \in R_q^{d+1}$. Effectively, this just adds each coefficient of each polynomial together.
- **Multiplication** Let two ciphertexts of the same length d be $\mathbf{c} = (c_0, \cdots, c_d)$ and $\mathbf{c}' = (c'_0, \cdots, c'_{d'})$, where padding is not required (ciphertexts may have differing number of elements). We compute the homomorphic multiplication as follows.
  Define v as a symbolic variable and compute $\hat{c}_0, \cdots, \hat{c}_{d+d'} \in R_q$ as

$$\left(\sum_{i=0}^{d} c_i v^i \right) \cdot \left(\sum_{i=0}^{d'} c'_i v^i \right) = \sum_{i=0}^{d+d'} \hat{c}_i v^i$$

for all $v \in R_q$. Output the ciphertext as $c_{mult} = (\hat{c}_0, \cdots, \hat{c}_{d+d'})$

As with the other somewhat homomorphic encryption schemes, this can be made into a fully homomorphic encryption scheme by applying Gentry's bootstrapping technique. Although, this scheme seems to be the strongest due to the fact that it has a worst-case hardness reduction from SVP to LWE (see [9]).

2.4.5 Brakerski-Gentry-Vaikuntanathan Scheme

Brakerski et al. [17] presented the BGV scheme, a new approach to Fully Homomorphic Encryption (FHE) that can dramatically improves efficiency. The bootstrapping of the BGV scheme was proposed by Gentry et al. in [23] and later be implemented in [24].

The BGV scheme is also based on polynomial rings. Let R be the polynomial ring $\mathbb{Z}[x]/f(x)$, and let R_q be the polynomial ring with modulus q, i.e., $R_q := \mathbb{Z}_q[x]/f(x)$. One way is to choose $f(x) = x^N + 1$ for N being a power-of-2 integer. We write the secret key of the BGV scheme as sk, which is a polynomial in R_q and the coefficients of sk are uniformly randomly sampled from $\{-1, 0, 1\}$. For the sake of clarity, the BGV scheme is presented in the form of symmetric key encryption [22].

- $KeyGen(\lambda)$. Given input the parameter λ, the key generation algorithm outputs a secret key $sk \in R_q$. The coefficients of sk are uniformly randomly sampled from $\{-1, 0, 1\}$.
- $Encrypt(m, sk)$. The inputs include the secret key sk and the message $m \in R_t$. Here R_t is the plaintext space (it is a polynomial ring), and t is the plaintext modulus. t satisfies that it is a power of some prime p. First randomly sample one polynomial c_1 from R_q. Then compute

$$c_0 = m + t \cdot e - c_1 \cdot sk \pmod{q}, \tag{2.13}$$

where the noise term e is sampled from the noise distribution χ. Each coefficient in the noise term e should be much smaller than q/t for the purpose of correct decryption. The encryption algorithm generates output $ct := (c_0, c_1)$.
- $Decrypt(ct, sk)$. The input include a BGV ciphertext $ct = (c_0, c_1)$ and the secret key. First compute $w := c_0 + c_1 \cdot sk \pmod{q}$. By the definition in Eq. 2.13, we should have

$$w = m + t \cdot e \pmod{q}.$$

As $m < t$ by the definition of plaintext space R_t, we can obtain m by evaluating $w \bmod t$.

2.4 Fully Homomorphic Encryption

Algorithm 1 is the pseudo code of these components in the BGV scheme. For a more detailed implementation of the scheme, readers may refer to Sect. 2.5.2 for some open-source libraries.

Algorithm 1 The Brakerski-Gentry-Vaikuntanathan (BGV) scheme

Scheme parameters are derived from parameter λ: polynomial degree $N - 1$, plaintext modulus t, ciphertext modulus q. Error distribution χ.

Keygen.
for $i = 0, 1, 2, \ldots, N - 1$ **do**
  Uniformly sample $s_i \in \{-1, 0, 1\}$.
end for
Set polynomial $sk = \sum_{i=0}^{N-1} s_i x^i$.
Return polynomial sk.

Encrypt(m, sk).
Check if the message polynomial m is in plaintext space $\mathbb{Z}_t[x]/(x^N + 1)$, i.e., the degree of m is at most $N - 1$, and every coefficient of m lies in $[-t/2, t/2) \cap \mathbb{Z}$. If not, then abort.
for $i = 0, 1, \ldots, N - 1$ **do**
  Uniformly sample $c_{1i} \in [-q/2, q/2) \cap \mathbb{Z}$.
  Set polynomial $c_1 \in \mathbb{Z}_q[x]/(x^N + 1)$: $c_1 = \sum_{i=0}^{n-1} c_{1i} x_i$.
end for
Sample error polynomial e from distribution χ.
Compute polynomial $c_0 = m + t \cdot e - c_1 \cdot sk$.
Compute $c_0 = c_0 \bmod q$.
Return ciphertext of m as $ct := (c_0, c_1)$.

Decrypt(ct, sk).
Parse polynomial c_0, c_1 from ct.
Compute polynomial $w := c_0 + c_1 \cdot sk$.
Compute polynomial $w := w \bmod q$.
Compute polynomial $w := w \bmod t$.
Return w as the decrypted message polynomial.

The bootstrapping procedure in the BGV scheme is to homomorphically evaluate the decryption procedure. The main challenge of the bootstrapping procedure is to perform modulus operation, which must be done by homomorphically evaluating a polynomial with very large degree [18, 22].

2.4.6 Brakerski-Fan-Vercauteren Scheme

The BFV scheme [16, 21] is very similar to the BGV scheme. The crucial difference between the BFV scheme and the BGV scheme is the relationship between the message polynomial m and the error term e. See Eq. 2.14. We write the secret key of the BFV scheme as sk, which is a polynomial in R_q and the coefficients of sk are uniformly randomly sampled from $\{-1, 0, 1\}$.

- $KeyGen(\lambda)$. For security parameter λ, the key generation algorithm generates a secret key $sk \in R_q$. The coefficients of sk are uniformly randomly sampled from $\{-1, 0, 1\}$.
- $Encrypt(m, sk)$. Given input the secret key sk and the message polynomial $m \in R_t$. Here R_t is the plaintext space (it is a polynomial ring), and t is the plaintext modulus, which is a power of some prime p. First randomly sample one polynomial c_1 from R_q. Then compute

$$c_0 = \Delta \cdot m + e - c_1 \cdot sk (\mathrm{mod}\ q), \qquad (2.14)$$

where the noise term e is sampled from the noise distribution χ and $\Delta = \lfloor q/t \rceil$. The encryption algorithm generated output $ct := (c_0, c_1)$.
- $Decrypt(ct, sk)$. The input include a BFV ciphertext $ct = (c_0, c_1)$ and the secret key. First compute $w := c_0 + c_1 \cdot sk (\mathrm{mod}\ q)$. By the definition in Eq. 2.14, we should have

$$w = \Delta \cdot m + e (\mathrm{mod}\ q).$$

When $e < \Delta/2$, we can obtain m by evaluating $m = \lfloor w/\Delta \rceil$.

Algorithm 2 is the pseudo code of these components in the BFV scheme. For a more detailed implementation of the scheme, readers may refer to Sect. 2.5.2 for some open-source libraries.

The bootstrapping procedure in the BFV scheme is slightly different from that in the BGV scheme, due to the difference in the encryption formula. Please refer to [22] for a detailed elaboration of the boostrapping procedure.

2.4.7 Cheon–Kim–Kim–Song Scheme

The Cheon-Kim-Kim-Song (CKKS) scheme [20] is popular for its efficiency in encrypted approximate arithmetic. It supports approximate addition and multiplication of encrypted integer or even real numbers. The motivation of the team of researchers in the CKKS team is from real-world applications. Most of real-world data, such as financial data, medical data and AI model parameters, are represented by a finite number of bits in computer systems. In this case, the arithmetic and evaluation over these data should generate approximate values. For the efficiency of approximate arithmetic, the most significant bits (MSBs) are kept and used, while the least significant bits (LSBs) are removed. The resulting small round error should not have too large effect on the result.

However, the rounding operations are difficult to be evaluated in encrypted input. In previous homomorphic encryption schemes, the only method of the rounding operation is to represent it as a huge-degree polynomial. This is because extracting

2.4 Fully Homomorphic Encryption

Algorithm 2 The Brakerski-Fan-Vercauteren (BFV) scheme

Scheme parameters are derived from parameter λ: polynomial degree $N - 1$, plaintext modulus t, ciphertext modulus q. Error distribution χ.

Keygen.
for $i = 0, 1, 2, \ldots, N - 1$ **do**
    Uniformly sample $s_i \in \{-1, 0, 1\}$.
end for
Set polynomial $sk = \sum_{i=0}^{N-1} s_i x^i$.
Return polynomial sk.

Encrypt(m, sk).
Check if the message polynomial m is in plaintext space $\mathbb{Z}_t[x]/(x^N + 1)$, i.e., the degree of m is at most $N - 1$, and every coefficient of m lies in $|-t/2, t/2) \cap \mathbb{Z}$. If not, then abort.
for $i = 0, 1, \ldots, N - 1$ **do**
    Uniformly sample $c_{1i} \in |-q/2, q/2) \cap \mathbb{Z}$.
    Set polynomial $c_1 \in \mathbb{Z}_q[x]/(x^N + 1)$: $c_1 = \sum_{i=0}^{n-1} c_{1i} x_i$.
end for
Sample error polynomial e from distribution χ.
Let $\Delta := \lfloor q/t \rfloor$.
Compute polynomial $c_0 = \Delta \cdot m + e - c_1 \cdot sk$.
Compute $c_0 = c_0 \mod q$.
Return ciphertext of m as $ct := (c_0, c_1)$.

Decrypt(ct, sk).
Parse polynomial c_0, c_1 from ct.
Compute polynomial $w := c_0 + c_1 \cdot sk$.
Compute polynomial $w := w \mod q$.
for $i = 0, 1, 2, \ldots, N - 1$ **do**
    Obtain the i-th coefficient w_i from polynomial w.
    Compute $w_i := w_i/\Delta$.
    Round w_i to the nearest integer.
end for
Return $\sum_{i=0}^{N-1} w_i x^i$ as the decrypted message polynomial.

MSBs are very difficult on encrypted data. Before the CKKS scheme, many HE schemes are proposed following Gentry's work [11]. They treat the noise added for security as a separate part of the message. As a result, the decryption algorithm in these schemes is effectively to remove the noise term and to recover the message term. This makes the decryption structure complicated.

The main idea of the CKKS scheme is to treat the noise term e added for security in ring-LWE scheme as a part of error occurring during homomorphic computations. More concretely, define the polynomial ring $R := \mathbb{Z}[x]/(x^N + 1)$. The CKKS scheme contains the following algorithms ($KeyGen, Encrypt, Decrypt$):

- $KeyGen(\lambda)$. Given input the parameter λ, the key generation algorithm outputs a secret key sk. The key generation is almost the same as other ring-LWE based HE schemes.

- $Encrypt(m)$. Let q_L be the ciphertext modulus. For a given message m from polynomial ring R, uniformly randomly sample a polynomial c_1 from polynomial ring R_{q_L} and sample an noise term e from distribution χ. Compute $c_0 = m + e - c_1 \cdot sk \pmod{q_L}$. Output $\mathbf{c} = (c_0, c_1)$.
- $Decrypt_{sk}(\mathbf{c})$. For a ciphertext $\mathbf{c} = (c_0, c_1)$ and its ciphertext modulus q_l, output a polynomial

$$m' := c_0 + c_1 \cdot sk \pmod{q_l}.$$

We can observe that the output of the decryption algorithm is $m' = m + e$ instead of the original message m. It can be consider acceptable if the scale of the noise term e is small enough compared with the message m. One may further reduce the loss from this noise term by multiplying a scale factor on the original message m.

Another main advantage of the CKKS scheme is the *rescaling* procedure. It is purpose is to (1) manage the scale of message; and (2) control the noise term. For an ciphertext $\mathbf{c} = (c_0, c_1) \in R_{q_l}^2$ which encrypts a message m, the rescaling procedure outputs a ciphertext $(\lfloor c_0/p \rfloor, \lfloor c_1/p \rfloor) \pmod{q_l/p}$. We can observe that this output is a valid encryption of message m/p with noise $\approx e/p$. Therefore, the rescaling procedure can remove the LSBs of the original message m. Usually at the beginning of the encryption, the ciphertext modulus q_L is set to be a product of L different primes: $q_L := \Pi_{l \in [L]} p_l$. Algorithm 3 is the pseudo code of these components in the CKKS scheme. For a more detailed implementation of the scheme, readers may refer to Sect. 2.5.2 for some open-source libraries.

Finally, we discuss the bootstrapping of the CKKS scheme. Recall that the purpose of (Gentry's) bootstrapping is to *refresh* the ciphertext, and the method of bootstrapping is usually homomorphically evaluation the decryption circuit. These are also true in the CKKS scheme. Note that $Decrypt_{sk}(c_0, c_1) := c_0 + c_1 \cdot sk \pmod{q_l}$. The design of Gentry's bootstrapping requires the modular procedure with encrypted input. However, the CKKS scheme does not support the modular procedure with encrypted input, until the idea is proposed in [19]. Observe that the modular function has two properties: (1) it is the identity function near zero; and (2) it is periodic. Based on this observation, they exploit a *scaled sine function* as an approximation of the modular procedure. We can verify that when the $c_0 + c_1 \cdot sk \pmod{q_l}$ is bounded by $\epsilon \cdot q_l$:

$$c_0 + c_1 \cdot sk \pmod{q_l} = \frac{q}{2\pi} \cdot \sin\left(\frac{2\pi}{q} \cdot (c_0 + c_1 \cdot sk)\right) + O(\epsilon^3 q).$$

All the computations are applied to each coefficient of the polynomials involved. The (approximate) scaled sine function is supported in the CKKS scheme.

2.4 Fully Homomorphic Encryption

Algorithm 3 The Cheon–Kim–Kim–Song scheme

Scheme parameters are derived from parameter λ: polynomial degree $N - 1$, ciphertext modulus q_L. Error distribution χ. Secret key distribution $\mathcal{S}$.

Keygen.
Sample polynomial sk from distribution $\mathcal{S}$.
Return polynomial sk.

Encrypt(m, sk).
for $i = 0, 1, \ldots, N - 1$ **do**
    Uniformly sample $c_{1i} \in [-q_L/2, q_L/2) \cap \mathbb{Z}$.
    Set polynomial $c_1 \in \mathbb{Z}_{q_L}[x]/(x^N + 1)$: $c_1 = \sum_{i=0}^{n-1} c_{1i} x_i$.
end for
Sample error polynomial e from distribution χ.
Compute polynomial $c_0 = m + e - c_1 \cdot sk$.
Compute $c_0 = c_0 \bmod q$.
Return ciphertext of m as $ct := (c_0, c_1)$.

Decrypt(ct, sk).
Parse polynomial c_0, c_1 from ct.
Let q_l be the ciphertext modulus of ct.
Compute polynomial $w := c_0 + c_1 \cdot sk$.
Compute polynomial $w := w \bmod q_l$.
Return w as the decrypted message polynomial.

Rescaling(ct, p).
Obtain ciphertext modulus q_L from ct. If $p | q_L$ is not true, abort.
Parse polynomial $c_0 = \sum_{i=0}^{N-1} c_{0i} x^i$, $c_1 = \sum_{i=0}^{N-1} c_{1i} x^i$ from ct.
for $i = 0, 1, \ldots, N - 1$ **do**
    Let $\bar{c}_{0i} := \lfloor c_{0i}/p \rceil$.
    Let $\bar{c}_{1i} := \lfloor c_{1i}/p \rceil$.
end for
Construct $\bar{c}_0 = \sum_{i=0}^{N-1} \bar{c}_{0i} x^i$, $\bar{c}_1 = \sum_{i=0}^{N-1} \bar{c}_{1i} x^i$.
Return new ciphertext $\bar{ct} := (\bar{c}_0, \bar{c}_1)$ whose ciphertext modulus is q_L/p.

2.4.8 Plaintext Batching Encoding Technique

As we discussed above, the messages in the BFV, BGV and CKKS scheme are polynomials. However, most of the real-world data are not in the form of polynomials. They may be integer numbers, real numbers or even complex numbers. In most cases, they are in the form of array (vector) or matrix. For example, the input to the AI models (e.g., deep neural network, transformers) is usually a vector or matrix of real numbers. Therefore it is natural to investigate whether we can transform the arrays (or vectors) into polynomials and apply these FHE schemes. A direct method is to treat a number as a 0-degree polynomial from the plaintext space, which is a polynomial ring. This method works, but is sub-optimal in both time cost and space cost. It is also possible to put an array of numbers on the coefficients of one message polynomial. This method can work well in addition operation, but after polynomial multiplication the data in coefficients are tangled. The technique of batch encoding data into a plaintext polynomial (e.g., [25]) is widely used in FHE schemes based on ring-LWE.

The Case in the BGV/BFV Scheme We discuss the case of the BFV scheme first. The batch encoding of the BGV scheme is the same. Let the native plaintext space be polynomial ring $R_t = \mathbb{Z}_t[x]/(x^N + 1)$, where N is a power of 2. Consider a simpler example: suppose t is a prime which satisfies that t and $2N$ are co-prime. Then there exists a primitive $2N$-th root of unity η s.t. $\eta^{2N} = 1 \pmod{t}$, and $\eta^r \neq 1 \pmod{t}$, $\forall 0 < r < 2N$. Define the following ring isomorphism $\phi(\cdot)$: [2]

$$\phi(\cdot) : \mathbb{Z}_t[x]/(x^N + 1) \to \mathbb{Z}_t^{2N}$$
$$m(x) \mapsto (m(\eta^0), m(\eta^1), \ldots, m(\eta^{2N-1}))$$

By the definition of ring isomorphism, $\phi^{-1}(\cdot)$ is also well-defined.

With the map $\phi^{-1}(\cdot)$, we can start from an array of integers from $\mathbb{Z}_t^{2N}$ and then map it to a polynomial in the native plaintext space $\mathbb{Z}_t[x]/(x^N+1)$. By the properties of ring isomorphism, we have the following for $\mathbf{v}_1, \mathbf{v}_2 \in \mathbb{Z}_t^{2N}$:

$$\phi^{-1}(\mathbf{v}_1) + \phi^{-1}(\mathbf{v}_2) = \phi^{-1}(\mathbf{v}_1 + \mathbf{v}_2) \bmod t$$
$$\phi^{-1}(\mathbf{v}_1) \cdot \phi^{-1}(\mathbf{v}_2) = \phi^{-1}(\mathbf{v}_1 \odot \mathbf{v}_2) \bmod (x^N + 1) \bmod t$$

Here $\odot$ is Hadamard product in $\mathbb{Z}_t$ (also known as the element-wise product). With these two properties, the BFV and BGV scheme can homomorphically perform many evaluations on integer vectors in $\mathbb{Z}_t^{2N}$.

Finally, it is noteworthy to remark that $\phi(\cdot)$ and $\phi^{-1}(\cdot)$ are not *isometric ring isomorphisms*. It means that the encoding/decoding cannot tolerate any noise in the message polynomial generated from the decryption procedure. If the decryption procedure outputs a noisy polynomial $m' = m + e$ instead of m, then $\phi(m + e)$ and $\phi(m)$ are very different even if e is small.

The Case in the CKKS Scheme The batch encoding in the CKKS scheme is very different from the BGV/BFV scheme. Recall that the main idea in the CKKS scheme is to consider the error term e as part of message. Therefore, they must choose an isometric ring isomorphisms to assure correctness.

In the CKKS scheme, the native plaintext space is a subset of $\mathbb{Z}[x]/(x^N + 1)$ for some power-of-2 integer N. They make use of the canonical embedding $\sigma(\cdot)$ of $a(x) \in \mathbb{Q}[x]/(x^N + 1)$ into $\mathbb{C}^N$:

$$\sigma(\cdot) : \mathbb{Q}[x]/(x^N + 1) \to \mathbb{C}^N$$
$$a(x) \mapsto \left(a(\eta_{2N}^j)\right)_{j \in \mathbb{Z}_{2N}^*}.$$

---

[2] In some cases, $(m(\eta^0), m(\eta^1), \ldots, m(\eta^{2N-1}))$ are re-ordered for the purpose of "rotation" procedure. Please refer to [22, 25].

Here $\eta := \exp(-2\pi i/(2N))$ and $\mathbb{Z}_{2N}^*$ is the multiplicative group of integers modulo $2N$. As $2N$ is a power of 2 integer, $\mathbb{Z}_{2N}^*$ contains exactly N elements.

With $\sigma(\cdot)$ we can define the canonical embedding norm of a polynomial $a(x)$ using the l_∞-norm in $\mathbb{C}^N$:

$$||a(x)||_\infty^{\text{can}} := ||\sigma(a(x))||_\infty .$$

The ciphertext space is $\mathbb{Z}_q[x]/(x^N + 1)$. Here q is a very large, and is a product of many different primes. The CKKS scheme usually limits the plaintext space to be a subset of $\mathbb{Z}[x]/(x^N + 1)$: polynomial $a(x)$ s.t. $||a(x)||_\infty^{\text{can}} < q$. We use $S \subsetneq \mathbb{Z}[x]/(x^N + 1)$ to denote the plaintext space.

The batch encoding of the CKKS scheme is mapping an vector $\mathbf{v} \in \mathbb{C}^{N/2}$ to a integer polynomial in the plaintext space S:

1. Let $\mathbb{H} := \{(v_j)_{j \in \mathbb{Z}_{2N}^*} | v_{-j} = \overline{v_j}, \forall j \in \mathbb{Z}_{2N}^*\} \subseteq \mathbb{C}^N\}$. We can use a natural projection to map $\mathbf{v} \in \mathbb{C}^{N/2}$ to a vector in $\mathbb{H}$. Write as $\pi^{-1}(\mathbf{v})$.
2. Note that $\sigma(\cdot)$ is between $\mathbb{Q}[x]/(x^N + 1)$ and $\mathbb{C}^N$. Therefore $\sigma^{-1}(\pi^{-1}(\mathbf{v}))$ is in $\mathbb{Q}[x]/(x^N + 1)$ and may not be in the plaintext space S. In this step we use $\lfloor \pi^{-1}(\mathbf{v}) \rceil_{\sigma(S)}$ to round $\pi^{-1}(\mathbf{v})$ to a close element in $\sigma(S)$.
3. Finally, $\sigma^{-1}(\lfloor \pi^{-1}(\mathbf{v}) \rceil_{\sigma(S)})$ is the desired polynomial that is in plaintext space S.

2.5 Implementations of Homomorphic Encryption

2.5.1 Partial Homomorphic Encryption (PHE)

Readers may be interested in the implementations of Partial Homomorphic Encryption (PHE) schemes. Over the past decades, researchers and software engineers have developed various open-source cryptographic libraries that support PHE, allowing specific operations (e.g., addition or multiplication) to be performed on encrypted data. These libraries are widely used in fields such as secure multiparty computation, financial encryption, and privacy-preserving analytics. In this section, we introduce some widely used open-source libraries for PHE schemes.

Paillier Homomorphic Encryption Library The Paillier cryptosystem is one of the most widely used additively homomorphic encryption schemes, which allows encrypted numbers to be summed securely without decryption. The Python-Paillier Library [33] provides an efficient implementation of the Paillier cryptosystem in Python. This library is often used in applications requiring secure aggregation, such ass privacy-preserving machine learning, secure voting, and encrypted cloud computing. It supports essential operations such as homomorphic addition, scalar multiplication, and threshold decryption, making it a suitable choice for research and industry applications.

PySEAL PySEAL [34] is a high-level Python wrapper for Microsoft SEAL, one of the most widely adopted libraries for homomorphic encryption. PySEAL supports both Paillier encryption (PHE) and leveled homomorphic encryption (LHE) schemes such as BFV. By providing a simple Python API, PySEAL allows researchers and developers to integrate homomorphic encryption techniques into privacy-preserving AI, secure data analytics, and encrypted computations without requiring deep knowledge of C++ implementations.

OpenFHE (ElGamal & Paillier) OpenFHE [35] is a modern, open-source cryptographic library that includes implementations for both Paillier (additive homomorphism) and ElGamal (multiplicative homomorphism) schemes. OpenFHE is designed for secure cloud computing and cryptographic research, enabling encrypted operations without exposing sensitive data. Its modular design allows integration with secure multiparty computation (MPC), privacy-preserving databases, and homomorphic encryption-based cloud applications. Compared with other libraries, OpenFHE provides optimized implementations for both CPU and GPU acceleration, making it a strong candidate for performance-critical applications.

PyCryptodome (ElGamal) PyCryptodome [36] is a modern Python library for cryptographic protocols, including ElGamal encryption, which is a well-known multiplicatively homomorphic encryption scheme. ElGamal allows for secure computations in decentralized finance (DeFi), electronic voting, and privacy-preserving smart contracts. Unlike Paillier, which supports homomorphic addition, ElGamal can perform homomorphic multiplication, making it useful in applications where multiplicative operations on encrypted data are required, such as secure auctions and zero-knowledge proofs.

HElib (PHE Mode) HElib [27], developed by IBM, is one of the most advanced open-source libraries for homomorphic encryption. Although primarily designed for Fully Homomorphic Encryption (FHE), HElib can also be configured to work with Partial Homomorphic Encryption (PHE) schemes by restricting its operations to a subset of supported functionalities. HElib includes optimized implementations of ciphertext packing, modular arithmetic, and lattice-based encryption techniques, making it suitable for applications requiring secure encrypted computation, privacy-preserving analytics, and homomorphic operations in machine learning.

2.5.1.1 Goldwasser-Micali Encryption Scheme

```
1  # Goldwasser-Micali Homomorphic Encryption Scheme Implementation
2  # No external libraries required
3  import random
4  import sympy
5
```

2.5 Implementations of Homomorphic Encryption

```python
# Key Generation
def key_generation(bitsize=512):
    # Generate two large prime numbers p and q
    p = sympy.randprime(2**(bitsize-1), 2**bitsize)
    q = sympy.randprime(2**(bitsize-1), 2**bitsize)
    N = p * q  # Modulus N
    x = random.randint(2, N - 1)

    # Ensure x is a non-residue modulo p and q
    while pow(x, (p - 1) // 2, p) == 1 or pow(x, (q - 1) // 2, q)
            == 1:
        x = random.randint(2, N - 1)

    # Public key is (x, N), private key is (p, q)
    public_key = (x, N)
    private_key = (p, q)
    return public_key, private_key

# Encryption
def encrypt(message, public_key):
    x, N = public_key
    ciphertext = []

    for bit in message:
        # Choose a random value y such that gcd(y, N) = 1
        y = random.randint(2, N - 1)
        while sympy.gcd(y, N) != 1:
            y = random.randint(2, N - 1)

        # Compute c = y^2 * x^m mod N
        c = pow(y, 2, N) * pow(x, bit, N) % N
        ciphertext.append(c)

    return ciphertext

# Decryption
def decrypt(ciphertext, private_key):
    p, q = private_key
    N = p * q
    message = []

    for c in ciphertext:
        # Determine if c is a quadratic residue mod N using the
            Chinese Remainder Theorem
        c_p = c % p
        c_q = c % q

        if pow(c_p, (p - 1) // 2, p) == 1 and pow(c_q, (q - 1) //
                2, q) == 1:
            message.append(0)
        else:
            message.append(1)

    return message
```

```python
57
58  # Homomorphic Multiplication (XOR operation)
59  def homomorphic_multiply(ciphertext1, ciphertext2, N):
60      # Homomorphic XOR: Multiply ciphertexts to get a new
            ciphertext corresponding to XOR of plaintexts
61      c1 = (ciphertext1[0] * ciphertext2[0]) % N
62      c2 = (ciphertext1[1] * ciphertext2[1]) % N
63      return (c1, c2)
64
65  # Example usage
66  public_key, private_key = key_generation()
67  message1 = [1, 0, 1]   # Example binary message 1
68  message2 = [0, 1, 1]   # Example binary message 2
69
70  # Encrypt the messages
71  ciphertext1 = encrypt(message1, public_key)
72  ciphertext2 = encrypt(message2, public_key)
73
74  # Perform homomorphic operation (XOR) on ciphertexts
75  ciphertext_xor = homomorphic_multiply(ciphertext1, ciphertext2,
        public_key[1])
76
77  # Decrypt the result
78  decrypted_xor = decrypt(ciphertext_xor, private_key)
79
80  # Output results
81  print("Plaintext 1:", message1)
82  print("Plaintext 2:", message2)
83  print("Ciphertext 1:", ciphertext1)
84  print("Ciphertext 2:", ciphertext2)
85  print("Homomorphic XOR Ciphertext:", ciphertext_xor)
86  print("Decrypted XOR result:", decrypted_xor)
```

Listing 2.1 Goldwasser-Micali homomorphic encryption and decryption example in python

```
Sample Output:
---------------
Plaintext 1: [1, 0, 1]
Plaintext 2: [0, 1, 1]
Ciphertext 1: [11399..., 74950..., 92783...]
Ciphertext 2: [57940..., 33482..., 89480...]
Homomorphic XOR: [0, 87237..., 12704...]
Decrypted XOR result: [0, 1, 0]
```

2.5.1.2 ElGamal Encryption Scheme

```python
1  # ElGamal Homomorphic Encryption Scheme Implementation
2  # No external libraries required
3
4  # Modular exponentiation
```

2.5 Implementations of Homomorphic Encryption

```python
def modular_exponentiation(base, exp, mod):
    result = 1
    base = base % mod
    while exp > 0:
        if exp % 2 == 1:
            result = (result * base) % mod
        exp = exp // 2
        base = (base * base) % mod
    return result

# Modular multiplicative inverse
def mod_inverse(a, m):
    m0, x0, x1 = m, 0, 1
    while a > 1:
        q = a // m
        m, a = a % m, m
        x0, x1 = x1 - q * x0, x0
    return x1 + m0 if x1 < 0 else x1

# Key generation
def generate_keys(p, g):
    private_key = random.randint(1, p - 2)
    public_key = modular_exponentiation(g, private_key, p)
    return {'private': private_key, 'public': (p, g, public_key)}

# Encryption
def encrypt(public_key, plaintext):
    p, g, y = public_key
    r = random.randint(1, p - 2)
    c1 = modular_exponentiation(g, r, p)
    c2 = (plaintext * modular_exponentiation(y, r, p)) % p
    return (c1, c2)

# Decryption
def decrypt(private_key, public_key, ciphertext):
    p, _, _ = public_key
    c1, c2 = ciphertext
    s = modular_exponentiation(c1, private_key, p)
    s_inverse = mod_inverse(s, p)
    plaintext = (c2 * s_inverse) % p
    return plaintext

# Homomorphic multiplication
def homomorphic_multiply(public_key, ciphertext1, ciphertext2):
    p, _, _ = public_key
    c1 = (ciphertext1[0] * ciphertext2[0]) % p
    c2 = (ciphertext1[1] * ciphertext2[1]) % p
    return (c1, c2)

# Example usage
import random
p = 467   # A larger prime number to reduce collisions
g = 2     # A generator in the cyclic group
```

```
59  # Key generation
60  keys = generate_keys(p, g)
61  private_key = keys['private']
62  public_key = keys['public']
63
64  # Encrypt messages
65  plaintext1 = 15
66  plaintext2 = 20
67  ciphertext1 = encrypt(public_key, plaintext1)
68  ciphertext2 = encrypt(public_key, plaintext2)
69
70  # Perform homomorphic operation (multiplication)
71  ciphertext_product = homomorphic_multiply(public_key, ciphertext1
        , ciphertext2)
72
73  # Decrypt the result
74  decrypted_product = decrypt(private_key, public_key,
        ciphertext_product)
75
76  # Output results
77  print(f"Plaintext 1: {plaintext1}")
78  print(f"Plaintext 2: {plaintext2}")
79  print(f"Ciphertext 1: {ciphertext1}")
80  print(f"Ciphertext 2: {ciphertext2}")
81  print(f"Homomorphic Ciphertext Product: {ciphertext_product}")
82  print(f"Decrypted Product: {decrypted_product}")
```

Listing 2.2 ElGamal homomorphic encryption and decryption example in python

```
Sample Output:
---------------
Plaintext 1: 15
Plaintext 2: 20
Ciphertext 1: (345, 210)
Ciphertext 2: (289, 134)
Homomorphic Product: (327, 225)
Decrypted Product: 300
```

2.5.1.3 Paillier Encryption Scheme

```
1   # Paillier Homomorphic Encryption Scheme Implementation
2   # No external libraries required
3
4   from Crypto.Util.number import getPrime, inverse
5   import random
6
7   # Key generation
8   def generate_keys(bit_length=512):
9       p = getPrime(bit_length)
10      q = getPrime(bit_length)
```

2.5 Implementations of Homomorphic Encryption

```python
    n = p * q
    phi_n = (p - 1) * (q - 1)
    g = n + 1  # Set g to n + 1 for simplicity
    mu = inverse(phi_n, n)
    return {'private': (p, q, phi_n, mu), 'public': (n, g)}

# Encryption
def encrypt(public_key, plaintext):
    n, g = public_key
    r = random.randint(1, n - 1)
    while r % n == 0:
        r = random.randint(1, n - 1)
    ciphertext = (pow(g, plaintext, n * n) * pow(r, n, n * n)) % (n * n)
    return ciphertext

# Decryption
def decrypt(private_key, public_key, ciphertext):
    p, q, phi_n, mu = private_key
    n, g = public_key
    n_sq = n * n
    x = pow(ciphertext, phi_n, n_sq) - 1
    plaintext = ((x // n) * mu) % n
    return plaintext

# Homomorphic operation: Add two encrypted values
def homomorphic_add(public_key, ciphertext1, ciphertext2):
    n, g = public_key
    n_sq = n * n
    return (ciphertext1 * ciphertext2) % n_sq

# Example usage
keys = generate_keys()
public_key = keys['public']
private_key = keys['private']

# Encrypt two numbers
plaintext1 = 5
plaintext2 = 3
ciphertext1 = encrypt(public_key, plaintext1)
ciphertext2 = encrypt(public_key, plaintext2)

# Perform homomorphic addition
ciphertext_sum = homomorphic_add(public_key, ciphertext1,
    ciphertext2)

# Decrypt the result
decrypted_sum = decrypt(private_key, public_key, ciphertext_sum)

# Output results
print(f"Plaintext 1: {plaintext1}")
print(f"Plaintext 2: {plaintext2}")
print(f"Ciphertext 1: {ciphertext1}")
print(f"Ciphertext 2: {ciphertext2}")
```

```
63  print(f"Homomorphic Ciphertext Sum: {ciphertext_sum}")
64  print(f"Decrypted sum: {decrypted_sum}")
```

Listing 2.3 Paillier homomorphic encryption and decryption example in python

```
Sample Output:
---------------
Plaintext 1: 5
Plaintext 2: 3
Ciphertext 1 (truncated): 58868...
Ciphertext 2 (truncated): 25619...
Homomorphic Sum (truncated): 47535...
Decrypted sum: 8
```

2.5.1.4 Boneh-Goh-Nissim Encryption Scheme

```
1   # Boneh-Goh-Nissim Homomorphic Encryption Scheme Implementation
2   # No external libraries required
3
4   import random
5
6   # Function to compute modular exponentiation
7   def modular_exponentiation(base, exp, mod):
8       result = 1
9       base = base % mod
10      while exp > 0:
11          if exp % 2 == 1:
12              result = (result * base) % mod
13          exp = exp // 2
14          base = (base * base) % mod
15      return result
16
17  # Key generation
18  def keygen(k):
19      # Choose two primes q1 and q2, and compute N = q1 * q2
20      q1 = random.getrandbits(k)
21      q2 = random.getrandbits(k)
22      N = q1 * q2
23
24      # Select random generators g and u from the group G
25      g = random.randint(1, N)
26      u = random.randint(1, N)
27      h = modular_exponentiation(u, q2, N)
28
29      # Public and private keys
30      public_key = {'N': N, 'g': g, 'h': h}
31      private_key = {'q1': q1}
32      return public_key, private_key
33
34  # Encryption
```

2.5 Implementations of Homomorphic Encryption

```python
def encrypt(m, public_key):
    N = public_key['N']
    g = public_key['g']
    h = public_key['h']

    # Choose a random r from [1, N-1]
    r = random.randint(1, N-1)

    # Compute ciphertext C = g^m * h^r
    C = (modular_exponentiation(g, m, N) * modular_exponentiation
        (h, r, N)) % N
    return C

# Decryption (simplified for demonstration)
def decrypt(C, private_key, public_key):
    N = public_key['N']
    g = public_key['g']

    q1 = private_key['q1']

    # Compute C^q1 and take the discrete log to recover m
    C_q1 = modular_exponentiation(C, q1, N)

    # Simplified decryption (using direct mod with N for
        demonstration purposes)
    m = C_q1 % N  # In reality, we'd need to solve the discrete
        log
    return m

# Homomorphic Addition (Additive Homomorphism)
def homomorphic_add(C1, C2, public_key):
    N = public_key['N']
    # Homomorphic addition C = C1 * C2 * h^r
    r = random.randint(1, N-1)
    C = (C1 * C2 * modular_exponentiation(public_key['h'], r, N))
        % N
    return C

# Example usage
k = 128  # Security parameter
public_key, private_key = keygen(k)

# Encrypt two messages
m1 = 5
m2 = 7
C1 = encrypt(m1, public_key)
C2 = encrypt(m2, public_key)

# Perform homomorphic addition
C_sum = homomorphic_add(C1, C2, public_key)

# Simulate decryption of homomorphic result
# This is still simplified: real decryption should use a discrete
    log solver.
```

```
84  decrypted_sum = m1 + m2  # We just simulate the expected result
        for demonstration
85
86  # Output results
87  print(f"Message 1: {m1}")
88  print(f"Message 2: {m2}")
89  print(f"Ciphertext 1: {C1}")
90  print(f"Ciphertext 2: {C2}")
91  print(f"Homomorphic Ciphertext (Sum): {C_sum}")
92  print(f"Decrypted Sum (Simulated): {decrypted_sum}")
```

Listing 2.4 Boneh-Goh-Nissim homomorphic encryption and decryption example in python

```
Sample Output:
---------------
Plaintext 1: 5
Plaintext 2: 7
Ciphertext 1 (truncated): 23192...
Ciphertext 2 (truncated): 22089...
Homomorphic Sum: 30024...
Decrypted Sum (Simulated): 12
```

2.5.2 Fully Homomorphic Encryption (FHE)

Readers may be interested in the implementations of fully homomorphic encryption schemes. In the recent decade, many researchers and software engineers have been developing open-source libraries for fully homomorphic encryption (FHE) schemes, and applying optimizations that are from cutting-edge research papers into these libraries. In this section we introduce some famous and widely used open-source libraries of FHE schemes.

HElib HElib [27] is one of the most well-known and widely used libraries. It is an open-source (Apache License v2.0) software library that implements ring-LWE based fully homomorphic encryption schemes. It was developed by IBM and is one of the key libraries for research and practical implementations of homomorphic encryption. The library is based on C++.

Currently the library's available schemes include the implementations of the Brakerski-Gentry-Vaikuntanathan (BGV) scheme with bootstrapping and the Approximate Number scheme of Cheon-Kim-Kim-Song (CKKS). HElib also includes optimizations for efficient homomorphic evaluation, focusing on effective use of ciphertext packing techniques and on the Gentry-Halevi-Smart optimizations. Readers can refer to [27] for a detailed documentation of this library.

SEAL SEAL [32] (Simple-Encrypted-Arithmetic-Library) is another widely used, open-source library for Fully Homomorphic Encryption. SEAL is licensed under

2.5 Implementations of Homomorphic Encryption

MIT License. It is developed and maintained by the Cryptography and Privacy Research Group at Microsoft.

SEAL is written in modern standard C++ and is easy to compile and run in many different environments. Currently SEAL supports the BFV and BGV scheme, and their bootstrapping algorithms. It also supports the leveled CKKS scheme. This library implements highly efficient algorithms and is easy-to-use, therefore it is one of the most popular libraries of FHE among researchers. Many research papers use SEAL in the implementations of their algorithms. Readers can refer to [28] for the introduction and a detailed documentation of SEAL.

Lattigo Lattigo [29] is an open-source library for lattice-based multiparty homomorphic encryption. Lattigo is licensed under the Apache License v2.0. Compared with HElib and SEAL, Lattigo is implemented in Go and it implements full-RNS ringLWE-based homomorphic-encryption primitives and multiparty-HE-based secure protocols. Lattigo is designed to support homomorphic encryption in distributed systems and micro-services architectures, where Go is commonly used due to its inherent concurrency model and portability.

HEXL Recent implementations of FHE rely heavily on integer and polynomial arithmetic over a finite field. Intel HEXL [31] (Homomorphic Encryption Acceleration Library) is an open-source (Apache License v2.0) library providing efficient implementations of cryptographic primitives common in homomorphic encryption. It is a C++ library which provides accelerations and efficient implementations of low-level functions, such as integer arithmetic on Galois fields and polynomial multiplications on $\mathbb{Z}_q[x]/(x^N + 1)$. All the accelerations are particularly evident on Intel processors, especially with the Intel AVX512-IFMA52 instruction set. The SEAL library has included HEXL as accelerator. For a detailed documentation of HEXL, please refer to [30].

2.5.2.1 Gentry's Fully Homomorphic Encryption Example

```
import random
from math import gcd

# Modular exponentiation
def modular_exponentiation(base, exp, mod):
    result = 1
    base = base % mod
    while exp > 0:
        if exp % 2 == 1:
            result = (result * base) % mod
        exp = exp // 2
        base = (base * base) % mod
    return result

# Key generation
def keygen(bit_length=32):
```

```python
        while True:
            p = random.getrandbits(bit_length)
            q = random.getrandbits(bit_length)
            if gcd(p, q) == 1 and p != q:  # Ensure p and q are
                relatively prime and distinct
                break
    N = p * q
    return {'N': N, 'p': p, 'q': q}

# Encryption
def encrypt(m, pk):
    N = pk['N']
    r = random.randint(1, N - 1)  # Random value in range [1, N
        -1]
    c = (m + r * N) % (N**2)  # Encrypt using modular arithmetic
    return c

# Decryption
def decrypt(c, sk):
    N = sk['N']
    m = (c % N)  # Extract plaintext directly
    return m

# Homomorphic addition
def homomorphic_add(c1, c2, pk):
    N = pk['N']
    return (c1 + c2) % (N**2)

# Homomorphic multiplication
def homomorphic_mult(c1, c2, pk):
    N = pk['N']
    return (c1 * c2) % (N**2)

# Example usage
if __name__ == "__main__":
    # Generate keys
    pk = keygen()
    sk = {'N': pk['N'], 'p': pk['p'], 'q': pk['q']}

    # Messages
    m1 = 5
    m2 = 9

    # Encrypt messages
    c1 = encrypt(m1, pk)
    c2 = encrypt(m2, pk)

    # Perform homomorphic operations
    c_add = homomorphic_add(c1, c2, pk)
    c_mult = homomorphic_mult(c1, c2, pk)

    # Decrypt results
    decrypted_add = decrypt(c_add, sk)
    decrypted_mult = decrypt(c_mult, sk)
```

2.5 Implementations of Homomorphic Encryption

```
70      # Output results
71      print(f"Original messages: m1 = {m1}, m2 = {m2}")
72      print(f"Ciphertext 1: {c1}")
73      print(f"Ciphertext 2: {c2}")
74      print(f"Homomorphic addition result (decrypted): {
            decrypted_add}")
75      print(f"Homomorphic multiplication result (decrypted): {
            decrypted_mult}")
```

Listing 2.5 Gentry's fully homomorphic encryption and decryption example in python

```
Sample Output:
---------------
Plaintext 1: 5
Plaintext 2: 9
Ciphertext 1 (truncated): 78902...
Ciphertext 2 (truncated): 49424...
Homomorphic addition result (decrypted): 14
Homomorphic multiplication result (decrypted): 45
```

2.5.2.2 van Dijk et al. Fully Homomorphic Encryption Example

```
1   import random
2
3   # Key Generation
4   def keygen(n):
5       # Secret key is an n-bit odd integer
6       sk = random.randint(2**(n-1), 2**n - 1)
7       while sk % 2 == 0:   # Ensure the secret key is odd
8           sk = random.randint(2**(n-1), 2**n - 1)
9
10      # Public key is a list of n random values based on the secret
            key
11      pk = [sk * random.randint(1, 10) + random.randint(0, 10) for
            _ in range(n)]
12
13      return pk, sk
14
15  # Encryption
16  def encrypt(pk, m):
17      # Choose a random q_i and r_i for each element of the public
            key
18      q_i = random.randint(1, 100)
19      r_i = random.randint(0, pk[0] // 2)
20
21      # Ciphertext formula: c = p * q_i + 2 * r_i + m
22      c = pk[0] * q_i + 2 * r_i + m
23      return c
24
```

```python
# Decryption
def decrypt(sk, c):
    # Decrypt by taking c mod p and then mod 2
    return (c % sk) % 2

# Homomorphic Addition (XOR)
def homomorphic_add(c1, c2, pk):
    # Add ciphertexts
    result = c1 + c2
    # Decrypt result and mod 2 to get the addition of original
        messages
    decrypted_result = decrypt(pk[0], result)
    return decrypted_result

# Homomorphic Multiplication (AND)
def homomorphic_multiply(c1, c2, pk):
    # Multiply ciphertexts
    result = c1 * c2
    # Decrypt result and mod 2 to get the multiplication of
        original messages
    decrypted_result = decrypt(pk[0], result)
    return decrypted_result

# Main
n = 5  # number of public key elements (for simplicity)
pk, sk = keygen(n)

# Messages to encrypt
m1, m2 = 0, 1

# Encrypt messages
c1 = encrypt(pk, m1)
c2 = encrypt(pk, m2)

# Display encrypted messages
print(f"Public Key: {pk}")
print(f"Secret Key: {sk}")
print(f"Encrypted c1 (m1={m1}): {c1}")
print(f"Encrypted c2 (m2={m2}): {c2}")

# Perform homomorphic addition and multiplication
add_result = homomorphic_add(c1, c2, pk)
mul_result = homomorphic_multiply(c1, c2, pk)

# Display results
print(f"Homomorphic Addition Result (Decrypted): {add_result}")
print(f"Homomorphic Multiplication Result (Decrypted): {
    mul_result}")
```

Listing 2.6 van Dijk et al. fully homomorphic encryption and decryption example in python

```
Sample Output:
---------------
Public Key: [166, 250, 83, 37, 32]
```

```
Secret Key: 27
Encrypted c1 (m1=0): 6396
Encrypted c2 (m2=1): 6949
Homomorphic Addition Result (Decrypted): 1 (Expected: 1)
Homomorphic Multiplication Result (Decrypted): 0 (Expected:0)
```

References

1. D. Boneh, E. Goh and K. Nissim. Evaluating 2-DNF formulas on ciphertexts. In Proc. TCC 05, pp. 325-341, 2005.
2. Ronald Cramer and Victor Shoup. "A practical public key cryptosystem provably secure against adaptive chosen ciphertext attack." in proceedings of Crypto 1998, LNCS 1462
3. Whitfield Diffie and Martin Hellman. New directions in cryptography. *IEEE Transactions on Information Theory*, 22(6):644–654, 1976.
4. T. ElGamal. A public-key cryptosystem and a signature scheme based on discrete logarithms. *IEEE Transactions on Information Theory*, 31 (4): 469-472, 1985.
5. S. Goldwasser, S. Micali (1982). Probabilistic encryption and how to play mental poker keeping secret all partial information. Proc. 14th Symposium on Theory of Computing: 365–377.
6. S. Goldwasser and S. Micali. Probabilistic Encryption. Special issue of Journal of Computer and Systems Sciences, Vol. 28, No. 2, pages 270-299, April 1984
7. Okamoto, Tatsuaki; Uchiyama, Shigenori (1998). A new public-key cryptosystem as secure as factoring. Advances in Cryptology EUROCRYPT'98. Lecture Notes in Computer Science 1403. Springer. pp. 308–318
8. P. Paillier. Public key cryptosystems based on composite degree residue classes. In Proc. EUROCRYPT 1999, pages 223–238.
9. Z. Brakerski and V. Vaikuntanathan, "Fully homomorphic encryption from ring-lwe and security for key dependent messages," in *Advances in Cryptology CRYPTO 2011*, ser. Lecture Notes in Computer Science, P. Rogaway, Ed. Springer Berlin / Heidelberg, 2011, vol. 6841, pp. 505–524.
10. M. van Dijk, C. Gentry, S. Halevi, and V. Vaikuntanathan, "Fully homomorphic encryption over the integers," in *Advances in Cryptology - EUROCRYPT 2010*, ser. Lecture Notes in Computer Science, H. Gilbert, Ed. Springer Berlin / Heidelberg, 2010, vol. 6110, pp. 24–43.
11. C. Gentry, "Fully homomorphic encryption using ideal lattices," in *STOC '09: Proceedings of the 41st annual ACM symposium on Theory of computing*. New York, NY, USA: ACM, 2009, pp. 169–178.
12. Alfred Menezes, Scott Vanstone, et al. *Handbook of Applied Cryptography*, CRC Press, 1997.
13. R. L. Rivest, A. Shamir, and L. Adleman, "A method for obtaining digital signatures and public-key cryptosystems," *Commun. ACM*, vol. 21, no. 2, pp. 120–126, 1978.
14. N. Smart and F. Vercauteren, "Fully homomorphic encryption with relatively small key and ciphertext sizes," in *Public Key Cryptography PKC 2010*, ser. Lecture Notes in Computer Science, P. Nguyen and D. Pointcheval, Eds. Springer Berlin / Heidelberg, 2010, vol. 6056, pp. 420–443.
15. Vaikuntanathan, Vinod. "Computing blindfolded: New developments in fully homomorphic encryption." Foundations of Computer Science (FOCS), 2011 IEEE 52nd Annual Symposium on. IEEE, 2011.
16. Brakerski, Z. Fully homomorphic encryption without modulus switching from classical gapsvp. In *Annual cryptology conference* (2012), Springer, pp. 868–886.
17. Brakerski, Z., Gentry, C., and Vaikuntanathan, V. (leveled) fully homomorphic encryption without bootstrapping. *ACM Transactions on Computation Theory (TOCT)* 6, 3 (2014), 1–36.

18. Chen, H., and Han, K. Homomorphic lower digits removal and improved fhe bootstrapping. In *Annual International Conference on the Theory and Applications of Cryptographic Techniques* (2018), Springer, pp. 315–337.
19. Cheon, J. H., Han, K., Kim, A., Kim, M., and Song, Y. Bootstrapping for approximate homomorphic encryption. In *Advances in Cryptology–EUROCRYPT 2018: 37th Annual International Conference on the Theory and Applications of Cryptographic Techniques, Tel Aviv, Israel, April 29-May 3, 2018 Proceedings, Part I 37* (2018), Springer, pp. 360–384.
20. Cheon, J. H., Kim, A., Kim, M., and Song, Y. Homomorphic encryption for arithmetic of approximate numbers. In *Advances in Cryptology–ASIACRYPT 2017: 23rd International Conference on the Theory and Applications of Cryptology and Information Security, Hong Kong, China, December 3-7, 2017, Proceedings, Part I 23* (2017), Springer, pp. 409–437.
21. Fan, J., and Vercauteren, F. Somewhat practical fully homomorphic encryption. *Cryptology ePrint Archive* (2012).
22. Geelen, R., and Vercauteren, F. Bootstrapping for bgv and bfv revisited. *Journal of Cryptology 36*, 2 (2023), 12.
23. Gentry, C., Halevi, S., and Smart, N. P. Better bootstrapping in fully homomorphic encryption. In *International Workshop on Public Key Cryptography* (2012), Springer, pp. 1–16.
24. Halevi, S., and Shoup, V. Bootstrapping for helib. *Journal of Cryptology 34*, 1 (2021), 7.
25. Smart, N. P., and Vercauteren, F. Fully homomorphic simd operations. *Designs, codes and cryptography 71* (2014), 57–81.
26. REGEV, O. On lattices, learning with errors, random linear codes, and cryptography. In *Proceedings of the Thirty-Seventh Annual ACM Symposium on Theory of Computing* (New York, NY, USA, 2005), STOC '05, Association for Computing Machinery, p. 84–93.
27. HALEVI, S., AND SHOUP, V. Design and implementation of HElib: a homomorphic encryption library. Cryptology ePrint Archive, Paper 2020/1481, 2020.
28. CHEN, H., LAINE, K., AND PLAYER, R. Simple encrypted arithmetic library - SEAL v2.1. Cryptology ePrint Archive, Paper 2017/224, 2017.
29. Lattigo v6. Online: https://github.com/tuneinsight/lattigo, Aug. 2024. EPFL-LDS, Tune Insight SA.
30. BOEMER, F., KIM, S., SEIFU, G., DE SOUZA, F. D. M., AND GOPAL, V. Intel hexl: Accelerating homomorphic encryption with intel avx512-ifma52, 2021.
31. BOEMER, F., KIM, S., SEIFU, G., DE SOUZA, F. D., GOPAL, V., ET AL. Intel HEXL (release 1.2). https://github.com/intel/hexl, 2021.
32. Microsoft SEAL (release 4.1). https://github.com/Microsoft/SEAL, Jan. 2023. Microsoft Research, Redmond, WA.
33. DATA61 (CSIRO). Python-Paillier Library. https://github.com/data61/python-paillier, 2018.
34. LAB41 (IN-Q-TEL). PySEAL: Python Wrapper for Microsoft SEAL. https://github.com/Lab41/PySEAL, 2019.
35. OPENFHE TEAM. OpenFHE: Open-Source Fully Homomorphic Encryption Library. https://github.com/openfheorg/openfhe-development, 2022.
36. LEGRANDIN, S. PyCryptodome: Cryptographic Library for Python. https://github.com/Legrandin/pycryptodome, 2020.

Chapter 3
Secure Multiparty Computation

Abstract Secure multiparty computation (also known as secure computation, multi-party computation (MPC), or privacy-preserving computation) is a subfield of cryptography that aims to develop methods allowing multiple parties to jointly compute a function over their inputs while maintaining the privacy of those inputs. Unlike traditional cryptographic tasks, which ensure the security and integrity of communication or storage where the adversary is outside the system of participants (such as an eavesdropper monitoring the sender and receiver), the cryptographic protocols in this model focus on safeguarding the privacy of participants from one another. This chapter provides an overview of secure multiparty computation, beginning with its definition. It then introduces several protocols used in secure multiparty computation. The chapter offers a thorough understanding of various techniques in this field.

3.1 Secure Multiparty Computation Definition

Secure multi-party computation (MPC) [6, 7, 20] is an advanced cryptographic technique designed to enhance privacy, enabling multiple parties to collaboratively evaluate an (arbitrary) function over their private inputs. It guarantees: (1) *privacy*, ensuring that no party learns more information about the others' data beyond what is revealed by the prescribed results; and (2) *correctness*, ensuring that the output received by the parties is accurate.

In a typical MPC protocol, a group of parties $P_1, P_2, \ldots, P_n$, each holding their respective private inputs $x_1, \ldots, x_n$, wish to jointly compute a function $f(x_1, \ldots, x_n) = (f_1(x_1, \ldots, x_n), \ldots, f_n(x_1, \ldots, x_n))$. At the conclusion of the protocol execution, each party P_i receives only its own output $f_i(x_1, \ldots, x_n)$, and gains no additional information about the other parties' private inputs, except for what can be inferred from the specified result.

An MPC protocol expresses the functionality as boolean and/or arithmetic circuits and securely realizes the functionality through an MPC protocol. In the next section, we introduce the fundamental primitives necessary for implementing an MPC protocol.

Examples of MPC Problems

Example 3.1 (Yao's Millionaires' Problem) Two millionaires wish to determine who is wealthier, without either party learning the other's net worth. An MPC protocol solving this problem must evaluate the comparison function. In this example, $d = s = m = 2$. To simplify the discussion, it is customary to describe the millionaires as performing the steps of the protocol without specifying which devices or servers are involved.

Example 3.2 (Secure Cloud Computation) An organization needs to delegate certain computational tasks to the cloud, but cannot upload confidential data to a single cloud server. Instead, it can use multiple independent cloud servers and organize secure MPC computation. All communication occurs exclusively between the data owner and the servers, with no server being connected to or aware of the others' details. In this example, $s > m = 1$ and all data is known only to the data owner.

Example 3.3 (Analysis of Medical Data) Hospitals or medical organizations can use secure MPC protocols for the statistical analysis of medical data owned by their patients. Multiple MPC servers, controlled by different and independent medical organizations, can be used to ensure privacy protection.

Example 3.4 (Electronic Voting) Electronic voting can be conducted by submitting votes to several MPC servers, which can perform private computations to tally the submitted votes and securely determine the winning candidate.

3.2 Security Requirements for MPC Protocols

An MPC protocol is said to have *information-theoretic* security, and its corresponding security requirements are considered to hold *unconditionally* if these security properties remain intact against adversaries with unlimited computing resources and time. On the other hand, if the security properties of a protocol are only guaranteed by the prohibitive computational cost of cryptanalysis, meaning the protocol can be broken by adversaries with unlimited computational resources, then the protocol is referred to as being *conditionally* or *computationally* secure.

The main security requirements that are crucial for the adoption of MPC protocols in practical applications are summarized as follows:

1. **Confidentiality**: It must be impossible for the participants to derive any additional information about each other's private data from the communications exchanged during the execution of the protocol, other than what can be inferred from the final output value provided by the protocol.

   To formalize this security requirement, consider the *trusted authority model*, often referred to as the *ideal world model*. In this model, the data owners send their private input data directly to a trusted external authority. The authority then computes the output function and returns the output value to all participants. The

3.2 Security Requirements for MPC Protocols

confidentiality requirement is considered satisfied by an MPC protocol if, upon completion, the participants cannot derive more information than they would have been able to deduce from the computation in the trusted authority model. The trusted authority model is also known as the ideal world model, in contrast to the real world, where no trusted authority is available. In the real-world scenario, the protocol participants should not be able to learn more about each other's private data than they could in the ideal world, with the trusted authority.

2. **Integrity**: Each honest participant must be guaranteed that if they receive an output value, it is correct. It must be impossible for adversaries to provide an invalid output to an honest participant without detection, even if they collude or perform incorrect computations.

   Secure MPC protocols can satisfy this requirement in two ways. The stricter definition applies to protocols that guarantee honest participants always receive the correct output. These protocols are said to possess *robust* integrity. In the second, more relaxed definition, a protocol can return an incorrect value to some honest participants, but it guarantees that they will be able to detect the error. In this case, the protocol satisfies security *with abort*.

   For example, in Example 3.3, the integrity of the protocol ensures that the correct statistical values describing aggregated medical data are returned. In Example 3.4, it guarantees that the correct candidate is chosen as the winner of the election, even if some servers are corrupted.

3. **Independence of Inputs and Computations of Honest Parties**: An MPC protocol must ensure that honest participants can choose their inputs and perform their computations independently of the actions of corrupted parties. This is essential because, if the computations of some participants depend on the actions of corrupted parties, then those participants cannot be considered fully honest.

4. **Guaranteed Honest Delivery**: An MPC protocol must ensure that corrupted parties cannot prevent honest participants from receiving their required output. In other words, adversaries should not be able to disrupt the computation by refusing to perform some steps of the protocol or by supplying incorrect values as their input.

5. **Interruption Fairness**: MPC protocols are often required to guarantee that malicious participants cannot gain an advantage by aborting their steps after receiving any output. The protocol must ensure that honest participants will receive at least the same information, or possibly more, even if an interruption occurs. Specifically, corrupted participants should not be able to prevent honest participants from receiving the information provided by the protocol. This property is crucial in scenarios like the use of MPC for signing official documents. If a malicious party were able to obtain a legitimate signed document while honest participants could not, due to the protocol being aborted by the malicious party, it could lead to significant security issues.

These requirements must be satisfied by all secure MPC protocols (cf. [4]).

3.3 Adversaries Against MPC Protocols

Various adversaries may attempt to influence the outcomes of an MPC protocol or gain access to confidential information. Some of the participants themselves may carry out attacks that subvert the protocol. The following are the main categories of adversaries that must be considered when designing MPC protocols.

1. **Honest-but-Curious, Semihonest, or Passive Adversaries**: *Honest-but-curious* adversaries, also known as *semihonest* or *passive* adversaries, follow all steps of the protocol correctly. However, they attempt to collect all intermediate data available to them during the execution of the process and seek to derive additional confidential information about the data of other owners. Certain groups of honest-but-curious adversaries may even collude and share information available to them in an attempt to gain insight into data stored by other owners.
   Let t be a positive integer. A protocol is considered *t-private* if it guarantees that, after the protocol's computation is completed, no set of at most t participants can obtain any more information than they could derive solely from their own private inputs and the protocol's output.
2. **Active or Malicious Adversaries**: *Active* or *malicious* adversaries may execute the protocol steps incorrectly. They can modify the values they transmit to other participants or even refuse to transmit some intermediate values that are necessary for other participants. When a malicious adversary attempts to disguise their behavior by pretending to be an honest participant, they typically perform many protocol steps correctly. Such malicious activity is referred to as a *Byzantine fault*.
   Let t be a positive integer. A protocol is considered *t-resilient* if it ensures that any participants who refuse to provide their inputs or are caught cheating are excluded from the protocol, and guarantees that no set of t or fewer malicious participants can influence the correctness of the protocol's outcome, denoted as $f(v_1, \ldots, v_d)$. Here, the function f is defined for the participants who provided their inputs and were not found to be cheating.
3. **External Adversaries**: *External* adversaries are individuals or organizations that do not officially participate in the protocol but can eavesdrop on the communication channels or attack some of the protocol's participants to gain additional information or coerce them into altering their actions. There are two main categories of external adversaries: honest-but-curious and malicious adversaries. If an honest-but-curious external adversary corrupts some participants, those participants behave like a group of colluding internal honest-but-curious adversaries. Similarly, if a malicious external adversary corrupts several participants, they behave like colluding malicious internal adversaries.
4. **Adaptive Adversaries**: *Adaptive adversaries* have the capability to corrupt participants of the protocol based on the information they have gathered during the earlier stages of the protocol's execution. Specifically, adaptive adversaries can thwart all MPC protocols that rely on the random selection of small participant sets for certain computations, as adaptive adversaries can always

corrupt the selected participants once they are identified. The key result from [5] demonstrates how to ensure secure communication even in the presence of adaptive adversaries.

3.4 Assumptions on MPC Protocols

In order to establish that an MPC protocol is secure, formal security proofs must be provided. These proofs depend on the corresponding security model, which includes the assumptions used to derive the proof. This section outlines the typical security assumptions that are employed in formal proofs for various MPC protocols.

1. **Assumptions on Security of Communication Channels**: MPC protocols can utilize the following two types of communication channels.
2. **Secure Communication Channels**: Formal MPC security models that assume secure communication channels are based on the premise that adversaries cannot access or modify any values transmitted over the communication channels between the participants and computation servers. These security models thus assume that all communications are completely secure.

    Open Communication Channels: Alternatively, formal MPC security models may assume that the communication channels are open, meaning that adversaries can access all values transmitted over the channels. In this case, participants can employ standard encryption techniques to ensure that their communications remain inaccessible to the adversaries. Additionally, [5] shows how encryption can be applied to secure communication channels even in the presence of adaptive adversaries.
3. **Assumption of Validity of the Private Data**: Every MPC protocol assumes that all participants, who do not perform detectable malicious actions, store valid private data. Specifically, it is assumed that the participants do not replace their original data with incorrect values before the protocol begins. A *replacement of original data* occurs when a participant alters their data before the protocol starts. Secure MPC protocols can detect incorrect steps in the computation, or identify inconsistent values or shares sent to the processors during the computation. Since the original data is private, the MPC protocol or other participants cannot directly access it.

    If a participant replaces their original data but behaves honestly in all other aspects of the protocol, this is considered a *concealed replacement of original data*. In such cases, it is impossible for other participants to detect the replacement. This hidden replacement can lead to incorrect protocol outcomes, as the invalid data may significantly affect the final result.

    In certain practical scenarios, it is assumed that data owners cannot perform concealed replacements. For example, in electronic elections, voters can make their vote invalid if they fail to follow the rules. However, if they follow the rules, their vote is automatically considered valid and the correct outcome is guaranteed. Similarly, in large-scale surveys, such as a national census, it is

Fig. 3.1 1-out-of-2 oblivious transfer

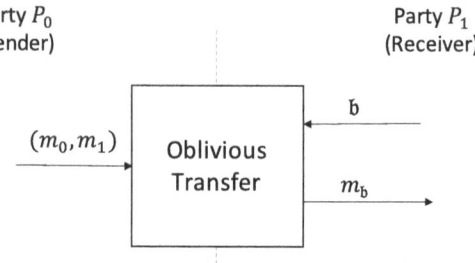

reasonable to assume that participants do not replace their original responses since statistical outcomes are not affected by a small number of incorrect entries. However, in some cases, a participant may be able to perform a concealed replacement of their data to produce an invalid outcome in the MPC protocol. For instance, in the Yao's Millionaires' Problem described in Example 3.1, a participant may intentionally submit incorrect data to influence the result, such as to boast about their wealth or conceal financial difficulties. In such situations, the protocol cannot detect the concealed data replacement, as the participant behaves honestly in all other aspects. This illustrates why it is common for secure MPC protocols to assume, either explicitly or implicitly, that participants do not perform concealed replacements of their original data.

3.5 Oblivious Transfer

Oblivious Transfer (OT) [37] is a fundamental cryptographic primitive that enables two parties to securely perform computations while remaining oblivious to each other's inputs. It serves as a key building block for various Multi-Party Computation (MPC) protocols, such as the Garbled Circuit (GC) protocol [6, 23] and the Goldwasser-Micali-Wigderson (GMW) protocol [7, 21, 38]. Figure 3.1 illustrates the basic concept of a 1-out-of-2 OT protocol. In this protocol, there are two parties: a *sender* P_0, who possesses a pair of binary strings $m_0, m_1 \in \{0, 1\}^\ell$, and a *receiver* P_1, who holds a choice bit $\mathfrak{b} \in \{0, 1\}$. The 1-out-of-2 OT protocol allows the receiver P_1 to obtain the string $m_\mathfrak{b}$, while ensuring that P_0 learns nothing about $\mathfrak{b}$, and P_1 does not gain any information about $m_{1-\mathfrak{b}}$.

3.5.1 Public Key-Based Oblivious Transfer

Naor and Pinkas [34] introduced one of the most prominent OT constructions in the semi-honest setting. This construction relies on public-key cryptography and assumes the Decisional Diffie-Hellman (DDH) problem as a security foundation.

3.5 Oblivious Transfer

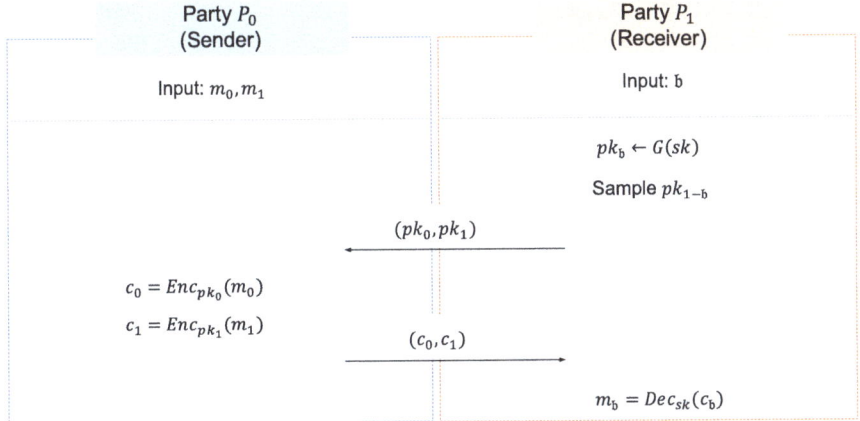

Fig. 3.2 Public key-based 1-out-of-2 semi-honest oblivious transfer

As depicted in Fig. 3.2, a 1-out-of-2 OT protocol based on this construction proceeds as follows: The receiver P_1 holds a secret choice bit b, while the sender P_0 has two secret messages, m_0 and m_1. The goal is to execute an OT protocol where the receiver P_1 learns the message m_b corresponding to their choice, while ensuring that the sender P_0 does not gain any information about b, and the receiver does not learn anything about m_{1-b}.

To begin the protocol, the receiver P_1 generates a public-private key pair, denoted as (pk_b, sk), and randomly selects a public key pk_{1-b}. These keys are all drawn from the same key space. If $b = 0$, the receiver sends the tuple (pk_b, pk_{1-b}) to the sender. Otherwise, if $b = 1$, the receiver sends (pk_{1-b}, pk_b) to the sender.

Upon receiving the tuple of public keys (pk_0, pk_1), the sender uses them to encrypt the two messages: $c_0 = Enc_{pk_0}(m_0)$ and $c_1 = Enc_{pk_1}(m_1)$. The sender then returns the ciphertexts (c_0, c_1) to the receiver.

Finally, the receiver P_1 decrypts the ciphertexts using their private key sk. Since sk corresponds to pk_b, the receiver can only decrypt the ciphertext c_b, thus obtaining the message m_b.

3.5.2 Oblivious Transfer Extension

The OT construction described above relies on public key cryptography, which limits its scalability. It requires three exponentiations for each execution of a 1-out-of-2 OT protocol, which can introduce substantial computational overhead when dealing with large-scale data transfers.

To address the performance challenges inherited from public-key cryptography, OT extensions [17, 25] have been proposed. These extensions decompose the

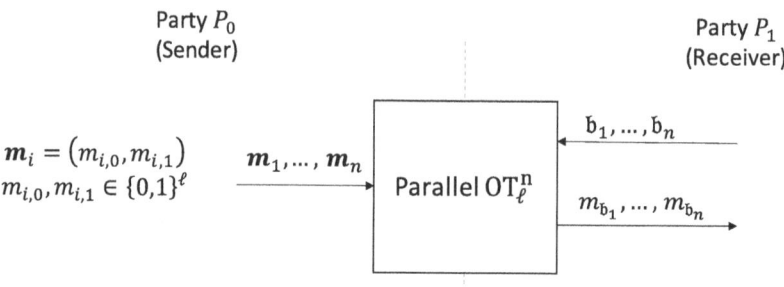

Fig. 3.3 Parallel 1-out-of-2 oblivious transfer for n ℓ-bit strings (OT_ℓ^n)

monolithic OT protocol into smaller OT units, using a small set of base OTs and leveraging symmetric cryptography (e.g., cryptographic hash functions) to efficiently handle a large number of OTs.

Parallel Oblivious Transfer To introduce the OT extension protocol, we first discuss a parallel 1-out-of-2 OT protocol that is executed n times, denoted as OT_ℓ^n. As shown in Fig. 3.3, this protocol allows the transfer of n ℓ-bit strings in parallel.

In this protocol, the sender P_0 holds n pairs of messages $\mathbf{m}_1, \ldots, \mathbf{m}_n$, where each pair consists of two bit strings $m_{i,0}, m_{i,1} \in \{0, 1\}^\ell$ for $i \in \{1, n\}$, and the receiver P_1 has a corresponding set of n choice bits $b_1, \ldots, b_n$, where each $b_i \in \{0, 1\}$.

Upon executing the OT_ℓ^n protocol, the receiver P_1 obtains the messages $m_{b_1}, \ldots, m_{b_n}$, and the sender P_0 learns nothing about the choice bits $b_1, \ldots, b_n$. Likewise, the receiver P_1 learns no information about the messages $m_{1-b_1}, \ldots, m_{1-b_n}$.

Oblivious Transfer Extension An OT extension [25] efficiently implements the parallel OT protocol by minimizing the need for expensive public-key cryptographic operations. The extension reduces the number of public key-based operations required and instead uses symmetric cryptography to perform the oblivious transfer.

To compute OT_ℓ^n, the OT extension protocol begins by selecting a small number of base OTs, which are used to transfer symmetric cryptographic keys. Suppose the base OTs are k-times OTs for n-bit strings, denoted as OT_n^k, where $k \ll n$. Each of the base OTs is realized by executing an OT_k^k protocol, using a pseudo-random generator to transfer the necessary messages.

The OT extension protocol uses the base OTs, OT_n^k, along with $3n$ cryptographic hash functions, to derive the OT_ℓ^n. Specifically, P_0 first samples a series of random choice bits $\mathbf{r} = r_1, \ldots, r_k \in_R \{0, 1\}$. Meanwhile, P_1 samples a $n \times k$ matrix $W = \mathbf{w}_1, \ldots, \mathbf{w}_k$, where each $\mathbf{w}_j \in \{0, 1\}^n$ for $j \in [1, k]$ and $i \in [1, n]$.

Next, P_0 and P_1 execute the OT_n^k protocol, swapping their roles: P_1 becomes the sender and P_0 the receiver. In this protocol, P_1 inputs the pairs $(\mathbf{w}_1, \mathbf{w}_1 \oplus \mathbf{b}), \ldots, (\mathbf{w}_k, \mathbf{w}_k \oplus \mathbf{b})$, where $\mathbf{b} = b_1, \ldots, b_n$ represents the choice bits held by P_1. P_0 inputs the random choice bit string $\mathbf{r}$.

3.6 Yao's Garbled Circuits

Upon receiving the matrix, P_0 computes a new matrix $T = \mathbf{t}_1, \ldots, \mathbf{t}_k$, where the j-th column is defined as $\mathbf{t}_j = r_j \cdot \mathbf{b} \oplus \mathbf{w}_j$, and the i-th row as $\mathbf{t}_i = \mathsf{b}_i \cdot \mathbf{r} \oplus \mathbf{w}_j$. Then, P_0 computes:

$$u_i^0 = m_i^0 \oplus H(i, \mathbf{t}_i)$$

and

$$u_i^1 = m_i^1 \oplus H(i, \mathbf{t}_i \oplus \mathbf{b}),$$

where H denotes a cryptographic hash function. P_0 then sends the tuples (u_i^0, u_i^1) to P_1.

Upon receiving these, P_1 computes the message $m_i^{\mathsf{b}_i}$ as follows:

$$m_i^{\mathsf{b}_i} = u_i^{\mathsf{b}_i} \oplus H(i, \mathbf{w}_i).$$

As a result, P_1 learns $m_1^{\mathsf{b}_1}, \ldots, m_n^{\mathsf{b}_n}$ based on its choice bits $\mathsf{b}_1, \ldots, \mathsf{b}_n$, while P_0 learns nothing.

Note that $H : \{0, 1\}^n \to \{0, 1\}^\ell$ is a random oracle.

3.5.3 Correlated Oblivious Transfer

Correlated Oblivious Transfer (COT) [12] is a specific variant of Oblivious Transfer (OT) that builds upon OT extension, facilitating communication between two parties in a manner similar to traditional OT but with enhanced practicality. The key distinction in COT lies in the fact that, instead of directly transmitting two messages, the sender constructs and inputs a *correlation function* $f_\Delta(\cdot)$ to link the two input messages, m_0 and m_1, such that m_0 is a random value and $m_1 = f_\Delta(m_0)$. The output of COT under the function $f_\Delta(\cdot)$ consists of the values m_0 and $m_\mathsf{b} \in \{m_0, m_1\}$. It is important to note that the function $f_\Delta(\cdot)$ serves as a correlation-robust random oracle $H : \{0, 1\}^\ell \to \{0, 1\}^\ell$ and can be instantiated using a one-way hash function. We denote the COT functionality described above as $COT(m_0, f_\Delta(\cdot); \mathsf{b})$. The n-times COT_ℓ (i.e., $n \times \text{COT}_\ell$) can be executed in parallel, where each COT_ℓ is responsible for transferring ℓ-bit strings.

3.6 Yao's Garbled Circuits

Garbled Circuits (GC) were first introduced by Yao in 1986 [6]. Since their introduction, numerous works (a few examples include [15, 19, 23, 27, 28, 32, 35, 36, 36, 41]) have been proposed to explore efficient GC constructions and optimization techniques within the semi-honest setting. Additional research [24, 26, 29–31, 39, 40] has sought to strengthen GC protocols with malicious security, though these efforts are beyond the scope of this thesis. In essence, a GC protocol allows two parties to securely evaluate an arbitrary function over their private inputs, ensuring that no information beyond the function's outputs is revealed. The fundamental idea behind a GC protocol is to represent a polynomial-time functionality f as a Boolean circuit C, and then to create a garbled circuit $\tilde{C}$ by encrypting all wires using symmetric encryption. In this section, we introduce the basic components and functions that constitute garbled circuits. We then proceed to present notable GC protocol constructions and optimization techniques aimed at improving the efficiency of GC implementations.

3.6.1 Components and Functions of Garbled Circuits

Basic Components of GC A typical garbled circuit (GC) construction consists of two fundamental components: *garbled labels* and *garbled tables*.

The garbled labels are cryptographic keys (symmetric encryption keys) randomly chosen to encode the truth values of the circuit's inputs and outputs. For clarity, we consider a logic gate g with input wires i, j and output wire k as part of the Boolean circuit being evaluated. To construct the garbled circuit for g, a pair of garbled labels (w^0, w^1) is randomly selected for each wire to encode the truth values 0 and 1, respectively. That is, the garbled labels (w_i^0, w_i^1), (w_j^0, w_j^1) are associated with the input wires i and j, and the labels (w_k^0, w_k^1) are associated with the output wire k. The labels w_i^0, w_j^0, and w_k^0 encode the truth values 0 for the wires i, j, and k, respectively. Similarly, the labels w_i^1, w_j^1, and w_k^1 encode the truth values 1 for their respective wires. Figure 3.4 provides a concrete illustration of the garbled labels of an AND gate and the truth values encoded by the labels.

Informally, the garbled table is an encrypted and permuted look-up table used to evaluate every gate within the Boolean circuit. For a single gate g, a look-up table of size 4 is constructed. Each entry in the look-up table is a ciphertext of the encrypted

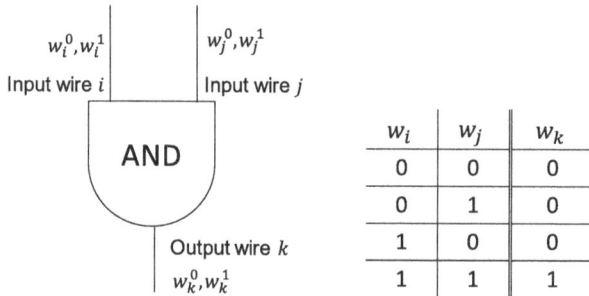

Fig. 3.4 A logic AND gate with its garbled labels and truth table

3.6 Yao's Garbled Circuits

output garbled label under two encryption keys: the garbled labels corresponding to the input wires. A garbled table contains the ciphertexts of all possible combinations of input values. Thus, when two input bits are provided, the garbled table facilitates the secure evaluation of the gate g by allowing only one entry to be decrypted, corresponding to the output bit of g.

Formally, suppose the logic gate g takes as input two bits b_i, b_j and outputs a bit b_k. The garbled table for g is constructed from the ciphertexts:

$$Enc_{w_i^{b_i}, w_j^{b_j}}(w_k^{g(b_i,b_j)}). \tag{3.1}$$

Here, the input garbled labels $w_i^{b_i}$ and $w_j^{b_j}$, which encode the input bits b_i and b_j, are used as encryption keys to encrypt the output garbled label $w_k^{g(b_i,b_j)}$, which encodes the output bit $g(b_i, b_j)$. An encrypted look-up table is produced with four ciphertexts, each corresponding to one of the possible input bit combinations:

$$T_g = \begin{pmatrix} Enc_{w_i^0, w_j^0}(w_k^{g(0,0)}) \\ Enc_{w_i^0, w_j^1}(w_k^{g(0,1)}) \\ Enc_{w_i^1, w_j^0}(w_k^{g(1,0)}) \\ Enc_{w_i^1, w_j^1}(w_k^{g(1,1)}) \end{pmatrix}. \tag{3.2}$$

During the secure evaluation, only the output garbled label $w_k^{g(b_i,b_j)}$, which encodes the result of the Boolean operation defined by the gate, can be decrypted. For example, if g is an AND gate, the table would be:

$$T_{\text{AND}} = \begin{pmatrix} Enc_{w_i^0, w_j^0}(w_k^0) \\ Enc_{w_i^0, w_j^1}(w_k^0) \\ Enc_{w_i^1, w_j^0}(w_k^0) \\ Enc_{w_i^1, w_j^1}(w_k^1) \end{pmatrix}. \tag{3.3}$$

If the input bits for the AND gate are $(0,1)$, then only the second entry, $Enc_{w_i^0, w_j^1}(w_k^0)$, can be correctly decrypted to obtain the label w_k^0 corresponding to the output $0 = 0 \wedge 1$ computed by the gate.

However, simply encrypting the look-up table reveals the position of the decrypted entry, which could lead to the deduction of the corresponding input truth values b_i and b_j. For example, decrypting the second entry of an AND gate reveals that the input bits are $(0,1)$. To prevent such information leakage, all entries in the encrypted look-up table are randomly permuted, so that the secure evaluation of the gate requires decrypting every entry. Since all entries are accessed, the position of the correctly decrypted entry remains hidden. Thus, the garbled table for the gate is securely constructed. Table 3.1 summarizes the basic components of a GC, using a garbled AND gate as an example, without loss of generality.

Table 3.1 Basic components of a garbled AND gate

Garbled labels				
Input wire i	Input wire j	Output wire k	Truth table	Garbled table
w_i^0	w_j^0	w_k^0	0	$Enc_{w_i^0, w_j^0}(w_k^0)$
w_i^0	w_j^1	w_k^0	0	$Enc_{w_i^0, w_j^1}(w_k^0)$
w_i^1	w_j^0	w_k^0	0	$Enc_{w_i^1, w_j^0}(w_k^0)$
w_i^1	w_j^1	w_k^1	1	$Enc_{w_i^1, w_j^1}(w_k^1)$

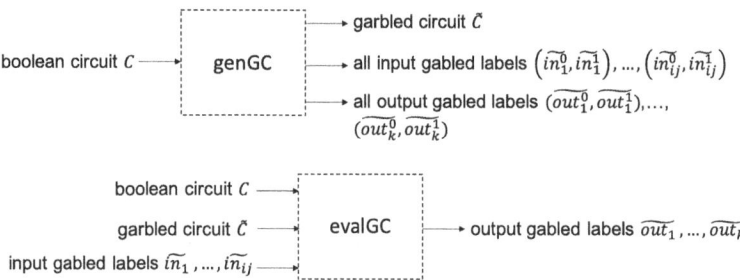

Fig. 3.5 Basic functions of a GC construction

3.6.2 Basic Functions of Garbled Circuits

The construction of a Garbled Circuit (GC) protocol consists of two key randomized functions: the GC generation function genGC, and the GC evaluation function evalGC, as illustrated in Fig. 3.5.

- The genGC function takes as input a boolean circuit C, and generates the corresponding garbled circuit $\tilde{C}$ along with all pairs of input and output garbled labels.
- The evalGC function accepts as input the garbled circuit $\tilde{C}$, the boolean circuit C (including its topology), and the garbled labels associated with the circuit's inputs. It then evaluates $\tilde{C}$ using the input labels to produce the garbled labels corresponding to the final output.

In detail, consider a boolean circuit $z = C(x, y)$ that represents the polynomial-time functionality $f(x, y)$. The inputs x and y of the circuit can be efficiently enumerated as all possible bit combinations $\{x_1, \ldots, x_i\}$ and $\{y_1, \ldots, y_j\}$. The circuit output z consists of all possible values $\{z_1, \ldots, z_k\}$ computed over all input combinations.

During the generation process, the genGC function produces a set of garbled labels $\{(\widetilde{in}^0, \widetilde{in}^1)\}_{1,\ldots,i \cdot j}$ for all possible input combinations, which are assigned to the input wires of the circuit C. For each logic gate in C, a pair of garbled labels is

3.6 Yao's Garbled Circuits

assigned to the intermediate output wire, from which a garbled table is generated. Finally, the garbled labels $\{(\tilde{out}^0, \tilde{out}^1)\}_{1,...,k}$ for all possible output combinations are produced and assigned to the output wires.

During evaluation, the evalGC function takes as input the garbled labels $\{\tilde{in}^{(x_i, y_j)}\}_{1,...,i \cdot j}$ specified for the inputs x and y. By evaluating the garbled circuit $\tilde{C}$ gate-by-gate according to the topological order of the circuit C, the evalGC function produces the garbled labels $\{\tilde{out}^{z_k}\}_{1,...,k}$ corresponding to the result z.

3.6.3 Yao's Protocol

Yao's protocol (also known as Yao's Garbled Circuits) [6] is a classical and widely-recognized GC protocol. It provides a general-purpose solution for secure two-party computation in the presence of semi-honest adversaries. Figure 3.6 outlines the workflow of Yao's GC protocol.

Suppose two parties, P_0 and P_1, each hold private inputs b_i and b_j, respectively, and wish to jointly compute the function $b_k = f(b_i, b_j)$. At the conclusion, only the result b_k should be known to both parties, with no additional information about each other's private inputs revealed.

To securely evaluate the function, P_0 (the *generator*) constructs a garbled boolean circuit C representing the function f. P_0 then runs the generation function genGC to create random garbled labels for all the circuit wires, along with the garbled circuit $\tilde{C}$.

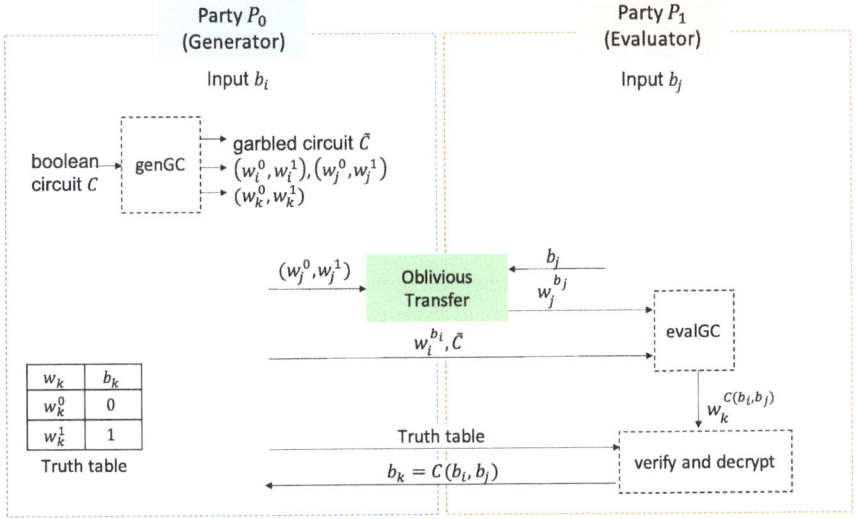

Fig. 3.6 Workflow of Yao's protocol

Given a security parameter λ, the garbled labels (w_i^0, w_i^1) are generated as follows:

$$w_i^0 = (w_i^0 \in_R \{0,1\}^\lambda, p_i^0 \in_R \{0,1\}),$$

$$w_i^1 = (w_i^1 \in_R \{0,1\}^\lambda, p_i^1 = 1 - p_i^0),$$

where p_i^0 and p_i^1 are color bits introduced by the point-and-permute technique [18], which efficiently identify which entry of the garbled table should be decrypted. Further details of the point-and-permute technique will be provided later.

The ciphertexts of the garbled circuit $\tilde{C}$ are computed as:

$$Enc_{w_i^{b_i}, w_j^{b_j}}(w_k^{C(b_i, b_j)}) = H(k_i^{b_i} || k_j^{b_j}) \oplus w_k^{C(b_i, b_j)},$$

where H is a random oracle, which can be instantiated as a cryptographic hash function in practice.

In addition, P_0 generates a truth table that maps the actual output bits b_k to the garbled output labels w_k, and maps its input b_i to a garbled label $w_i^{b_i}$. After constructing the garbled circuit, P_0 invokes a 1-out-of-2 oblivious transfer (OT) protocol with P_1, the *evaluator*, in order to obliviously assign a garbled label $w_j^{b_j}$ to P_1's private input b_j.

Once the OT is complete, P_0 sends the garbled circuit $\tilde{C}$, the garbled label $w_i^{b_i}$, and the truth table to P_1. Upon receipt, P_1 evaluates the circuit using **evalGC** to obtain the garbled label corresponding to the output $w_k^{C(b_i, b_j)}$.

Finally, either P_0 or P_1 (or both) can learn the result by communicating the truth table. P_1 can then verify and decrypt the garbled label to obtain the computed output b_k.

3.6.4 Efficient GC Constructions with Optimizations

In this subsection, we briefly introduce standard optimization techniques that enhance the efficiency of garbled circuits evaluation.

Point-and-Permute The point-and-permute technique [18], introduced by Beaver, alleviates the need for the evaluator to decrypt all four entries of a garbled table. It introduces a pair of color bits, p^0 and p^1, which are associated with the garbled labels w^0 and w^1. Given $p^b \in \{p^0, p^1\}$, the color bit p^b is generated using the expression $b \oplus r$, where $r \in \{0, 1\}$ is a randomly chosen bit used to mask the truth value $b \in \{0, 1\}$. This ensures that the two color bits, p^0 and p^1, are distinct, revealing no information about the underlying truth value b. For wires i, j, and k, the corresponding labels are constructed as follows:

$$(w_i^0 || p_i^0, w_i^1 || p_i^1), (w_j^0 || p_j^0, w_j^1 || p_j^1), (w_k^0 || p_k^0, w_k^1 || p_k^1).$$

3.6 Yao's Garbled Circuits

The garbled table for gate g is generated as:

$$T_g = \begin{pmatrix} Enc_{w_i^{p_i^0}, w_j^{p_j^0}}(w_k^{g(p_i^0, p_j^0)} \| g(p_i^0, p_j^0) \oplus p_k^{g(0,0)}) \\ Enc_{w_i^{p_i^0}, w_j^{p_j^1}}(w_k^{g(p_i^0, p_j^1)} \| g(p_i^0, p_j^1) \oplus p_k^{g(0,1)}) \\ Enc_{w_i^{p_i^1}, w_j^{p_j^0}}(w_k^{g(p_i^1, p_j^0)} \| g(p_i^1, p_j^0) \oplus p_k^{g(1,0)}) \\ Enc_{w_i^{p_i^1}, w_j^{p_j^1}}(w_k^{g(p_i^1, p_j^1)} \| g(p_i^1, p_j^1) \oplus p_k^{g(1,1)}) \end{pmatrix},$$

where the color bits are used to permute the entries in the garbled table. Upon receiving the garbled table, the evaluator decrypts only the entry identified by the color bits. For example, when the evaluator receives $w_i^{p_i^0} \| 0$, $w_j^{p_j^0} \| 0$, it decrypts the first entry in the garbled table. Thus, the color bits determine which ciphertext is decrypted, without revealing the actual truth value.

Free-XOR The Free-XOR optimization, introduced by Kolesnikov and Schnider [28], enables the secure evaluation of XOR operations "for free", meaning that a garbled XOR gate can be evaluated efficiently without the need to construct a garbled table. The technique involves selecting a common secret value $\Delta \in_R \{0, 1\}^\lambda$ and applying it to all wires of the garbled circuit. This results in a fixed correlation between the garbled labels:

$$w^1 = w^0 \oplus \Delta,$$

which allows the output garbled labels of the XOR gate for any pair of input bits to be computed as follows:

$$w_k^0 = w_i^0 \oplus w_j^0,$$
$$w_k^1 = w_k^0 \oplus \Delta.$$

Thus, the garbled XOR gate is generated by setting $w_k^0 = w_i^0 \oplus w_j^0$ during the circuit construction. The evaluation simply involves computing $w_k = w_i^0 \oplus w_j^0$, since w_k^1 is naturally correlated with w_k^0.

The correctness of this correlation is demonstrated in the following way: Consider a garbled XOR gate $\tilde{g}$, where the output garbled label is $w_k = \tilde{g}(w_i, w_j)$.

- *Input bits (0,0)*: The XOR gate evaluates these input bits to yield an output bit $0 = 0 \oplus 0$. The corresponding output garbled label is $w_k^0 = w_i^0 \oplus w_j^0$.
- *Input bits (0,1)*: The XOR gate evaluates these input bits to yield an output bit $1 = 0 \oplus 1$. The corresponding output garbled label is $w_k^1 = w_i^0 \oplus w_j^1 = (w_i^0 \oplus w_j^0) \oplus \Delta = w_k^0 \oplus \Delta$.

- *Input bits (1,0)*: The **XOR** gate evaluates these input bits to yield an output bit $1 = 1 \oplus 0$. The corresponding output garbled label is $w_k^1 = w_i^1 \oplus w_j^0 = (w_i^0 \oplus w_j^0) \oplus \Delta = w_k^0 \oplus \Delta$.
- *Input bits (1,1)*: The **XOR** gate evaluates these input bits to yield an output bit $0 = 1 \oplus 1$. The corresponding output garbled label is $w_k^0 = w_i^1 \oplus w_j^1 = (w_i^0 \oplus w_j^0) \oplus (\Delta \oplus \Delta) = w_i^0 \oplus w_j^0$.

This optimization is particularly useful in applications involving secure neural network inference on resource-constrained devices.

Fancy Garbling Ball et al. [15] introduced a specialized version of the garbled circuit protocol (BMR16), which represents the functionality as arithmetic circuits. This was later expanded into a comprehensive framework, known as fancy-garbling [14]. The protocol is specifically designed to efficiently support secure neural network applications, where most computations involve arithmetic operations, such as addition and multiplication. The proposed garbled circuit protocol builds upon Yao's protocol with the Free-XOR optimization as its foundation. By customizing garbled labels with an addition correlation (similar to the **XOR** correlation in Yao's protocol), this approach enables efficient evaluation of arithmetic circuits over integers, including *free* addition and scalar multiplication by a public constant. We now briefly outline the core concept of the fancy-garbling protocol.

In an arithmetic circuit, each wire is associated with an integer in the ring $x \in \mathbb{Z}_m$ with modulus m, and a pair of garbled labels w^0, w^x. For each wire, the fancy-garbling protocol randomly generates a garbled label as an element of the ring

$$w^0 \in_R \mathbb{Z}_m,$$

and a common secret value $\Delta \in_R \mathbb{Z}_m$. The integer x is encoded in the other garbled label as

$$w^x = w^0 + x \cdot \Delta.$$

It is noteworthy that when $m = 2$, the correlation between w^0 and w^x is equivalent to the **XOR** relationship found in Yao's protocol.

Given two input values x and y with garbled labels (w_i^0, w_i^x) on wire i and (w_j^0, w_j^y) on wire j, and an output z with labels (w_k^0, w_k^z) on wire k, the output labels are generated as follows:

Fancy-garbling *freely* supports the secure evaluation of certain arithmetic operations:

- Modular addition $z = x + y$ is securely computed as $w_i^x + w_j^y = (w_i^0 + w_j^0) + (x + y) \cdot \Delta = w_k^0 + z \cdot \Delta = w_k^z$.
- Multiplication by a public constant $z = cx$ is securely computed as $c \cdot w_i^x = c \cdot (w_i^0 + x \cdot \Delta) = c \cdot w_i^0 + (c \cdot x) \cdot \Delta = w_k^0 + z \cdot \Delta = w_k^z$.

3.7 Secure Multiparty Computation with Secret Sharing

3.7.1 Arithmetic Secret Sharing

Given an ℓ-bit value x, Arithmetic sharing [13] generates shares $\langle x \rangle_0^A, \langle x \rangle_1^A \in \mathbb{Z}_{2^\ell}$ such that $\langle x \rangle_0^A + \langle x \rangle_1^A \equiv x \pmod{2^\ell}$. Unless stated otherwise, *all operations are performed under modulo 2^ℓ*.

Sharing To share $\mathsf{Shr}_i^A(x)$ between party P_i, where $i \in \{0, 1\}$, party P_i generates $r \in_R \mathbb{Z}_{2^\ell}$ randomly and sends it to P_{1-i}. Party P_i then sets $\langle x \rangle_i^A = x - r$, and party P_{1-i} sets $\langle x \rangle_{1-i}^A = r$.

Reconstruction To reconstruct $\mathsf{Rec}_i^A(x)$, party P_{1-i} sends $\langle x \rangle_{1-i}^A$ to party P_i, who then computes $x = \langle x \rangle_0^A + \langle x \rangle_1^A$.

Secure Addition Given the shares of two secret values x and y, secure addition can be computed between two parties, denoted as P_0 and P_1, each holding shares of the values. Addition or subtraction over shares ($\langle z \rangle_i^A = \langle x \rangle_i^A \pm \langle y \rangle_i^A$), and multiplication by a public value ($\langle z \rangle_i^A = \eta \cdot \langle x \rangle_i^A$), can be efficiently performed by each party P_i locally without any interaction.

Secure Multiplication To securely multiply two shares ($\langle z \rangle^A = \langle x \rangle^A \cdot \langle y \rangle^A$), a pre-computed secret-shared multiplication triple (a, b, c) [16] is required, where $c = a \cdot b$. With the secret-shared triple, secure multiplication is supported as follows: Each party P_i first computes:

$$\langle e \rangle_i^A = \langle x \rangle_i^A - \langle a \rangle_i^A,$$

$$\langle f \rangle_i^A = \langle y \rangle_i^A - \langle b \rangle_i^A.$$

Then both parties interact to reconstruct e and f. Finally, each party P_i computes:

$$\langle z \rangle_i^A = i \cdot e \cdot f + f \cdot \langle a \rangle_i^A + e \cdot \langle b \rangle_i^A + \langle c \rangle_i^A.$$

Figure 3.7 provides an illustration of this secure multiplication procedure.

Beaver's Multiplication Triples An efficient approach for secure multiplication over two secret shares involves splitting the multiparty computation (MPC) protocol into a preprocessing phase and an online phase. The preprocessing phase occurs before the parties' private inputs are available and generates correlated secret shares, which are used in the online phase for secure multiplication. Beaver [16] introduced the concept of multiplication triples, a highly efficient method for generating such correlated secret shares. A Beaver multiplication triple consists of secret shares $\langle a \rangle^A, \langle b \rangle^A, \langle c \rangle^A$ such that $c = a \cdot b$. These triples can be generated offline and shared obliviously via correlated oblivious transfer extension (COT) [12, 22]. In

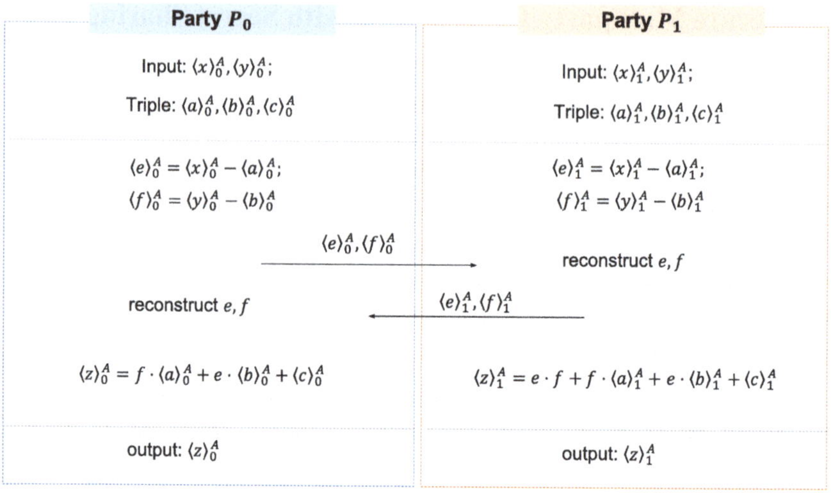

Party P_0	Party P_1
Input: $\langle x \rangle_0^A, \langle y \rangle_0^A$;	Input: $\langle x \rangle_1^A, \langle y \rangle_1^A$;
Triple: $\langle a \rangle_0^A, \langle b \rangle_0^A, \langle c \rangle_0^A$	Triple: $\langle a \rangle_1^A, \langle b \rangle_1^A, \langle c \rangle_1^A$
$\langle e \rangle_0^A = \langle x \rangle_0^A - \langle a \rangle_0^A$;	$\langle e \rangle_1^A = \langle x \rangle_1^A - \langle a \rangle_1^A$;
$\langle f \rangle_0^A = \langle y \rangle_0^A - \langle b \rangle_0^A$	$\langle f \rangle_1^A = \langle y \rangle_1^A - \langle b \rangle_1^A$
$\xrightarrow{\langle e \rangle_0^A, \langle f \rangle_0^A}$	reconstruct e, f
reconstruct e, f	$\xleftarrow{\langle e \rangle_1^A, \langle f \rangle_1^A}$
$\langle z \rangle_0^A = f \cdot \langle a \rangle_0^A + e \cdot \langle b \rangle_0^A + \langle c \rangle_0^A$	$\langle z \rangle_1^A = e \cdot f + f \cdot \langle a \rangle_1^A + e \cdot \langle b \rangle_1^A + \langle c \rangle_1^A$
output: $\langle z \rangle_0^A$	output: $\langle z \rangle_1^A$

Fig. 3.7 Secure multiplication over arithmetic shares

essence, the relationship is as follows:

$$\langle c \rangle^A = (\langle a \rangle_0^A + \langle a \rangle_1^A) \cdot (\langle b \rangle_0^A + \langle b \rangle_1^A) \quad (3.4)$$
$$= \underbrace{\langle a \rangle_0^A \langle b \rangle_0^A}_{\text{term of } P_0} + \underbrace{\langle a \rangle_1^A \langle b \rangle_1^A}_{\text{term of } P_1} + \underbrace{\langle a \rangle_0^A \langle b \rangle_1^A + \langle a \rangle_1^A \langle b \rangle_0^A}_{\text{mixed terms}}.$$

Let P_i randomly select $\langle a \rangle_i^A, \langle b \rangle_i^A \in_R \mathbb{Z}_{2^\ell}$, so that $\langle a \rangle_i^A \cdot \langle b \rangle_i^A$ can be computed locally. However, computing $\langle a \rangle_i^A \cdot \langle b \rangle_{1-i}^A$ directly, P_i can use $\langle a \rangle_{1-i}^A$ to recover the private input of the other party. Thus, the mixed terms are obliviously shared via COT_ℓ^ℓ.

Given $s_{i,0} \in_R \mathbb{Z}_{2^\ell}$ from P_0 and a bit-value $\mathsf{b} = \langle b \rangle_1^A[i]$ from P_1, the correlation function is defined as

$$f_i(s_{i,1}) = (\langle a \rangle_0^A \cdot 2^i - s_0) \bmod 2^\ell,$$

where $i \in [1, \ell]$. Note that $s_{i,0}$ and $s_{i,1}$ correspond to the cases when b equals 0 and 1, respectively. Finally, P_0 sets its share as

$$\langle u \rangle_0^A = (\Sigma_{i=1}^\ell s_{i,0}) \bmod 2^\ell,$$

while P_1 sets its share as

$$\langle u \rangle_1^A = (\Sigma_{i=1}^\ell s_{i,\mathsf{b}}) \bmod 2^\ell.$$

3.7 Secure Multiparty Computation with Secret Sharing

Likewise, P_0 and P_1 compute $\langle v \rangle^A = \langle \langle a \rangle_1^A \langle b \rangle_0^A \rangle^A$. Ultimately, each party P_i computes the triple item as

$$\langle c \rangle_i^A = \langle a \rangle_i^A \langle b \rangle_i^A + \langle u \rangle_i^A + \langle v \rangle_i^A.$$

3.7.2 Boolean Sharing MPC

Boolean sharing models the evaluated functionality as Boolean circuits and constructs XOR-based secret shares of a secret value. It can be viewed as arithmetic sharing over $\mathbb{Z}_2$. In this section, we employ the GMW protocol [7] to evaluate the function represented as a Boolean circuit.

Given a bit x, Boolean sharing produces two Boolean shares, $[\![x]\!]_0, [\![x]\!]_1 \in \mathbb{Z}_2$, between two parties, P_0 and P_1, such that $[\![x]\!]_0 \oplus [\![x]\!]_1 = x$.

Sharing When sharing $\mathsf{Shr}_i^B(x)$ of a secret bit x, party P_i ($i \in \{0, 1\}$), holding its secret x, generates a random value $r \in_R \{0, 1\}$ and sends it to party P_{1-i}. Party P_i sets $[\![x]\!]_i = x \oplus r$, while party P_{1-i} sets $[\![x]\!]_{1-i} = r$.

Reconstruction To reconstruct x, party P_{1-i} sends its share to party P_i, who computes $x = [\![x]\!]_0 \oplus [\![x]\!]_1$.

Secure XOR The addition operation over arithmetic shares in $\mathbb{Z}_2$ is reformulated as the XOR operation ($\oplus$), which is computed locally by each party P_i ($[\![z]\!]_i = [\![x]\!]_i \oplus [\![y]\!]_i$).

Secure AND Similarly, the multiplication operation is replaced by the bitwise AND operation ($\wedge$) over Boolean shares and is computed with the assistance of precomputed Boolean AND Triples, $[\![c]\!] = [\![a]\!] \wedge [\![b]\!]$. As with multiplication triples, the AND triple can be generated via Random OT (ROT) [12, 33] or by a trusted third party.

3.7.3 Replicated Secret Sharing

This section introduces 2-out-of-3 replicated secret sharing (RSS) as described by Araki et al. [11]. Let P_1, P_2, P_3 denote the three parties involved in the computation. For brevity, P_{i-1}, P_{i+1} represent the previous and succeeding parties of party P_i, where $i \in \{1, 2, 3\}$. Given a secret $x \in \mathbb{Z}_{2^\ell}$, we describe the sharing and reconstruction processes below.

Sharing To share a secret, x is split by uniformly sampling three random values, $\langle x \rangle_1, \langle x \rangle_2, \langle x \rangle_3 \in \mathbb{Z}_{2^\ell}$, such that $\langle x \rangle_1 + \langle x \rangle_2 + \langle x \rangle_3 \equiv \langle x \rangle \pmod{2^\ell}$. These replicated secret shares are distributed as pairs: P_1 holds $(\langle x \rangle_1, \langle x \rangle_2)$, P_2 holds $(\langle x \rangle_2, \langle x \rangle_3)$, and P_3 holds $(\langle x \rangle_3, \langle x \rangle_1)$.

Reconstruction To reveal the secret, P_i sends its share $\langle x \rangle_i$ to P_{i+1}, and each party locally adds the three shares to reconstruct x.

Secure Addition Arithmetic operations are evaluated by RSS as follows. Given two secret values x and y shared among three parties, the addition $\langle z \rangle_i = \langle x \rangle_i + \langle y \rangle_i$ (mod 2^ℓ) can be locally computed by each party. The combined share is given by $\langle z \rangle = \langle x + y \rangle := (\langle x \rangle_1 + \langle y \rangle_1, \langle x \rangle_2 + \langle y \rangle_2, \langle x \rangle_3 + \langle y \rangle_3)$.

Secure Multiplication Multiplication over two shares, $\langle z \rangle = \langle x \rangle \cdot \langle y \rangle$ (mod 2^ℓ), is computed interactively as follows. Each party P_i first locally computes 3-out-of-3 secret shares $\hat{z}_i = \langle x \rangle_i \langle y \rangle_i + \langle x \rangle_{i+1} \langle y \rangle_i + \langle x \rangle_i \langle y \rangle_{i+1}$ such that $\hat{z}_1 + \hat{z}_2 + \hat{z}_3 = xy$. Then, the parties jointly perform a *resharing* scheme to maintain the invariance of 2-out-of-3 sharing.

Given $\alpha_1, \alpha_2, \alpha_3 \in \mathbb{Z}_{2^\ell}$ as the *zero sharing* terms, subject to $\alpha_1 + \alpha_2 + \alpha_3 = 0$, each party P_i uses α_i to mask $\hat{z}_i$ as $\langle z \rangle_i = \hat{z}_i + \alpha_i$. P_i then sends $\langle z \rangle_i$ to P_{i-1}. Parties obtain their shares as $(\langle z \rangle_i, \langle z \rangle_{i+1})$. Note that the zero sharing terms [11, 33] can be sampled locally by each party using a pseudorandom function (PRF) with a common seed shared among the parties during preprocessing.

Boolean RSS When $\ell = 2$, the RSS for Boolean secret shares is defined as $[\![x]\!] \equiv [\![x]\!]_1 \oplus [\![x]\!]_2 \oplus [\![x]\!]_3$, where XOR $\oplus$ and AND $\wedge$ are computed as equivalent to addition and multiplication over $\mathbb{Z}_2$.

3.7.4 Shamir's Secret Sharing

Shamir's secret sharing (SSS) was first introduced in [1] (see also [3]). This section outlines the use of SSS in secure Multi-Party Computation (MPC), as proposed in [2].

3.7.4.1 Initialisation of the Protocol

To establish the MPC protocol based on SSS, the participants define several constant values that remain unchanged throughout all rounds of the protocol. First, they select a finite field E with $|E| > s$ and a primitive s-th root of unity, $\omega \in E$, such that $\omega^s = 1$. All numerical data processed by the protocol is represented as elements of the finite field E.

Next, they compute the elements $\xi_1 = \omega^0, \xi_2 = \omega^1, \ldots, \xi_s = \omega^{s-1}$ in E. The values $\xi_1, \xi_2, \ldots, \xi_s$ are distinct and nonzero, and these values are known to all participants of the protocol.

Then, the participants compute the constant $s \times s$ Vandermonde matrix $B = [b_{i,j}]$ with entries $b_{i,j} = \xi_j^i$ for $i, j = 1, \ldots, s$, as well as the constant $s \times s$ linear projection matrix $P = [p_{i,j}]$ with entries

3.7 Secure Multiparty Computation with Secret Sharing

$$p_{i,j} = \begin{cases} 1 & \text{if } i = j \leq t+1, \\ 0 & \text{otherwise.} \end{cases} \quad (3.5)$$

It is well-established that the Vandermonde matrix is non-singular for pairwise distinct $\xi_1, \ldots, \xi_s$. Therefore, the inverse matrix B^{-1} always exists.

Finally, the processors compute the $s \times s$ matrix

$$A = B^{-1}PB. \quad (3.6)$$

The matrix A is constant and depends only on the values $\xi_1, \ldots, \xi_s$ in E. Thus, it is publicly available, meaning it is known to all participants of the algorithm.

3.7.4.2 Creating Private Shares of Secret Values

For each $i \in [1:m]$ and $j \in [a_i, b_i]$, to represent a secret value v_j, the data owner $\mathcal{O}_i$ generates s private shares $\mu_1(v_j), \ldots, \mu_s(v_j)$, one share for each processor $\mathcal{P}_1, \ldots, \mathcal{P}_s$. To do this, the owner selects t random elements $h_{j,1}, \ldots, h_{j,t} \in E$ and defines the following polynomial

$$h_j(x) = v_j + h_{j,1}x + h_{j,2}x^2 + \cdots + h_{j,t}x^t, \quad (3.7)$$

then computes the private shares

$$\mu_\ell(v_j) = h_j(\xi_\ell), \quad (3.8)$$

for all $\ell \in [1:s]$.

Thus, the secret value v_j is represented by the collection of all private shares

$$\mu_1(v_j), \mu_2(v_j), \ldots, \mu_s(v_j). \quad (3.9)$$

The owner then sends each share (3.8) to the corresponding processor $\mathcal{P}_\ell$ for all $\ell \in [1:s]$. The set (3.9) consists of t uniformly distributed random variables over E (cf. [1]).

3.7.4.3 Recovering a Value from its Private Shares

The Lagrange interpolation formula can be used to recover the value v_j from the collection of private shares (3.9) (if $t < s$). Recall that for pairwise distinct values $\xi_1, \ldots, \xi_{t+1}$ and any polynomial $g(x) = g_0 + g_1 x + \cdots + g_t x^t$, if the values

$$\gamma_1 = g(\xi_1), \gamma_2 = g(\xi_2), \ldots, \gamma_{t+1} = g(\xi_{t+1}) \quad (3.10)$$

are known, then the following formula uniquely determines $g(x)$:

$$g(x) = \sum_{k=1}^{t+1} \left(\gamma_k \frac{\prod_{\ell \in [1:t+1], \ell \neq k}(x - \xi_\ell)}{\prod_{\ell \in [1:t+1], \ell \neq k}(\xi_k - \xi_\ell)} \right). \tag{3.11}$$

Equation (3.11) is known as the *Lagrange interpolation formula*. It follows from the well-known fact that there exists a unique polynomial of degree t that satisfies all $t + 1$ Eq. (3.10).

By applying (3.11) to the polynomial (3.7), we obtain

$$h_j(x) = \sum_{k=1}^{t+1} \left(\mu_k(v_j) \frac{\prod_{\ell \in [1:t+1], \ell \neq k}(x - \xi_\ell)}{\prod_{\ell \in [1:t+1], \ell \neq k}(\xi_k - \xi_\ell)} \right). \tag{3.12}$$

This formula can be used to recover $h_j(x)$ from the set of private shares (3.9). Once $h_j(x)$ is restored, the secret value v_j can be recovered since $v_j = h_j(0)$. Therefore, the processors $\mathscr{P}_1, \ldots, \mathscr{P}_s$ must keep their private shares confidential to prevent the exposure of the secret value v_j. At least $t + 1$ public shares are required to recover v_j. However, if t or fewer honest-but-curious processors collude and combine their shares, they will be unable to reconstruct the secret value.

3.7.4.4 Computing Linear Combinations with Integer Coefficients

This section explains how the protocol can compute the sum

$$w = v_1 + v_2 + \cdots + v_d \tag{3.13}$$

of the secret values. A straightforward modification of this procedure allows the computation of any linear combination with integer coefficients.

As described earlier, for $i \in [1 : m]$, $j \in [a_i, b_i]$, the owner $\mathscr{O}_i$ sends each share (3.8) to the corresponding processor $\mathscr{P}_\ell$ for all $\ell \in [1 : s]$. The processor $\mathscr{P}_\ell$ receives the private shares $\mu_\ell(v_1), \mu_\ell(v_2), \ldots, \mu_\ell(v_d)$ and computes their sum

$$w_\ell = \mu_\ell(v_1) + \mu_\ell(v_2) + \cdots + \mu_\ell(v_d). \tag{3.14}$$

The values $w_1, \ldots, w_s$ are the values of the polynomial $g(x) = g_1(x) + \cdots + g_d(x)$ evaluated at the points $\xi_1, \ldots, \xi_s$, i.e., $w_\ell = g(\xi_\ell)$ for all ℓ.

It follows that each w_ℓ is equivalent to the private share $\mu_\ell(w) = g(\xi_\ell)$ of the sum w corresponding to the polynomial $g(x)$. Therefore, the values $w_1, \ldots, w_s$ represent the private shares of the processors $\mathscr{P}_1, \ldots, \mathscr{P}_s$ encoding the sum $w = g(0)$. To return the result to the data owners, each processor $\mathscr{P}_\ell$ sends the private share w_ℓ back to all data owners. Every owner can then recover the sum w from its shares $\mu_1(w), \ldots, \mu_s(w)$ using Lagrange interpolation.

Since every linear combination with integer coefficients can be derived using only addition and subtraction of the secret values, we can state the following theorem, which directly follows from the results in [2].

Theorem 3.1 ([2]) *For every $t < s$, the MPC protocol based on Shamir's secret sharing can t-privately compute every linear combination of secret values with integer coefficients.*

3.7.4.5 Computing Linear Combinations with Public Constant Coefficients

Now suppose that $c_1, \ldots, c_d \in E$ are public constants known to all participants in the protocol. This section describes how the processors can privately compute the linear combination

$$c_1 v_1 + c_2 v_2 + \cdots + c_d v_d. \tag{3.15}$$

Since the procedure for computing sums has already been explained, we now focus on how the processors can privately compute the constant products of a private value and a public constant. To demonstrate this, we consider the product $c_1 v_1$.

Suppose that the secret value v_1 is encoded as a polynomial $h_1(x)$. Clearly, the polynomial $g(x) = c_1 \cdot h_1(x)$ encodes the product $c_1 \cdot v_1$. To compute $c_1 v_1$ privately, each processor $\mathscr{P}_\ell$ computes the private share $g(\xi_\ell) = c_1 \cdot h_1(\xi_\ell)$ and returns it to the data owners. Each data owner then applies the Lagrange interpolation formula to recover the value $c_1 v_1$ from its private shares $g(\xi_1), \ldots, g(\xi_s)$.

3.7.4.6 Multiplication of a Public Constant Matrix and a Matrix with Secret Values

Consider the matrix product of a public constant matrix and a matrix containing secret values. Each entry in this matrix product is a linear combination of the form (3.15), where $(c_1, c_2, \ldots, c_d)$ is a row vector from the public constant matrix, and $(x_1, x_2, \ldots, x_d)$ is a column vector from the matrix containing secret values. This linear combination can be computed as explained earlier.

Therefore, the processors can securely compute all entries of the matrix product by following the procedures outlined above.

3.7.4.7 Multiplication of Two Secret Values

In this section, we explain how the processors can privately compute the product of two secret values, y_1 and y_2, encoded as polynomials $f(x)$ and $g(x)$. For $i \in [1:d]$, each processor $\mathscr{P}_i$ computes the secret share $v_i = f(\xi_i)g(\xi_i)$. We then define the vector

$$V = [v_1, \ldots, v_s], \qquad (3.16)$$

and the processors privately compute the matrix product

$$W = [w_1, \ldots, w_s] = VA, \qquad (3.17)$$

where A is defined by (3.6). To perform this computation, the processors must encode each secret value in V as a new polynomial, communicate the secret shares of this polynomial, and compute all the linear combinations involved in VA. After this, for each $i \in [1 : d]$, each value w_i is revealed to only one processor, $\mathscr{P}_i$. To achieve this, all processors communicate their private shares of w_i to $\mathscr{P}_i$. Processor $\mathscr{P}_i$ then applies the Lagrange interpolation formula to recover the polynomial corresponding to these shares and determines w_i as the constant term of that polynomial.

It has been shown in [2, p. 3] that the values $w_1, \ldots, w_s$ are the secret shares of a polynomial $w(x)$ of degree $\leq t$, which is equal to the truncation of the product $f(x)g(x)$. It follows that $w(x)$ has the same constant term as $f(x)g(x)$. Since the constant term of $f(x)g(x)$ is equal to $y_1 y_2$, we conclude that $w(x)$ encodes the product $y_1 y_2$. Thus, the processors have privately computed the product of y_1 and y_2.

We denote the private shares $f(\xi_i)$, $g(\xi_i)$, and w_i discussed in the previous paragraph by the new symbols $\Psi_i(y_1)$, $\Psi_i(y_2)$, and $\Psi_i(y_1, y_2)$, respectively.

To prevent probabilistic attacks, an additional randomization step is required before finalizing the shares $\Psi_i(y_1, y_2)$. The reason is that the polynomial $w(x)$ defined in the previous paragraph is not random. If this polynomial were used to hold the result, a probabilistic attack could be arranged. To avoid such attacks, the processors ensure that the final polynomial representing the outcome is completely random. Each data owner $\mathscr{O}_i$ selects a random polynomial of degree d with the constant term set to zero, and distributes its shares among the processors. The processors then add up all the secret shares of these random polynomials to the outcome and use this to represent the final value.

It is well known that every function over a finite field can be represented as a polynomial. A simple proof of this fact follows from the Lagrange interpolation formula (3.11), which can be used to define a polynomial that coincides with any given function over a finite field. Therefore, the algorithms presented above lead to the following theorem, which was originally derived in [2].

Theorem 3.2 ([2]) *For every $t < s/2$, the MPC protocol based on Shamir's secret sharing can t-privately compute any public function f.*

This theorem cannot be improved, as demonstrated by the following result.

Theorem 3.3 ([2]) *There are functions for which no $s/2$-private protocols exist.*

3.7 Secure Multiparty Computation with Secret Sharing

3.7.4.8 Correction of Byzantine Faults Induced by Malicious Participants

Consider the scenario where, during the execution of a multi-party computation (MPC) protocol, one of the processors needs to recover a value v from its secret shares. The processor responsible for recovering v is denoted as $\mathscr{P}_\ell$, where $1 \leq \ell \leq s$. Let the secret shares encoding v be represented as:

$$\mu_1(v), \mu_2(v), \ldots, \mu_s(v), \tag{3.18}$$

where $\mu_i(v)$ is the secret share provided by the processor $\mathscr{P}_i$. The security model assumes that at most t participants may be corrupted simultaneously, where $t < s$. This implies that at most t of the values in (3.18) can be erroneous.

The processor $\mathscr{P}_\ell$ recovers the value v by applying Lagrange's interpolation formula (3.11), as defined in Sect. 3.7.4.3.

The definition in (3.8) shows that the secret shares encoding v are precisely the values in (3.18), which are equal to:

$$g(\xi_1), g(\xi_2), \ldots, g(\xi_s), \tag{3.19}$$

where $\xi_1 = \omega^0, \xi_2 = \omega^1, \ldots, \xi_s = \omega^{s-1}$, and $g(x) = a_0 + a_1 x + \cdots + a_t x^t$ with $a_0 = v$ represents the polynomial encoding v, as explained in Sect. 3.7.4.2.

Let us define additional zero values as $a_{t+1} = a_{t+2} = \cdots = a_{s-1} = 0$. Therefore, the sequence in (3.19) becomes a Discrete Fourier Transform [10, Definition 3.2.11] of the sequence $a_0, a_1, \ldots, a_{s-1}$.

Consider the following polynomial:

$$\widehat{g} = \mu_1(n) + \mu_2(n)x + \cdots + \mu_s(n)x^{s-1}. \tag{3.20}$$

Using the well-known formula for the Inverse Fourier Transform [10, Lemma 3.2.12], we obtain: $a_i = \frac{1}{s}\widehat{g}(\omega^{-i})$ for all $i \in [0 : s-1]$. In particular,

$$\widehat{g}(\omega^{-i}) = 0, \text{ for } i \in [t+1 : s-1]. \tag{3.21}$$

Since $\omega^d = 1$, we have $\omega^{-i} = \omega^{s-i}$. Substituting (3.20) into (3.21), we derive the following system of linear equations satisfied by the value $\mu_i(v)$ for each i:

$$\sum_{i=0}^{s-1} \omega^{r \cdot i} \cdot \mu_i(n) = 0 \text{ for } r \in [1 : 2t]. \tag{3.22}$$

These linear equations imply that the values $\omega, \omega^2, \ldots, \omega^{2t}$ are all roots of the polynomial in (3.20). Since $\omega^s = 1$ and $s = 3t + 1$, the set of these roots coincides with the set $\omega^{t+1}, \omega^{t+2}, \ldots, \omega^{s-1}$.

Let $M_{\omega^i}(x)$ denote the minimal polynomial of ω^i. It follows that the polynomial:

$$h(x) = \text{lcm}\{M_{\omega^{t+1}}(x), \ldots, M_{\omega^{s-1}}(x)\} \tag{3.23}$$

divides (3.20). In coding theory terms, this implies that the sequence in (3.18) is a codeword in the cyclic code of length d generated by $h(x)$. The choice of the generator polynomial $h(x)$ means that this cyclic code is the well-known BCH code with a designed distance of $2t + 1$ [10, Definition 3.2.7]. It has a straightforward error-correcting procedure capable of correcting up to t errors [10, §3.3].

By applying this error-correction procedure, processor D_ℓ can correct all t errors introduced by the Byzantine faults in the sequence (3.18). Afterward, D_ℓ can recover the value v as outlined earlier. This leads to the following theorem:

Theorem 3.4 ([2]) *For every $t < s/3$, the MPC protocol based on Shamir's secret sharing can compute any public function f both t-privately and t-resiliently.*

This theorem cannot be improved, as demonstrated by the following result from [2]:

Theorem 3.5 ([2]) *There exist functions for which no $s/3$-resilient protocols exist.*

3.7.5 Function Secret Sharing

A two-party function secret sharing (FSS) [8, 9] divides a function f into compact shares of the function. This method ensures that no individual function share reveals any information about f. When these shares are evaluated at a given point x, the combined results correspond to $f(x)$.

We present two constructions for FSS: (1) distributed point functions (DPF [8]) and (2) arithmetic FSS [9]. The formal definitions of these constructions are as follows.

DPF for Equality Test DPF.Equa($[\![x]\!]$) is designed to evaluate whether a secret value x, when input into a point function, equals 0. Specifically, we have:

$$\mathsf{Eval}^=(k_0^=, x') + \mathsf{Eval}^=(k_1^=, x') = \begin{cases} y = 1 & \text{if } x' = \gamma \\ 0 & \text{otherwise.} \end{cases}$$

In the key generation phase, the client selects a random value $\gamma \in \mathbb{Z}_{2^\ell}$ and runs $\mathsf{KeyGen}^=(1^\lambda, \alpha = \gamma, \beta = 1)$ to produce k_b for two parties.

During evaluation, a secret value $[\![x']\!]_b$ is constructed as the additive masking of $\gamma \in \mathbb{Z}_{2^\ell}$, i.e., $[\![x']\!]_b = [\![\gamma]\!]_b + [\![x]\!]_b$.

The parties exchange $[\![x']\!]_b$ to reveal the masked value of x'. Each party P_b then evaluates the DPF $\mathsf{Eval}^=(k_b, x')$ to generate the secret share $[\![y]\!]_b$. This share is reconstructed to result in $y = 1$ if and only if $x' = \gamma$ (i.e., $x = 0$).

DPF for Comparison DPF.Comp($[\![x]\!]$) operates similarly, but in this case, the function compares whether x is less than or equal to 0. The construction works as follows:

3.7 Secure Multiparty Computation with Secret Sharing

$$\text{Eval}^<(k_0^<, x') + \text{Eval}^<(k_1^<, x') = \begin{cases} y = 1 & \text{if } x' \leq \gamma \\ 0 & \text{if } x' > \gamma \end{cases}$$

In this case, the client runs DPF.Comp's key generation on $\alpha = \gamma$ and $\beta = 1$ to generate k_b. During the DPF evaluation, each party P_b evaluates the function $\text{Eval}(k_b, x')$ on the masked value x', producing the shared secret y. If $x' \leq \gamma$ (i.e., $x \leq 0$), then $y = 1$; otherwise, if $x' > \gamma$ (i.e., $x > 0$), y is evaluated to 0.

It is important to note that in both constructions, the key generation for the DPF is independent of the actual computation. Therefore, it can be performed offline, randomly generated, and stored in a key pool for later use.

For simplicity in this demonstration, we omit the security parameter from the key generation procedure KeyGen.

Algorithm 1 FSS multiplication $\text{FSS.Mul}(x_1', x_2')$

Multiplication Gate ($\text{KeyGen}^\times$, $\text{Eval}^\times$)
$\text{KeyGen}^\times(1^\lambda, r_1^{in}, r_2^{in}, r^{out})$

(1) Randomly sample $\tau_0, \tau_1 \in \mathbb{Z}_{2^\ell}^{in}$ such that $\tau_0 + \tau_1 = r_1^{in}$.
(2) Randomly sample $\rho_0, \rho_1 \in \mathbb{Z}_{2^\ell}^{in}$ such that $\rho_0 + \rho_1 = r_2^{in}$.
(3) Randomly sample $\xi_0, \xi_1 \in \mathbb{Z}_{2^\ell}^{in}$ such that $\xi_0 + \xi_1 = r_1^{in} \cdot r_2^{in} + r^{out}$.
(4) Define $k_b = \tau_b \parallel \rho_b \parallel \xi_b$, where $b \in \{0, 1\}$.
Return: $(k_0^\times, k_1^\times)$.
$\text{Eval}^\times(b, k_b^\times, x_1', x_2')$

(1) Parse $k_b^\times = \tau_b \parallel \rho_b \parallel \xi_b$.
Return: $b \cdot (x_1' \cdot x_2') - x_1' \cdot \rho_b - x_2' \cdot \tau_b + \xi_b$.

FSS Multiplication Gate The FSS multiplication gate is designed to securely compute the arithmetic function $g^\times(x_1, x_2) := x_1 \cdot x_2$ with the use of masks r_1^{in} and r_2^{in}, i.e., $(x_1 + r_1^{in}) \cdot (x_2 + r_2^{in})$. The two key routines in the FSS multiplication gate are denoted as $\text{KeyGen}^\times$ and $\text{Eval}^\times$. The evaluation process of the FSS multiplication gate is formally defined as $[\![x_1 \cdot x_2 + r_{out}]\!]_b \leftarrow \text{FSS.Mul}(x_1', x_2')$. The detailed algorithm for this procedure is presented in Algorithm 1.

Algorithm 2 FSS addition $\text{FSS.Add}(x_1', x_2')$

Addition Gate (KeyGen^+, Eval^+)
$\text{KeyGen}^+(1^\lambda, r_1^{in}, r_2^{in}, r^{out})$

(1) Randomly sample $\xi_0, \xi_1 \in \mathbb{Z}_{2^\ell}^{in}$ such that $\xi_0 + \xi_1 = r^{out} - (r_1^{in} + r_2^{in})$.
(2) Define $k_b^+ = \xi_b$, where $b \in \{0, 1\}$.
Return: (k_0^+, k_1^+).
$\text{Eval}^+(b, k_b^+, x_1', x_2')$

(1) Parse $k_b^+ = \xi_b$.
Return: $x_1' + \xi_0$ and $x_2' + \xi_1$.

FSS Addition Gate The FSS addition gate is designed to securely compute the masked function $g^+(x_1, x_2) := x_1 + x_2$ using the masks r_1^{in} and r_2^{in}, i.e., $(x_1 + r_1^{in}) + (x_2 + r_2^{in})$, in a non-interactive manner. The two core routines in the FSS addition gate are referred to as KeyGen^+ and Eval^+. We define the FSS addition evaluation process as follows: $[\![x_1 + x_2 + r_{out}]\!]_b \leftarrow \mathsf{FSS.Add}(x_1', x_2')$. The formal algorithm is provided in Algorithm 2.

3.8 Secure Multiparty Computation Implementations

3.8.1 Overview of SMC Libraries

Over the years, researchers and engineers have developed various open-source libraries to support secure computations among multiple parties, enabling cryptographic techniques such as secret sharing and garbled circuits. We introduce some widely used MPC libraries as follows.

MP-SPDZ MP-SPDZ [42] is a comprehensive framework that supports multiple MPC protocols, including Yao's Garbled Circuits (GC), BMR, Additive Secret Sharing, Replicated Secret Sharing, and Shamir's Secret Sharing, with various trust settings such as semi-honest and malicious adversaries. MP-SPDZ is licensed under the BSD 3-Clause License, allowing redistribution and use in source and binary forms, with or without modification, under certain conditions.

Although MP-SPDZ is implemented in C++, it provides Python bindings for accessibility. Its compiler optimizes computation by minimizing communication rounds for secret-sharing-based protocols and leveraging AES-NI pipelining for garbled circuits. Many research papers have been built on MP-SPDZ to develop secure computation applications [43, 44].

MPyC MPyC [45] is an implementation of Shamir Secret Sharing, enabling secure multiparty computation while tolerating a dishonest minority of up to t passively corrupted parties. MPyC is licensed under the MIT License, which allows unrestricted use, modification, distribution, and sublicensing, provided that the original copyright and permission notices are retained. It is implemented in Python, making it accessible for rapid prototyping and academic research.

CrypTen CrypTen [46] is an implementation of Replicated Secret Sharing, supporting a three-party model under a semi-honest trust setting. Licensed under the MIT License, CrypTen is implemented in Python and is specifically designed for Privacy-Preserving Machine Learning (PPML). It integrates seamlessly with PyTorch, allowing researchers to develop encrypted models without modifying existing PyTorch-based deep learning workflows.

Several research papers have leveraged CrypTen for efficient MPC-based deep learning frameworks, such as secure transformer inference [47] and GPU-accelerated privacy-preserving ML [48] (Table 3.2).

3.8 Secure Multiparty Computation Implementations

Table 3.2 Comparison of MPC libraries

Libraries	Yao's GC (2PC)	BMR (2PC)	Additive SS (2PC)	Replicated SS (3PC)	Shamir SS (nPC)
MP-SPDZ	*	*	*	*	*
MPyC					*
CrypTen				*	

∗ indicates the supported protocols of an MPC library

3.8.2 Example of Yao's Garbled Circuit Implementation

```
1  from cryptography.hazmat.primitives.ciphers import Cipher,
       algorithms, modes
2  from cryptography.hazmat.primitives import padding
3  from cryptography.hazmat.backends import default_backend
4  import os
5  import random
6  # The code below is implemented based on Yakoubov S. A gentle
       introduction to Yao garbled circuits [J]. preprint on webpage
       at https://web.mit.edu/sonka89/www/papers/2017ygc.pdf, 2017.
7
8
9  def aes_encrypt(key, plaintext):
10     """
11     Encrypt plaintext using AES in CBC mode.
12
13     :param key: A 16-byte (128-bit) encryption key
14     :param plaintext: The plaintext to encrypt
15     :return: A tuple of (IV, ciphertext)
16     """
17     if len(key) != 16:
18         raise ValueError("Key must be 16 bytes (128 bits) long.")
19
20     # Generate a random IV (Initialization Vector)
21     iv = os.urandom(16)
22
23     # Pad plaintext to a multiple of the block size (16 bytes for
           AES)
24     padder = padding.PKCS7(128).padder()
25     padded_plaintext = padder.update(plaintext.encode()) + padder
           .finalize()
26
27     # Initialize AES cipher in CBC mode
28     cipher = Cipher(algorithms.AES(key), modes.CBC(iv), backend=
           default_backend())
29     encryptor = cipher.encryptor()
30
31     # Encrypt the plaintext
32     ciphertext = encryptor.update(padded_plaintext) + encryptor.
           finalize()
33
34     return iv, ciphertext
```

```python
def aes_decrypt(key, iv, ciphertext):
    """
    Decrypt ciphertext using AES in CBC mode.

    :param key: A 16-byte (128-bit) decryption key
    :param iv: The initialization vector used during encryption
    :param ciphertext: The ciphertext to decrypt
    :return: The decrypted plaintext as a string
    """
    if len(key) != 16:
        raise ValueError("Key must be 16 bytes (128 bits) long.")
    if len(iv) != 16:
        raise ValueError("IV must be 16 bytes (128 bits) long.")

    # Initialize AES cipher in CBC mode
    cipher = Cipher(algorithms.AES(key), modes.CBC(iv), backend=
        default_backend())
    decryptor = cipher.decryptor()

    # Decrypt the ciphertext
    padded_plaintext = decryptor.update(ciphertext) + decryptor.
        finalize()

    # Unpad the plaintext
    unpadder = padding.PKCS7(128).unpadder()
    plaintext = unpadder.update(padded_plaintext) + unpadder.
        finalize()

    return plaintext.decode()

# An example of AES encryption
key = os.urandom(16)

# Plaintext to encrypt
plaintext = "This is a secret message."

# Encrypt the plaintext
iv, ciphertext = aes_encrypt(key, plaintext)
print(f"Ciphertext: {ciphertext.hex()}")

# Decrypt the ciphertext
decrypted_plaintext = aes_decrypt(key, iv, ciphertext)
print(f"Decrypted Plaintext: {decrypted_plaintext}")

key_a_0 = os.urandom(16)
len(key_a_0)

# Alice

# Alice picks four random keys. For example, key_b_0 corresponds
    to the event that Bob's input is 0.
key_a = os.urandom(16)
key_b = os.urandom(16)
```

3.8 Secure Multiparty Computation Implementations

```python
85
86  key_a_0, key_a_1 = key_a[:8], key_a[8:]
87  key_b_0, key_b_1 = key_b[:8], key_b[8:]
88
89  # garbled output
90
91  iv, ciphertext = aes_encrypt(key, plaintext)
92
93  # XOR operation
94  plaintext_table = {
95      (0, 0): '0',
96      (0, 1): '1',
97      (1, 0): '1',
98      (1, 1): '0',
99  }
100
101 #Alice send the garbled_output to Bob and its input key. For
        example, if Alice holds the bit of 1, it will send key_a_1 to
        Bob.
102 garbled_output = [
103     aes_encrypt(key_a_0 + key_b_0, plaintext_table[(0,0)]),
104     aes_encrypt(key_a_0 + key_b_1, plaintext_table[(0,1)]),
105     aes_encrypt(key_a_1 + key_b_0, plaintext_table[(1,0)]),
106     aes_encrypt(key_a_1 + key_b_1, plaintext_table[(1,1)]),
107 ]
108
109 # Alice shuffle the garbled table
110 random.shuffle(garbled_output)
111
112 # Bob
113
114 # Bob receive the Alice's input and the shuffled garbled_output.
        For example, key_a_0.
115 # However, by only getting key_a_0, which is a random value, Bob
        can not learn what's Alice's true input.
116 print(key_a_0)
117
118 # Bob knows nothing to the permuted_garbled_output
119 print(garbled_output)
120
121 # Bob holds input of 1, by Bob and Alice running a oblivious
        transfer, Bob will learn key_b_1
122 print(key_b_0)
123
124 key = key_a_0 + key_b_0
125
126 # Bob start to open the value it wants
127 for i, (iv, ciphertext) in enumerate(garbled_output):
128     try:
129        decrypted_plaintext = aes_decrypt(key, iv, ciphertext)
130        print("The", i, "-th value can be decrypted as",
               decrypted_plaintext)
131     except ValueError:
132        print("The", i, "-th value can not be decrypted")
```

3.8.3 Example of SMC with Secret Sharing Implementation

Two-party Arithmetic Secret Sharing

```python
import random

def float_to_fixed(value, f, N):
    return int(value * (2**f)) % N

def fixed_to_float(value, f):
    return value / (2**f)

def share_secret(secret, num_parties, N):
    shares = [random.randint(0, N - 1) for _ in range(num_parties
         - 1)]
    final_share = (secret - sum(shares)) % N
    shares.append(final_share)
    return shares

def reconstruct_secret(shares, N):
    return sum(shares) % N

def addition (shares_0, shares_1, num_parties, N):
    new_shares = [(shares_0[i] + shares_1[i]) % N for i in range(
        num_parties)]
    return new_shares

# Example usage
if __name__ == "__main__":
    # Integer field for modular arithmetic
    N = 2**32

    # Number of fractional bits for fixed-point representation
    f = 16

    # Number of parties involved in secret sharing
    num_parties = 2

    # Example secret value
    secret_float = 123.456

    # Convert the secret to fixed-point representation
    secret_fixed = float_to_fixed(secret_float, f, N)

    # Split the fixed-point secret into shares
    shares = share_secret(secret_fixed, num_parties, N)

    print(f"Original Secret (float): {secret_float}")
    print(f"Original Secret (fixed-point): {secret_fixed}")
    print(f"Shares: {shares}")

    # Reconstruct the fixed-point secret
    reconstructed_fixed = reconstruct_secret(shares, N)
```

```
48
49        # Convert back to floating-point representation
50        reconstructed_float = fixed_to_float(reconstructed_fixed, f)
51        print(f"Reconstructed Secret (fixed-point): {
              reconstructed_fixed}")
52        print(f"Reconstructed Secret (float): {reconstructed_float}")
53
54        # Alice's secret value (floating-point)
55        secret_float_0 = 123.456
56        # Convert Alice's secret to fixed-point representation
57        secret_fixed_0 = float_to_fixed(secret_float_0, f, N)
58        # Create additive shares for Alice's secret
59        shares_0 = share_secret(secret_fixed_0, num_parties, N)
60
61        # Bob's secret value (floating-point)
62        secret_float_1 = 789.123
63        # Convert Bob's secret to fixed-point representation
64        secret_fixed_1 = float_to_fixed(secret_float_1, f, N)
65        # Create additive shares for Bob's secret
66        shares_1 = share_secret(secret_fixed_1, num_parties, N)
67
68        # Perform addition of the shares
69        shares_added = addition(shares_0, shares_1, num_parties, N)
70
71        # Reconstruct the sum of the shared secrets and convert back
              to floating-point
72        reconstructed_fixed_result = fixed_to_float(
              reconstruct_secret(shares_added, N), f)
73        print(f"Reconstructed Secret (float): {
              reconstructed_fixed_result}")
74
75        # Compute the ground-truth value by directly adding the
              floating-point numbers
76        print(f"Ground-truth value: {secret_float_0+secret_float_1}")

     Original Secret (float): 123.456
     Original Secret (fixed-point): 8090812
     Shares: [1420876667, 2882181441]
     Reconstructed Secret (fixed-point): 8090812
     Reconstructed Secret (float): 123.45599365234375
     Ground-truth value: 912.5790000000001
     Reconstructed Secret (float): 912.5789794921875
```

References

1. Shamir, A. How to share a secret. *Communications of the ACM*, 22:612–613, 1979.
2. Ben-Or, M., Goldwasser, S., and Wigderson, A. Completeness Theorems for Non-Cryptographic Fault-Tolerant Distributed Computation. In *Proc. 20th Annual ACM Symposium on Theory of Computing (STOC'88)*, pages 1–10, 1988.

3. Dawson, E., and Donovan, D. The breadth of Shamir's secret-sharing scheme. *Computers & Security*, 13:69–78, 1994.
4. Lindell, Y. Secure multiparty computation. *Communications of the ACM*, 64:86–96, 2021.
5. Canetti, R., Feige, U., Goldreich, O., and Naor, M. Adaptively Secure Multi-party Computation. In *Proc. 28th ACM Symposium on Theory of Computing (STOC 96)*, pages 639–648, 1996.
6. Yao, A. C.-C. How to generate and exchange secrets. In *27th Annual Symposium on Foundations of Computer Science (Sfcs 1986)*, pages 162–167, IEEE, 1986.
7. Goldreich, O., Micali, S., and Wigderson, A. How to play any mental game, or a completeness theorem for protocols with honest majority. In *Providing Sound Foundations for Cryptography: On the Work of Shafi Goldwasser and Silvio Micali*, pages 307–328, 2019.
8. E. Boyle, N. Gilboa, and Y. Ishai, "Function secret sharing: Improvements and extensions," in *Proceedings of the 2016 ACM SIGSAC Conference on Computer and Communications Security*, 2016, pp. 1292–1303.
9. E. Boyle, N. Chandran, N. Gilboa, D. Gupta, Y. Ishai, N. Kumar, and M. Rathee, "Function secret sharing for mixed-mode and fixed-point secure computation," in *Annual International Conference on the Theory and Applications of Cryptographic Techniques*, Springer, 2021, pp. 871–900.
10. Kelbert, M., Suhov, Y.: Information Theory and Coding by Example. Cambridge University Press, Cambridge (2013) Kelbert, M., and Suhov, Y. *Information Theory and Coding by Example*. Cambridge University Press, Cambridge, 2013.
11. T. Araki, J. Furukawa, Y. Lindell, A. Nof, and K. Ohara. High-throughput semi-honest secure three-party computation with an honest majority. In *Proceedings of the 2016 ACM SIGSAC Conference on Computer and Communications Security*, pages 805–817, 2016.
12. G. Asharov, Y. Lindell, T. Schneider, and M. Zohner. More efficient oblivious transfer and extensions for faster secure computation. In *Proc. of ACM CCS*, 2013.
13. M. Atallah, M. Bykova, J. Li, K. Frikken, and M. Topkara. Private collaborative forecasting and benchmarking. In *Proc. of WPES*, 2004.
14. M. Ball, B. Carmer, T. Malkin, M. Rosulek, and N. Schimanski. Garbled neural networks are practical. *IACR Cryptol. ePrint Arch.*, 2019:338, 2019.
15. M. Ball, T. Malkin, and M. Rosulek. Garbling gadgets for boolean and arithmetic circuits. In *Proc. of ACM CCS*, 2016.
16. D. Beaver. Efficient multiparty protocols using circuit randomization. In *Proc. of Crypto*, 1991.
17. D. Beaver. Correlated pseudorandomness and the complexity of private computations. In *Proceedings of the twenty-eighth annual ACM symposium on Theory of computing*, pages 479–488, 1996.
18. D. Beaver, S. Micali, and P. Rogaway. The round complexity of secure protocols. In *Proceedings of the twenty-second annual ACM symposium on Theory of computing*, pages 503–513, 1990.
19. M. Bellare, V. T. Hoang, S. Keelveedhi, and P. Rogaway. Efficient garbling from a fixed-key blockcipher. In *2013 IEEE Symposium on Security and Privacy*, pages 478–492. IEEE, 2013.
20. M. Ben-Or, S. Goldwasser, and A. Wigderson. Completeness theorems for non-cryptographic fault-tolerant distributed computation. In *Proceedings of the twentieth annual ACM symposium on Theory of computing*, pages 1–10. ACM, 1988.
21. S. G. Choi, K.-W. Hwang, J. Katz, T. Malkin, and D. Rubenstein. Secure multi-party computation of boolean circuits with applications to privacy in on-line marketplaces. In *Cryptographers' Track at the RSA Conference*, pages 416–432. Springer, 2012.
22. D. Demmler, T. Schneider, and M. Zohner. Aby-a framework for efficient mixed-protocol secure two-party computation. In *Proc. of NDSS*, 2015.
23. Y. Huang, D. Evans, J. Katz, and L. Malka. Faster secure two-party computation using garbled circuits. In *Proc. of USENIX Security*, 2011.
24. Y. Huang, J. Katz, and D. Evans. Efficient secure two-party computation using symmetric cut-and-choose. In *Annual Cryptology Conference*, pages 18–35. Springer, 2013.

References

25. Y. Ishai, J. Kilian, K. Nissim, and E. Petrank. Extending oblivious transfers efficiently. In *Annual International Cryptology Conference*, pages 145–161. Springer, 2003.
26. J. Katz, S. Ranellucci, M. Rosulek, and X. Wang. Optimizing authenticated garbling for faster secure two-party computation. In *Annual International Cryptology Conference*, pages 365–391. Springer, 2018.
27. V. Kolesnikov, P. Mohassel, and M. Rosulek. Flexor: Flexible garbling for xor gates that beats free-xor. In *Annual Cryptology Conference*, pages 440–457. Springer, 2014.
28. V. Kolesnikov and T. Schneider. Improved garbled circuit: Free xor gates and applications. In *Proc. of ICALP*. Springer, 2008.
29. B. Kreuter, A. Shelat, and C.-H. Shen. Billion-gate secure computation with malicious adversaries. In *21st {USENIX} Security Symposium ({USENIX} Security 12)*, pages 285–300, 2012.
30. Y. Lindell. Fast cut-and-choose-based protocols for malicious and covert adversaries. *Journal of Cryptology*, 29(2):456–490, 2016.
31. Y. Lindell and B. Pinkas. Secure two-party computation via cut-and-choose oblivious transfer. *Journal of cryptology*, 25(4):680–722, 2012.
32. D. Malkhi, N. Nisan, B. Pinkas, Y. Sella, et al. Fairplay-secure two-party computation system. In *Proc. of USENIX Security*, 2004.
33. P. Mohassel and P. Rindal. Aby3: A mixed protocol framework for machine learning. In *Proc. of ACM CCS*, 2018.
34. M. Naor and B. Pinkas. Efficient oblivious transfer protocols. In *SODA*, volume 1, pages 448–457, 2001.
35. M. Naor, B. Pinkas, and R. Sumner. Privacy preserving auctions and mechanism design. In *Proceedings of the 1st ACM Conference on Electronic Commerce*, pages 129–139, 1999.
36. B. Pinkas, T. Schneider, N. P. Smart, and S. C. Williams. Secure two-party computation is practical. In *International conference on the theory and application of cryptology and information security*, pages 250–267. Springer, 2009.
37. M. O. Rabin. How to exchange secrets with oblivious transfer. *Harvard University, Tech. Rep. TR-81*, 1981.
38. T. Schneider and M. Zohner. Gmw vs. yao? efficient secure two-party computation with low depth circuits. In *International Conference on Financial Cryptography and Data Security*, pages 275–292. Springer, 2013.
39. C.-h. Shen et al. Two-output secure computation with malicious adversaries. In *Annual International Conference on the Theory and Applications of Cryptographic Techniques*, pages 386–405. Springer, 2011.
40. X. Wang, A. J. Malozemoff, and J. Katz. Faster secure two-party computation in the single-execution setting. In *Annual International Conference on the Theory and Applications of Cryptographic Techniques*, pages 399–424. Springer, 2017.
41. S. Zahur, M. Rosulek, and D. Evans. Two halves make a whole. In *Proc. of Eurocrypt*, 2015.
42. KELLER, M. MP-SPDZ: A versatile framework for multi-party computation. In CCS, 2020.
43. XU, Z., LAI, S., LIU, X., ABUADBBA, A., YUAN, X., AND YI, X. OblivGNN: Oblivious inference on transductive and inductive graph neural networks. In USENIX Security, 2024.
44. PANG, Q., YUAN, Y., AND WANG, S. MPCDiff: Testing and repairing MPC-hardened deep learning models. In NDSS, 2024.
45. SCHOENMAKERS, B. MPyC: Secure multiparty computation in Python. https://github.com/lschoe/mpyc.
46. FACEBOOK RESEARCH. CrypTen: Secure multi-party computation for ML. https://github.com/facebookresearch/CrypTen.
47. LI, D., WANG, H., SHAO, R., GUO, H., XING, E., AND ZHANG, H. MPCFormer: Fast, performant, and private transformer inference with MPC. In ICLR, 2023.
48. TAN, S., KNOTT, B., TIAN, Y., AND WU, D. J. CryptGPU: Fast privacy-preserving machine learning on the GPU. In IEEE S&P, 2021.

Chapter 4
Differential Privacy

Abstract In this chapter, we introduce Differential Privacy (DP), a mathematical framework designed to preserve individual privacy in data analysis by incorporating carefully calibrated noise into computations. We provide an intuitive understanding of its principles and delve into its formal definitions, fundamental mechanisms, and composition theorems. Additionally, we discuss its variations, including Relaxed Differential Privacy—such as Rényi Differential Privacy (RDP) and Zero-Concentrated Differential Privacy (zCDP)—which offer greater flexibility and tighter bound for iterative algorithms and complex data analysis tasks. Furthermore, we examine Local Differential Privacy (LDP), a decentralized model that ensures privacy at the data source without the need for a trusted central curator, and highlight its key mechanisms. Finally, we explore widely used tools and libraries for implementing differential privacy and provided examples to illustrate their practical application.

4.1 Introduction to Differential Privacy

This section introduces the foundation for understanding differential privacy, presenting its core concepts, mechanisms, and theoretical guarantees.

4.1.1 Intuition Behind Differential Privacy

Differential privacy tackles the challenge of gaining meaningful insights about a population while protecting the privacy of any individual in the dataset. At its core, DP operates on the principle of indistinguishability: the results of any analysis on a dataset should remain nearly identical whether or not any single individual's data is included. This ensures that adversaries cannot infer information about a specific individual, even with auxiliary knowledge or repeated queries. To achieve this, DP introduces carefully calibrated random noise into the computation.

Imagine a survey is conducted to study the average income in a city. Without privacy protections, an individual's income might be identifiable from the survey's results, especially if the dataset is small or the individual's income is unique. Differential privacy ensures that the published average income has noise added to it, making it impossible to infer any specific individual's income. For instance, even if an adversary knows certain external details about someone, the randomized noise prevents them from determining whether that person's data was included or excluded. At the same time, the noisy average remains accurate enough to provide useful insights about the overall income distribution of the city.

Differential privacy guarantees that the insights derived from a dataset, such as the finding that regular exercise reduces the risk of heart disease, remain consistent regardless of whether a specific individual's data is included or excluded. More precisely, it ensures that the likelihood of any particular output is nearly the same, whether or not an individual's data is part of the dataset. This "near equality" is determined by probabilities arising from the randomness introduced by the privacy mechanism, which is managed by the data curator. The degree of similarity in these probabilities is controlled by a privacy parameter ϵ. A smaller ϵ provides stronger privacy at the cost of reduced data accuracy, while a larger ϵ allows greater utility with weaker privacy guarantees.

4.1.2 Formal Definition

Differential privacy ensures that the inclusion or exclusion of any single individual's data in a dataset does not significantly affect the outcome of an analysis.

Definition 4.1 (Differential Privacy [1]) A randomized algorithm $\mathcal{M}$ with domain $\mathbb{N}^{|X|}$ is (ϵ, δ)-differential privacy if for all $\mathcal{S} \subseteq \text{Range}(\mathcal{M})$ and for all $x, y \in \mathbb{N}^{|X|}$ such that $||x - y||_1 \leq 1$:

$$Pr[\mathcal{M}(x) \in \mathcal{S}] \leq exp(\epsilon) Pr[\mathcal{M}(y) \in \mathcal{S}] + \delta$$

(ϵ, δ)-differential privacy is also known as approximate differential privacy, which is an extension of the standard differential privacy framework. It introduces an additional parameter, δ, which allows for a small probability of a stronger privacy loss than ϵ would permit. The pure ϵ-differential privacy is a special case of the approximate differential privacy when $\delta = 0$. DP ensures that the output of a mechanism is nearly identical regardless of whether an individual's data is included in the dataset. This makes it impossible for an attacker to confidently infer any single individual's participation or contribution. Unlike traditional methods, such as k-anonymity and l-diversity, DP guarantees hold even if an attacker has access to external information. This robustness makes DP particularly valuable in real-world scenarios. The parameters ϵ and δ provide a clear, quantifiable measure of privacy, allowing practitioners to balance privacy and utility based on their needs.

4.1 Introduction to Differential Privacy

Differential privacy is immune to post-processing. It means that once data has been processed through a DP mechanism, any additional operations or transformations applied to the output cannot degrade the original privacy guarantees. In other words, the privacy protection provided by the DP mechanism is preserved regardless of how the privatized results are subsequently used or manipulated, as long as the raw data is not revisited.

Let $\mathcal{M}$ be a DP mechanism, and let f be any post-processing function. For all subsets T in the output space of f:

$$Pr[f(\mathcal{M}(D)) \in T] = \sum_{S \subseteq \mathcal{M}(D): f(S) \in T} Pr[\mathcal{M}(D) \in S] \qquad (4.1)$$

The probabilities remain governed by the original DP guarantee:

$$Pr[\mathcal{M}(D) \in S] \leq e^{\epsilon} Pr[\mathcal{M}(D') \in S] + \delta \qquad (4.2)$$

Thus:

$$Pr[f(\mathcal{M}(D)) \in T] \leq e^{\epsilon} Pr[f(\mathcal{M}(D')) \in T] + \delta \qquad (4.3)$$

This demonstrates that the output of f preserves the same privacy guarantees as the original mechanism. We can see that the privacy guarantee is entirely determined by the randomness added by the DP mechanism $\mathcal{M}$ at the time of its interaction with the dataset. Post-processing functions have no access to the original data D or D' and it operate only on the already privatized output of $\mathcal{M}$. Since f cannot alter the internal randomness or the indistinguishability properties established by $\mathcal{M}$, the privacy guarantee remains intact.

4.1.3 Basic Mechanisms

4.1.3.1 The Laplace Mechanism

The Laplace mechanism is a fundamental tool for achieving differential privacy. It achieves it by adding random noise drawn from a Laplace distribution to the query result. The scale of the noise is calibrated based on the sensitivity of the query function, which measures the maximum possible change in the output caused by altering one individual's data, and the privacy parameter ϵ, which controls the trade-off between privacy and accuracy. By introducing this noise, the Laplace mechanism guarantees that the output is indistinguishable within defined bounds for neighboring datasets, providing strong privacy protection while preserving the utility of aggregate information. This mechanism is particularly suited for numeric queries, such as sums or averages, where noise can be added directly to the result to achieve DP compliance.

Definition 4.2 (ℓ_1—Sensitivity) Let $f : D \to \mathbb{R}^k$ be a query function, where D is the domain of the datasets, and let D and D' be two neigbouring dataset that differ by at most one record, the ℓ_1- sensitivity of f, denoted as Δf, is defined as:

$$\Delta f = \max_{D,D'} ||f(D) - f(D')||_1 \tag{4.4}$$

where $|| \cdot ||$ represents the ℓ_1—norm.

The ℓ_1-sensitivity of a query function measures the maximum possible change in the function's output when a single record in the dataset is added, removed, or altered. It quantifies the maximum potential impact a single individual's data can have on the output of a query. This relationship is crucial for determining the amount of noise that must be added to the query result to ensure privacy guarantees.

For the Laplace mechanism, the noise added to the query result is sampled from the Laplace distribution. The Laplace distribution is characterized by its probability density function shown as follows.

$$P(x|b) = \frac{1}{2b} exp(-\frac{|x|}{b}), \tag{4.5}$$

where b is the scale parameter of the distribution and is set based on the sensitivity of the query and the desired privacy budget. x is the random variable representing the noise. We will use Lap(b) to represent the Laplace distribution with scale b.

The Laplace mechanism operates by first computing the result of the query on the dataset and then adding noise to each dimension of the result to ensure privacy. The noise is sampled from a Laplace distribution, which is centered at zero and scaled by b, where $b = \frac{\Delta f}{\epsilon}$. Δf represents the sensitivity of the query and ϵ is the privacy budget controlling the trade-off between privacy and accuracy. The formal definition is shown as follows.

Definition 4.3 (The Laplace Mechanism) Let $f : D \to \mathbb{R}^k$ be a query function that maps a dataset $x \in D$ to a k—dimensional real—valued vector. The Laplace mechanism is defined as:

$$\mathcal{M}_L(x, f, \epsilon) = f(x) + (Y_1, Y_2, \cdots, Y_k), \tag{4.6}$$

where Y_i are i.i.d. random variables drawn from the Laplace distribution Lap($\Delta f/\epsilon$)

The Laplace mechanism satisfied ϵ—differential privacy, because for any two neighboring datasets x and x' and for any possible output $O \subseteq \mathbb{R}^k$:

$$Pr[\mathcal{M}_L(x, f, \epsilon) \in O] \le e^\epsilon Pr[\mathcal{M}_L(x', f, \epsilon) \in O] \tag{4.7}$$

The added noise ensures that the outputs for neighboring datasets are indistinguishable within a factor of e^ϵ, preserving individual privacy.

4.1.3.2 The Gaussian Mechanism

The Gaussian mechanism is a widely used approach to ensure (ϵ, δ)—differential privacy, particularly in applications requiring approximate privacy guarantees. It achieves this by adding noise sampled from a Gaussian (normal) distribution, characterized by its mean of zero and a carefully determined standard deviation. The amount of noise depends on the sensitivity of the query function, which represents the maximum change in the result due to modifying one individual's data, and the privacy parameters ϵ and δ, which define the privacy level and allowable probability of a privacy breach, respectively. This approach ensures that outputs for neighboring datasets are statistically similar with high probability, balancing privacy and utility. The Gaussian mechanism is especially useful for complex queries or machine learning models, where distributing noise across multiple dimensions is necessary for robust privacy guarantees.

Definition 4.4 (ℓ_2 Sensitivity) Let $f : D \rightarrow \mathbb{R}^k$ be a query function, where D is the domain of datasets and $\mathbb{R}^k$ is the range of the query output. For any two neighbouring datasets D and D' (differing by at most one record), the ℓ_2—sensitivity of f, denoted as $\Delta_2 f$, is defined as:

$$\Delta_2 f = \max_{D, D'} ||f(D) - f(D')||_2, \tag{4.8}$$

where $||\cdot||_2$ is the ℓ_2—norm.

The Gaussian mechanism adds noise drawn from a normal distribution, which is defined by its mean (centered at zero) and standard deviation σ. The Gaussian distribution's tail probabilities depend on the Euclidean distance, that is ℓ_2-norm, between outputs. This makes ℓ_2-sensitivity a natural measure for calibrating the noise. Besides, the ℓ_2—norm measures the overall distance between two points in Euclidean space, making it well-suited for functions with multidimensional outputs. In high-dimensional spaces, ℓ_2-norm provides a more compact and realistic sensitivity measure by accounting for changes across all dimensions simultaneously, unlike ℓ_1-sensitivity, which aggregates the absolute changes per dimension. The ℓ_2—sensitivity plays a fundamental role in the Gaussian mechanism, it directly informs how much Gaussian noise is required to obfuscate this maximum change and ensure privacy.

For the Gaussian mechanism, the noise added to the query result is sampled from the Gaussian distribution. The Gaussian distribution is defined by its probability density function:

$$P(x|\mu, \sigma^2) = \frac{1}{\sqrt{2\pi\sigma^2}} exp(-\frac{(x-\mu)^2}{2\sigma^2}), \tag{4.9}$$

where μ represents the mean of the distribution, set to zero in the Gaussian mechanism, and σ^2 is the variance, which determines the scale of the noise added.

The Gaussian distribution is denoted as $\mathcal{N}(0, \sigma^2)$, where the standard deviation σ is proportional to the ℓ_2—sensitivity of the query, which is given by:

$$\sigma = \frac{\Delta_2 f \cdot \sqrt{2ln(1.25/\delta)}}{\epsilon}, \tag{4.10}$$

This formula ensures that the added noise sufficiently masks the largest potential change in the query result, as measured by ℓ_2-sensitivity. Next, we give the formal definition of the Gaussian mechanism.

Definition 4.5 (The Gaussian Mechanism) Let $f : D \to \mathbb{R}^k$ be a query fucntion mapping a dataset D to a k-dimensinoal vector. The Gaussian mechanism is defined as:

$$\mathcal{M}_G(x, f, \epsilon, \delta) = f(x) + (\eta_1, \eta_2, \cdots, \eta_k), \tag{4.11}$$

where η_i are independent noise sampled from a Gaussian distribution $\eta_i \sim \mathcal{N}(0, \sigma^2)$.

The Gaussian mechanism satisfies (ϵ, δ)—differential privacy for any two neighboring datasets x and x', ensuring that:

$$Pr[\mathcal{M}_G(x, f, \epsilon, \delta) \in O] \leq e^\epsilon Pr[\mathcal{M}_G(x', f, \epsilon, \delta) \in O] + \delta \tag{4.12}$$

for any possible output $O \subseteq \mathbb{R}^k$.

4.1.3.3 The Exponential Mechanism

The Exponential Mechanism is a powerful method for achieving differential privacy, particularly when the query output is categorical or non-numeric. Instead of adding random noise to the result itself, the Exponential Mechanism modifies the probability of selecting each possible output based on a utility function. This utility function quantifies the quality or relevance of each output relative to the dataset and ensures that outputs with higher utility are more likely to be chosen. The probability of selecting an output is calibrated according to the sensitivity of the utility function, which measures the maximum possible change in utility caused by altering one individual's data, and the privacy parameter ϵ, which controls the trade-off between privacy and the quality of the output. By probabilistically selecting outputs in this way, the Exponential Mechanism ensures that outputs remain indistinguishable within defined bounds for neighboring datasets, providing strong privacy guarantees. This mechanism is particularly suited for tasks such as selecting optimal features, categories, or other discrete outputs where direct noise addition is impractical.

The Exponential Mechanism relies on a utility function $u(x, r)$ that evaluates how desirable or relevant a possible output r is for the dataset x. Outputs with

4.1 Introduction to Differential Privacy

higher utility values are more desirable, and the mechanism ensures that they are more likely to be selected while still preserving privacy. To ensure privacy, the Exponential Mechanism introduces randomness into the selection process by probabilistically choosing outputs. The probability of selecting an output r is proportional to $exp(\frac{\epsilon \cdot u(x,r)}{2\Delta u})$, where Δu is the sensitivity of the utility function (i.e., the maximum change in $u(x, r)$ caused by altering one individual's data). This formula ensures that outputs with higher utility are exponentially more likely to be chosen, but the privacy parameter ϵ controls how much more likely they are compared to lower-utility outputs. The formal definition of the Exponential Mechanism is shown as follows.

Definition 4.6 (The Exponential Mechanism) Let D be the dataset, $\mathcal{R}$ be the set of all possible outputs, and Δu be the sensitivity of the utility function, the Exponential Mechanism selects an output $r \in \mathcal{R}$ with probability:

$$Pr[\mathcal{M}_E(D) = r] \propto exp(\frac{\epsilon \cdot u(D, r)}{2\Delta u}) \tag{4.13}$$

Imagine you want to select the most frequent word in a dataset of text documents. Without privacy, you might simply pick the word with the highest count. However, if the dataset contains sensitive information, directly revealing the most frequent word could leak details about an individual. The Exponential Mechanism instead assigns a probability to each word based on its frequency (utility), ensuring that more frequent words are still likely to be chosen but not deterministically, preserving privacy.

The Exponential Mechanism satisfies ϵ—differential privacy. For any two neighboring datasets D and D', and for any subset of outputs $S \subseteq \mathcal{R}$, the mechanism ensures:

$$Pr[\mathcal{M}_E(D) \in S] \leq e^{\epsilon} Pr[\mathcal{M}_E(D') \in S] \tag{4.14}$$

The Exponential Mechanism ensures differential privacy by carefully balancing the likelihood of selecting higher-utility outputs with the need to obscure the influence of any single individual's data. By making outputs with similar utilities nearly indistinguishable, it protects privacy while still favoring results that are useful and relevant for the given task. The Exponential Mechanism does not require numeric outputs and can handle any discrete output space.

4.1.4 Composition Theorems

The composition properties of differential privacy describe how privacy guarantees degrade when multiple differentially private mechanisms are applied to the same dataset. These theorems provide a framework for analyzing the cumulative privacy loss, enabling practitioners to design complex systems while maintaining privacy guarantees.

Theorem 4.1 (Parallel Composition [1]) *Suppose that a method includes m independent randomized functions* $M = \{M_1, M_2, \ldots, M_m\}$, *each* M_i *provides* ϵ_i-*differential privacy guarantee. If each* M_i *is performed on a disjointed record of the entire dataset, then* M *is* $\max\{\epsilon_1, \cdots, \epsilon_m\}$-*differentially private.*

Parallel composition ensures that the total privacy loss is determined by the most privacy-leaking mechanism applied to any subset. It allows the same privacy budget to be used independently across multiple disjoint subsets, enabling more complex analyses without additional privacy costs. The subsets D_i must be disjoint. If there is overlap, the standard composition theorems (basic or advanced) apply, leading to cumulative privacy loss.

Example of Parallel Composition Suppose a hospital dataset is divided into subsets based on geographic regions (e.g., data from different cities). If a differentially private mechanism is used to compute the average patient age for each city. Each city's data is treated as a disjoint subset. The privacy loss for analyzing the entire dataset is determined by the maximum ϵ used for any single city.

Parallel composition is particularly beneficial for large-scale datasets where disjoint partitions can be processed independently, such as federated learning or regional statistics. It allows for modular privacy-preserving systems, where differentially private algorithms are applied independently to subsets of data.

Theorem 4.2 (Sequential Composition [1]) *Suppose that a method includes m independent randomized functions* $M = \{M_1, M_2, \ldots, M_m\}$, *each* M_i *provides* (ϵ_i, δ_i)-*differential privacy guarantee. If these functions are performed on the same dataset sequentially, then* M *is* $(\sum_{i=1}^{m} \epsilon_i, \sum_{i=1}^{m} \delta_i)$-*differentially private.*

The basic sequential composition theorem considers the biggest value of the privacy loss. The total privacy loss is the sum of the individual privacy losses ϵ_i. This makes it straightforward to calculate the cumulative impact of multiple mechanisms. It applies when mechanisms operate on the same dataset or overlapping parts of the dataset. This includes iterative algorithms, such as gradient descent in machine learning, where the same data is queried repeatedly. Sequential composition holds regardless of whether the mechanisms are identical or different, as long as each satisfies differential privacy.

Example of Sequential Composition A government agency aims to publish insights from a sensitive income dataset while preserving privacy using differential privacy. They release three statistics: average income, median income, and income variance, each protected by a privacy parameter $\epsilon = 0.5$. Noise is added to each query independently, ensuring privacy for individual results. However, since all queries are applied to the same dataset, the sequential composition theorem dictates that the total privacy loss accumulates, resulting in an overall privacy guarantee of $\epsilon_{total} = 1.5$.

Sequential composition highlights the cost of repeated queries: the more mechanisms applied to the same dataset, the higher the total privacy loss, potentially requiring more noise to preserve privacy, which can reduce utility. In systems

4.1 Introduction to Differential Privacy

with limited privacy budgets, careful allocation of ϵ-values is necessary to balance privacy and utility across multiple queries. Sequential composition is a limitation for large-scale systems with many queries or iterations. Advanced composition can mitigate this by providing tighter bounds on privacy loss.

Theorem 4.3 (Advanced Composition [1]) *For all $\epsilon, \delta, \delta' \geq 0$, the class of (ϵ, δ)-differentially private mechanisms satisfies $(\epsilon', m\delta + \delta')$-differential privacy under m-fold adaptive composition for:*

$$\epsilon' = \epsilon\sqrt{2m \log(1/\delta')} + m\epsilon(e^\epsilon - 1)$$

Advanced composition provides a tighter analysis by leveraging the probabilistic nature of privacy loss. It uses inequalities like Chernoff [2] or Hoeffding [3] bounds to analyze the distribution of cumulative privacy loss. These bounds show that the probability of large deviations (extreme privacy breaches) diminishes exponentially. This allows advanced composition to bound the total privacy loss more precisely, reflecting realistic behavior. Advanced composition demonstrates that the cumulative privacy loss grows sublinearly with the number of mechanisms. Instead of assuming all privacy loss instances are independent and maximal, advanced composition incorporates dependencies and tail behavior, ensuring that the total privacy loss estimate is both tight and practical.

While advanced composition provides sublinear growth of privacy loss, it still accumulates significantly for a large number of queries. This makes it less effective for highly iterative processes. Later, Abadi et al. [4] proposed concept moments accountant to track the privacy loss, which reduces the upper bound further. The general idea is that instead of only considering the expectation, they explore the higher-order moments to bound the tail of the privacy loss variable.

Definition 4.7 (Moments Accountant [4]) Given a random mechansim $\mathcal{M}$, auxiliary input aux, and neighbouring datasets $d, d \in D$, the moment account is defined as:

$$\alpha(\lambda) \triangleq \max_{aux,d,d'} \alpha_\mathcal{M}(\lambda; aux, d, d')$$

where $\alpha_\mathcal{M}(\lambda; aux, d, d')$ is the λth moment, which is defined as:

$$\alpha_\mathcal{M}(\lambda; aux, d, d') \triangleq \log E_{o \sim \mathcal{M}(aux,d)}[exp(\lambda c(o; \mathcal{M}, aux, d, d'))]$$

The moment accountant is a powerful technique used to track and analyze the cumulative privacy loss of composed mechanisms in differential privacy, particularly in iterative algorithms like private stochastic gradient descent (SGD). The moment accountant tracks the log moments of the privacy loss random variable over multiple iterations or queries. By analyzing the higher-order moments of the privacy loss, it provides tighter cumulative bounds on privacy loss under composition.

4.2 Relaxed Differential Privacy

This section introduces the concept of relaxed differential privacy, which extends the strict guarantees of pure ϵ-differential privacy to provide more flexibility and practicality in real-world applications. Relaxed definitions, such as (ϵ, δ)-Differential Privacy, Rényi Differential Privacy (RDP), and Zero-Concentrated Differential Privacy (zCDP), allow for a controlled probability of privacy loss or focus on alternative measures of privacy leakage. These relaxations are particularly useful for complex queries, iterative algorithms, and high-dimensional data, where strict differential privacy may require excessive noise, significantly reducing utility. By allowing small, well-defined relaxations of privacy guarantees, these frameworks balance the trade-off between privacy protection and the usability of the data, making them essential for scalable and efficient privacy-preserving systems.

4.2.1 Rényi Differential Privacy

Rényi Differential Privacy is a relaxed framework for measuring privacy loss that generalizes the standard notion of differential privacy using Rényi divergence, a measure of the distance between two probability distributions. Unlike ϵ-differential privacy, which enforces a strict bound on the ratio of probabilities for neighboring datasets, RDP focuses on the cumulative privacy loss over multiple executions, making it particularly suited for iterative processes like machine learning model training.

Definition 4.8 ((α, ϵ)-**RDP** [5]) A randomized mechansim $\mathcal{M} : \mathbb{D} \to \mathbb{R}$ is said to have ϵ-Rényi differential privacy of order α, or (α, ϵ)-RDP for short, if for any adjacent $D, D' \in \mathbb{D}$ it holds that

$$D_\alpha(\mathcal{M}(D)||\mathcal{M}(D')) \leq \epsilon$$

where $D_\alpha(\mathcal{M}(D)||\mathcal{M}(D'))$ is the Renyi Divergence of order $\alpha > 1$, specifically,

$$D_\alpha(\mathcal{M}(D)||\mathcal{M}(D')) = \frac{1}{\alpha - 1} \log E_{x \sim \mathcal{M}(D)}[(\frac{Pr[\mathcal{M}(D) = x]}{Pr[\mathcal{M}(D') = x]})^{\alpha-1}] \qquad (4.15)$$

RDP is parameterized by an order $\alpha > 1$ and a privacy budget ϵ, which quantifies the divergence at order α. This approach offers a more accurate measure of cumulative privacy loss and tighter bounds when multiple queries are composed. By parameterizing privacy loss with Rényi divergence, RDP enables a smoother and more controlled tradeoff between privacy and utility, making it easier to achieve an optimal balance [6]. It simplifies the accumulation of privacy loss, reducing the complexity of maintaining privacy guarantees throughout multiple stages of data analysis.

4.2 Relaxed Differential Privacy

Connection to (ϵ, δ)—DP An $(\alpha, \epsilon_\alpha)$—RDP mechanism can be converted to (ϵ, δ)—DP. The relationship between RDP and (ϵ, δ)—DP is established by bounding the tail probabilities of the likelihood ratio:

$$\Lambda(x) = \frac{M_D(x)}{M_{D'}(x)} \tag{4.16}$$

For (ϵ, δ)—DP, we are interested in the probability that $\Lambda(x) > e^\epsilon$:

$$Pr[\Lambda(x) > e^\epsilon] \leq \delta \tag{4.17}$$

RDP provides a bound on the higher moments of $\Lambda(x)$, which can be related to the tail probabilities via Markov's inequality [7]. Specifically:

$$Pr[\Lambda(x) > e^\epsilon] \leq \inf_{\alpha > 1} exp((\alpha - 1)\epsilon - \alpha \epsilon_\alpha) \tag{4.18}$$

This inequality shows that for a given RDP guarantee $(\alpha, \epsilon_\alpha)$, the tail probability can be minimized by choosing an optimal order α. For a given Rényi divergence order, the conversion ensures that the probability of exceeding a privacy loss ϵ is bounded by δ:

$$\delta \leq e^{(\epsilon_\alpha - \epsilon)(\alpha - 1)} \tag{4.19}$$

Rearranging, the privacy loss ϵ for a given δ is:

$$\epsilon = \epsilon_\alpha + \frac{ln(1/\delta)}{\alpha - 1} \tag{4.20}$$

This conversion shows that (ϵ, δ)—DP guarantees can be derived from RDP parameters, providing flexibility in balancing ϵ and δ.

Composition Bounds RDP provides tighter composition bounds than standard (ϵ, δ)—DP. When multiple mechanisms are composed, their cumulative privacy loss under RDP can be calculated using the additivity of Rényi divergence:

$$\epsilon_\alpha^{total} = \sum_i \epsilon_\alpha^i, \tag{4.21}$$

where ϵ_α^i represents the RDP budget for the i-th mechanism. This additivity allows more efficient use of privacy budgets, especially in iterative algorithms like stochastic gradient descent.

4.2.2 Zero-Concentrated Differential Privacy

Zero-Concentrated Differential Privacy is a refined privacy framework that provides a balance between the interpretability of (ϵ, δ)—differential privacy and the mathematical rigor of Rényi Differential Privacy. zCDP measures privacy loss using a privacy parameter ρ, which bounds the Rényi divergence between the output distributions of neighboring datasets.

Definition 4.9 (Zero-Concentrated Differential Privacy [8]) A randomized mechanism $\mathcal{M} : \mathcal{X}^n \to \mathcal{Y}$ is ρ-zero-concentrated differentially private if, for all $x, x' \in \mathcal{X}^n$ differing on a single entry and all $\alpha \in (1, \infty)$,

$$D_\alpha(M(x)||M(x')) \leq \rho\alpha$$

where $D_\alpha(M(x)||M(x'))$ is the α- Rényi divergence between the distribution of $M(x)$ and $M(x')$.

zCDP uses concentration inequalities to provide a more refined measure of privacy loss. This allows for tighter control over the distribution of privacy loss, leading to more efficient privacy guarantees. One of the major advantages of zCDP is its simpler and more efficient composition properties. zCDP allows for straightforward addition of privacy loss terms when composing multiple queries.

Conversion to (ϵ, δ)—DP zCDP inherently bounds privacy loss in a way that can be interpreted probabilistically through (ϵ, δ)—DP. zCDP bounds the tail probabilities using Rényi divergence, which controls the likelihood ratio between the distributions of $\mathcal{M}(x)$ and $\mathcal{M}(x')$. Using a concentration inequality (Chernoff or Markov bounds), we derive the probability that the privacy loss exceeds a given threshold.

For any $\epsilon > 0$, the probability of privacy loss exceeding ϵ can be bounded as:

$$\delta = Pr[\text{Privacy Loss} > \epsilon] \leq exp(-\frac{(\epsilon - \rho)^2}{4\rho}) \qquad (4.22)$$

Rearranging, we solve for ϵ in terms of ρ and δ:

$$\epsilon = \rho + 2\sqrt{\rho ln(1/\delta)} \qquad (4.23)$$

This conversion ensures that mechanisms designed with zCDP are compatible with the more common (ϵ, δ)-DP framework while benefiting from zCDP's tighter privacy accounting.

Composition Bound ρ-zCDP provides a more direct measure of privacy loss and composes additively under sequential queries.

$$\rho_{total} = \sum_{i=1}^{k} \rho_i, \qquad (4.24)$$

where ρ_i is the zCDP parameter for the i-th mechanism. This is simpler and tighter than directly composing (ϵ, δ)-DP.

4.3 Local Differential Privacy

4.3.1 Introduction to Local Differential Privacy

Local Differential Privacy (LDP) is a privacy framework where individuals perturb their data locally (on their own devices) before sharing it. This ensures that the raw data remains private and never leaves the individual's control, offering strong privacy guarantees even in untrusted or decentralized environments.

Definition 4.10 (Local Differential Privacy [9]) A randomization mechanism $\mathcal{M}$ satisfies ϵ—local differential privacy if, for any two possible input x and x' and any possible output y, the following condition holds:

$$Pr[\mathcal{M}(x) = y] \leq e^{\epsilon} Pr[\mathcal{M}(x') = y], \tag{4.25}$$

The output of the randomization mechanism is almost equally likely regardless of the individual's actual data, ensuring privacy protection at the source.

LDP Versus CDP Traditional differential privacy, also known as Centralized differential privacy, is denoted as CDP for brevity. Figure 4.1 illustrates a comparison between the models of CDP and LDP, highlighting their structural differences. In the CDP model, the raw data from individuals is collected and stored on a centralized server. Privacy protection is applied by adding noise to the aggregated statistics at the server side before releasing results to the public users. This requires trusting the server to securely handle the raw data. In contrast, the LDP model applies privacy protection at the client side, where each individual perturbs their data locally before sending it to the server. The server only has access to the perturbed dataset, ensuring that raw data is never exposed. This decentralized approach eliminates the need for

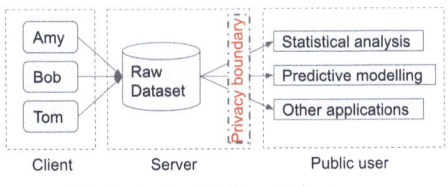

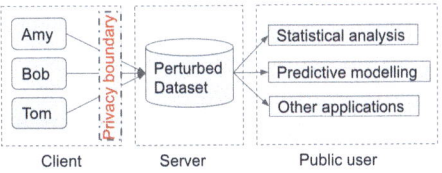

Fig. 4.1 Comparison between CDP and LDP models. (**a**) Centralized differential privacy. (**b**) Local differential privacy

Table 4.1 Comparison between centralized differential privacy (CDP) and local differential privacy (LDP)

Aspect	Centralized differential privacy (CDP)	Local differential privacy (LDP)
Definition	Privacy is guaranteed by perturbing the data after it has been collected by a trusted curator.	Privacy is guaranteed by perturbing data locally on the user's device before it is shared.
Trust model	Requires a trusted curator to handle raw data securely and apply the DP mechanism.	Does not rely on a trusted curator; privacy is ensured independently at the source.
Mechanism location	Noise is added centrally, after the raw data is collected.	Noise is added locally on the user's device, before data is transmitted.
Data accessibility	The trusted curator has access to the raw, unperturbed data.	No access to raw data; only the perturbed data is shared.
Privacy parameter	Typically uses (ϵ, δ)-DP, allowing small probability δ of privacy violation.	Uses pure ϵ-DP, with stricter guarantees.
Utility	Higher utility due to the use of unperturbed raw data during analysis.	Lower utility because noise is added locally to each individual's data.
Scalability	Suitable for large-scale data analysis where the curator aggregates data centrally.	Suitable for decentralized systems or cases where data collection is distributed.
Applications	Statistical data analysis, machine learning with private models, census data.	User telemetry (e.g., Google Chrome, Apple), surveys, federated learning.
Resilience to attacks	Vulnerable if the curator is compromised, as raw data may be exposed.	Strong privacy guarantees since raw data is never collected.

a trusted server but introduces more noise into the analysis, potentially reducing utility. Table 4.1 summaries the differences between LDP and CDP in more details.

Advantages of LDP LDP has several advantages compare with traditional differential privacy.

- *Strong Privacy Guarantees.* LDP ensures that privacy protection is applied locally, directly on the user's device, before any data is shared. This means that the raw data never leaves the user's control, offering strong privacy guarantees even in untrusted or adversarial environments. Users can confidently share their perturbed data without worrying about exposure of sensitive information.
- *No Trusted Curator Required.* Unlike CDP, LDP does not require a trusted third party or curator to process raw data. This eliminates the risk of breaches or misuse of raw data by central servers, making LDP highly suitable for distributed systems and scenarios where a trusted curator cannot be assumed.
- *Scalable and Decentralized.* LDP is inherently decentralized, allowing it to work well in large-scale, distributed systems such as user telemetry collection

4.3 Local Differential Privacy

or federated learning. Each user independently perturbs their data, making the approach naturally scalable for environments with millions of users.
- *Resistance to Server-Side Attacks.* Since raw data is never transmitted to a central server, LDP is highly resistant to server-side data breaches or attacks. Even if the server is compromised, attackers only gain access to already perturbed data, which offers limited utility for malicious purposes.

Disadvantages of LDP Simultaneously, several disadvantages have been identified.
- *Reduced Utility.* The primary drawback of LDP is the significant amount of noise added to individual data points, which can lead to reduced accuracy of aggregated results. Since noise is applied locally, before aggregation, the utility of data diminishes, especially for small datasets or analyses requiring fine-grained insights.
- *High Communication Overhead.* In some cases, LDP mechanisms require a larger volume of data to compensate for the reduced utility. For example, achieving accurate results for population-level statistics may require collecting more data or additional rounds of communication, which can increase system overhead.
- *Complex Implementation.* Implementing LDP mechanisms often requires designing algorithms tailored to specific data types or use cases (e.g., frequency estimation, histogram generation). This complexity can make it challenging for developers to adopt LDP without specialized expertise in privacy-preserving techniques.
- *Aggregation Challenges.* Since noise is applied locally, the aggregated results often require more sophisticated post-processing techniques to extract meaningful insights. This is particularly problematic for high-dimensional data or analyses that require correlations between multiple attributes.

LDP offers robust privacy guarantees by perturbing data at the source, making it highly effective in untrusted environments and decentralized systems. Its advantages include strong privacy, scalability, and resistance to server-side breaches. However, these benefits come with trade-offs in terms of reduced data utility, increased complexity, and challenges in achieving fine-grained insights. As a result, LDP is best suited for use cases where strong individual privacy is a priority, such as user telemetry collection, federated learning, and privacy-sensitive surveys, but it may require careful algorithmic design and data collection strategies to balance privacy and utility effectively.

4.3.2 Local Differential Privacy Mechanisms

This section introduces several fundamental mechanisms of LDP, which serve as the foundation for more complex algorithms. While there are numerous LDP algorithms tailored for specific applications such as statistical queries or data analysis, we focus on the basic mechanisms commonly used as building blocks. These mechanisms are versatile and widely applicable, providing the essential tools needed to design privacy-preserving systems for tasks like range query, heavy hitter identification, and more.

4.3.2.1 Frequency Estimation

Frequency estimation is a fundamental problem in data analysis that involves determining the number of frequency of specific items or values in a dataset. It provides essential insights into the distribution of values within a dataset, which can then be used to construct or support more advanced analyses. For example, frequency estimation helps in understanding the popularity of items, detecting trends, and forming the basis for histograms, contingency tables, and categorical data analysis. By accurately estimating frequencies in a privacy-preserving manner, it becomes possible to derive additional statistics or build more sophisticated machine learning models. In an LDP setting, the goal is to estimate the true frequency of items in the population based on the noisy, perturbed responses received from individuals. To achieve this, several LDP mechanisms have been developed, offering a different trade-off between privacy and utility.

Let $U = \{u_1, \cdots, u_n\}$ be a set of users and $A = \{I_1, \cdots, I_d\}$ be a set of possible items that each user may hold. The aim of the data aggregator is to find the frequency of each item in A. More specifically, it aims to estimate $(f_1, \cdots, f_d)$ where for $i = 1, \cdots, d$, $f_i = \frac{c(I_i)}{n}$ with $c(I_i) = \left|\left\{j \in \{1, \cdots, n\} : u_j \text{ holds } I_i\right\}\right|$. In general, frequency estimation methods can be divided into 4 processes.

- *Encoding.* Suppose that for $i = 1, \cdots, n$, u_i holds $v_i \in A$. Each user u_i encodes v_i using a predefined coding scheme to t_i, which can either be a value or a binary vector to be used as an input to a perturbation mechanism.
- *Perturbation.* Having t_i, each user u_i uses a perturbation mechanism $\mathcal{M}$ satisfying local differential privacy to perturb t_i to $\hat{t}_i$ under the guarantee that the distribution of $\hat{t}_i$ is statistically indistinguishable to the case when u_i holds any other value.
- *Aggregation.* The data aggregator receives all the perturbed values $\{\hat{t}_1, \cdots, \hat{t}_n\}$ from n users and aggregates them accordingly.
- *Estimation.* The aggregator performs post-processing to calculate an estimation $(\hat{f}_1, \cdots, \hat{f}_d)$ of $(f_1, \cdots, f_d)$ according to the perturbation strategy to ensure that the estimate is unbiased. Such post-processing techniques can also be done to improve estimation accuracy.

In the following, we discuss several typical local differentially private perturbation mechanism.

Randomized Response (RR) Let $p \in [0, 1]$, $\mathcal{X}$ be the domain of the user's data with size 2, $t \in \mathcal{X}$ be a user's private value and $\hat{t} \in \mathcal{X}$ be its perturbed response

4.3 Local Differential Privacy

outputted by the randomized response mechanism with probability p. Then $\hat{t}$ is a random variable such that for $v \in \mathcal{X}$,

$$Pr[\hat{t} = v] = \begin{cases} p, & \text{if } t = v \\ 1 - p, & \text{if } t \neq v \end{cases}. \tag{4.26}$$

Holohan et al. [10] showed that the above mechanism provides ϵ-LDP if $p \in \left[\frac{1}{1+e^\epsilon}, \frac{e^\epsilon}{1+e^\epsilon}\right]$ and among all such choices of p, choosing $p = \frac{e^\epsilon}{e^\epsilon+1}$ yields a mechanism with the minimum expected error. Note that the RR mechanism is only defined when $|\mathcal{X}| = 2$. It is then natural to extend it to a larger domain, which is defined by *generalized randomized response (GRR)*, which was proposed in [11].

Generalized Randomized Response (GRR) Suppose that $|\mathcal{X}| = d$ for some positive integer $d \geq 2$ and $p \in [0, 1]$. Suppose that a user u holds a value $t \in \mathcal{X}$ and let $\hat{t} \in \mathcal{X}$ be its perturbed response outputted by the generalized randomized response mechanism. Then $\hat{t}$ is defined to be a random variable such that for any $v \in \mathcal{X}$,

$$Pr[\hat{t} = v] = \begin{cases} p, & \text{if } t = v \\ \frac{1-p}{d-1}, & \text{if } t \neq v \end{cases}. \tag{4.27}$$

Similar to the case of RR, the above mechanism provides ϵ-LDP if $p \in \left[\frac{1}{1+(d-1)e^\epsilon}, \frac{e^\epsilon}{e^\epsilon+d-1}\right]$ and among all such choices of p, choosing $p = \frac{e^\epsilon}{e^\epsilon+d-1}$ yields a mechanism with the minimum expected error. It is also easy to see that when $d = 2$, the generalized randomized response mechanism reduces back to the RR mechanism defined above.

Optimized Unary Encoding (OUE) [12] Suppose that $|A| = d$ and for a user u_i that holds $v_i = I_{j_i} \in A$, the encoding of v_i is done by defining a binary vector $t_i = (t_{i,1}, \cdots, t_{i,d})$ of length d, where only $t_{i,j} = 1$, all other values equal to 0. The perturbation is then done for each bit independently. More specifically, given $t_{i,j}$, the j-th entry of the output $\hat{t}_{i,j}$ is generated using RR with possible outputs $\{0, 1\}$ where

$$p = \begin{cases} \frac{1}{2}, & \text{if } t_{i,j} = 1 \\ \frac{e^\epsilon}{e^\epsilon+1}, & \text{if } t_{i,j} = 0 \end{cases}. \tag{4.28}$$

By optimizing the perturbation probabilities, OUE is more efficient than Randomized Response for domains with a large number of categories, as it reduces the amount of noise needed to achieve privacy.

RR, GRR, and OUE are three fundamental building blocks in LDP. These mechanisms are not only essential for advanced frequency estimation methods but also serve as the foundation for more complex applications. For instance, typical frequency estimation algorithms like RAPPOR [13] and OLH [12] incorporate hash functions to reduce communication costs, yet their core perturbation mechanisms are still based on RR or GRR. OUE is widely utilized for various applications, including heavy hitter identification [14], joint distribution estimation [15], and range queries [16], among others.

4.3.2.2 Mean Value Estimation

Mean value estimation is a fundamental task in data analysis that involves calculating the average value of a numeric dataset. Two traditional numerical data perturbation mechanisms, the Laplace mechanism and the Gaussian mechanism, can be applied locally to users' data to provide Local Differential Privacy. In addition to these, two widely adopted schemes specifically designed for mean value estimation are commonly used in LDP applications.

Mechanism Proposed by Duchi et al. Duchi et al. proposed a method [17] to perturb the numeric value under LDP, which has been widely used for numerical statistics. Each user's data point is first normalized to ensure it lies within a unit sphere $||t|| \leq 1$. The algorithm returns a perturbed $\hat{t}$ that equals either $\frac{e^\epsilon+1}{e^\epsilon-1}$ or $-\frac{e^\epsilon+1}{e^\epsilon-1}$, with the following probabilities:

$$Pr[\hat{t}_i = x | t_i] = \begin{cases} \frac{e^\epsilon-1}{2e^\epsilon+2} \cdot t_i + \frac{1}{2}, if x = \frac{e^\epsilon+1}{e^\epsilon-1} \\ -\frac{e^\epsilon-1}{2e^\epsilon+2} \cdot t_i + \frac{1}{2}, if x = -\frac{e^\epsilon+1}{e^\epsilon-1} \end{cases} \quad (4.29)$$

$\hat{t}_i$ is an unbiased estimator of the input value t_i. Relative to other existing solutions, the solution by Duchi et al. outputs an estimate with relatively small variance when ϵ is small. However, their scheme has a characteristic that regardless of the value of ϵ, the statistical variance of the estimate of each user's data t_i is $\left(\frac{e^\epsilon+1}{e^\epsilon-1}\right)^2$, which is always larger than 1. Hence, compared to other protocols, which generally can reduce the statistical variance arbitrarily close to 0 by increasing ϵ, such solution by Duchi et al. will perform worse when ϵ is sufficiently large.

Piecewise Mechanism (PM) Piecewise Mechanism [18] is another typical LDP mechanism designed for mean value estimation of numerical data. The Piecewise Mechanism divides the input domain $[-1, 1]$ into distinct intervals (or "pieces") and applies different perturbation probabilities within each interval. By tailoring the perturbation strategy to the data distribution, it reduces the noise introduced into the mean estimation process, thus improving utility. Specifically, let $t_i \in [-1, 1]$ be the input value, the PM will output a perturbed value $\hat{t}_i$ in $[-C, C]$, where

$$C = \frac{exp(\epsilon/2) + 1}{exp(\epsilon/2) - 1} \quad (4.30)$$

The probability density function of $\hat{t}_i$ is:

4.3 Local Differential Privacy

$$pdf(\hat{t}_i = x|t_i) = \begin{cases} p, & \text{if } x \in [\ell(t_i), r(t_i)], \\ \frac{p}{exp(\epsilon)}, & \text{if } x \in [-C, \ell(t_i)) \cup (r(t_i), C] \end{cases} \quad (4.31)$$

where

$$p = \frac{exp(\epsilon) - exp(\epsilon/2)}{2exp(\epsilon/2) + 2}, \quad (4.32)$$

$$\ell(t_i) = \frac{C+1}{2} \cdot t_i - \frac{C-1}{2}, \text{ and} \quad (4.33)$$

$$r(t)i = \ell(t_i) + C - 1. \quad (4.34)$$

The Piecewise Mechanism achieves significantly lower statistical variance when the privacy budget ϵ is large. Combining the mechanism proposed by Duchi et al. with PM can enhance the utility of statistics across the entire privacy regime, making it a practical choice for real-world applications.

High-Dimensional Mean Estimation In real-world scenarios, users may possess multiple values or high-dimensional records, which makes mean value estimation more challenging. A common approach is to split the privacy budget across dimensions to ensure rigorous privacy guarantees. Alternatively, sampling a subset of values for reporting can help conserve the privacy budget while still providing meaningful aggregate statistics. These strategies aim to balance the trade-off between privacy and utility in complex data scenarios. Besides, several other methods might also be applied.

- *Dimensionality reduction.* Methods like random projection or principal component analysis (PCA) can be adapted to reduce the dimensionality of data before applying mean estimation mechanisms. This approach reduces the effective privacy budget required while retaining most of the relevant information for aggregation.
- *Adaptive Privacy Allocation.* In real-world scenarios, not all dimensions of a high-dimensional dataset are equally important. Adaptive privacy allocation strategies dynamically distribute the privacy budget ϵ based on the importance or sensitivity of each attribute. For example, more budget can be allocated to dimensions that contribute more significantly to the mean estimation task, thereby improving utility without increasing the overall privacy cost.
- *Domain Knowledge Incorporation.* Leveraging domain-specific knowledge can significantly enhance the utility of mean estimation under LDP. For instance, prior information about the expected range or distribution of the data (e.g., uniform, normal, or skewed) can guide the design of the perturbation strategy.

In addition to being fundamental tools in high-dimensional statistics, both Duchi's mechanism and the Piecewise Mechanism are widely utilized and integrated into more complex algorithms across various domains, including statistical aggregation, machine learning, and beyond.

4.4 Differential Privacy Implementations

4.4.1 Overview of Tools and Libraries

Python is widely used across industries and academia for data analysis. It is an ideal programming language for implementing differential privacy due to its extensive ecosystem of libraries and frameworks, as well as its user-friendly design. Its intuitive and readable syntax significantly reduces the complexity of implementing mathematical concepts like noise addition, sensitivity calculations, and privacy accounting. Core libraries such as NumPy and SciPy provide robust support for mathematical and statistical computations, simplifying the implementation of DP mechanisms, including Laplace and Gaussian noise addition, as well as advanced techniques. Python's dominance in the AI/ML ecosystem, supported by powerful frameworks like TensorFlow, PyTorch, and Scikit-learn, enables seamless integration of DP into model training and evaluation processes, making it particularly valuable in machine learning applications. Additionally, Python's portability across platforms ensures that privacy-preserving systems can be deployed reliably in diverse environments, from local machines to large-scale cloud infrastructures. A summary of some popular tools and libraries is provided below.

Google's Differential Privacy Library

Google's Differential Privacy Library (https://github.com/google/differential-privacy) is a powerful open-source framework designed to provide strong privacy guarantees for data analysis by applying differential privacy techniques. The library supports a variety of statistical queries and aggregations, such as counts, sums, averages, and histograms, making it versatile for diverse analytical needs. It provides tools for managing the privacy budget, allowing developers to balance privacy and utility effectively. The library is suitable for enterprise-level applications, including cloud-based systems. While primarily implemented in C++, it also offers Java bindings and Python support via PyDP, making it accessible to a broader range of developers.

PyDP

PyDP (https://github.com/OpenMined/PyDP), developed by OpenMined in 2020, is a Python wrapper for Google's differential privacy library, providing an easy-to-use interface for implementing differential privacy techniques in Python. It integrates the robustness and scalability of Google's C++ library with the simplicity of Python, making it an accessible tool for researchers, data scientists, and practitioners. It offers a range of ϵ-differentially private algorithms for securely analyzing sensitive numeric datasets. It supports computations like BoundedSum, BoundedMean, Median, and more. PyDP currently focuses on numeric aggregate computations and uses Laplace noise as its noise mechanism. Future updates are expected to enhance its capabilities with additional noise mechanisms and features.

4.4 Differential Privacy Implementations

Diffprivlib

Diffprivlib (https://github.com/IBM/differential-privacy-library) is an open-source Python library developed by IBM specifically for implementing differential privacy in machine learning and data analytics. The library provides differentially private versions of standard Scikit-learn models, including Logistic Regression, Naive Bayes, and Support Vector Machines (SVMs). It is fully compatible with Scikit-learn pipelines, enabling users to seamlessly integrate DP-enhanced models into their existing workflows. Additionally, it offers tools for calculating private means, medians, histograms, and covariance matrices, ensuring privacy-preserving computations in data analysis. The library includes built-in mechanisms for managing and tracking privacy budgets across computations, allowing users to configure privacy parameters to strike an optimal balance between privacy guarantees and model utility.

TensorFlow Privacy

TensorFlow Privacy (https://github.com/tensorflow/privacy) is an open-source library developed by Google that extends TensorFlow to include differential privacy capabilities, enabling the creation of privacy-preserving machine learning models. It provides key features like Differentially Private Stochastic Gradient Descent (DP-SGD) for noise addition and gradient clipping, robust privacy accounting tools for tracking privacy budgets (ϵ and δ), and flexible APIs that are compatible with both TensorFlow and Keras. With customizable parameters and extensive documentation, TensorFlow Privacy enables developers to balance privacy guarantees and model utility, making it ideal for privacy-preserving machine learning on sensitive data, such as in healthcare and federated learning scenarios.

Opacus

Opacus (https://opacus.ai/) is an open-source library developed by Meta (formerly Facebook) that enables differential privacy in PyTorch-based machine learning models. It provides an efficient implementation of DP-SGD, allowing developers to add noise to gradients and clip them during training to protect individual data points in a dataset. Opacus is designed for scalability and minimal overhead, leveraging PyTorch's dynamic computation graph to efficiently compute per-sample gradients without significantly impacting training performance. It includes built-in tools for privacy budget accounting, enabling users to monitor privacy loss throughout the training process. With its easy integration into PyTorch workflows and focus on efficiency, Opacus is particularly suited for training privacy-preserving deep learning models on sensitive datasets.

SmartNoise

SmartNoise (https://smartnoise.org/) is an open-source Differential Privacy (DP) platform developed by Microsoft in collaboration with OpenDP. SmartNoise offers tools for securely analyzing sensitive data in structured datasets, with support for SQL-based queries. It integrates seamlessly with data platforms such as databases and data lakes, enabling efficient and scalable privacy-preserving analysis. The platform provides both centralized and local differential privacy options, making it suitable for diverse use cases, from data custodians managing large datasets to individual data contributors. SmartNoise includes features for tracking and managing privacy budgets and allows users to configure noise levels and sensitivity parameters to strike an optimal balance between privacy and utility. While its primary focus is on data query and analysis, rather than direct support for DP in machine learning.

OpenDP

OpenDP (https://github.com/opendp/opendp) is an open-source initiative and community-driven platform dedicated to advancing the adoption and development of Differential Privacy (DP) tools and technologies. Founded in collaboration with Microsoft and led by researchers from Harvard University, OpenDP provides a suite of modular libraries and frameworks that enable developers, researchers, and practitioners to build privacy-preserving systems. The project emphasizes flexibility and transparency, offering tools that support a wide range of DP applications, including statistical analysis, data sharing, and machine learning. With a strong focus on academic rigor and practical usability, OpenDP fosters collaboration among privacy experts, data scientists, and policymakers, aiming to make state-of-the-art DP methods accessible to both technical and non-technical audiences.

In addition to the differential privacy libraries discussed, there are other libraries available, though they are less commonly used. Table 4.2 provides a comparison of these most widely used libraries, detailing their distinctive features and key differences.

4.4.2 *Examples for Differential Privacy Implementation*

In this section, we provide detailed examples of utilizing differential privacy libraries for implementing data analysis tasks, including statistical computations and machine learning applications.

Example 1: Using the Diffprivlib Library to Calculate a Differentially Private Mean and Histogram

- Install Diffprivlib. First, ensure you have the Diffprivlib library installed. You can install it using pip:

```
1  pip install diffprivlib
```

- Import Necessary Libraries. Start by importing the required modules:

```
1  import numpy as np
2  from diffprivlib.tools import mean, histogram
```

- Prepare the Data. We will simulate a dataset for this example. For instance, let's assume we have sensitive age data:

```
1  # Simulated dataset
2  data = np.array([23, 45, 31, 34, 29, 67, 41, 39, 56, 49])
3  print("Original Data:", data)
```

4.4 Differential Privacy Implementations

Table 4.2 Comparison between differential privacy libraries

Library	Primary use	Focus area	Mechanisms supported	Characteristics
Google's differential privacy library	General DP computations for aggregate queries	Aggregate data analysis	Laplace, Gaussian	High performance, scalable, and customizable.
PyDP	Python wrapper for Google's DP library	Aggregate data analysis	Laplace (current), Gaussian (future plans)	Dependent on Google's library with limited mechanisms.
Diffprivlib	Privacy-preserving machine learning and statistics	Statistical analysis and ML models	Laplace, Gaussian	Focuses primarily on Scikit-learn models.
TensorFlow privacy	Differential privacy for tensorflow-based models	Deep learning with TensorFlow	DP-SGD	Seamless integration with TensorFlow/Keras.
Opacus	Differential privacy for Pytorch-based models	Deep learning with PyTorch	DP-SGD	Optimized for PyTorch with minimal overhead.
SmartNoise	Privacy-preserving data query and analysis	Data query and SQL-based analysis	Laplace, Gaussian	SQL-based query support, scalable. Focused on query, less ML support.
OpenDP	Modular tools for broad DP applications	General DP tools and community-driven development	Varies (modular and flexible)	Extensive academic backing, modular design.

- Calculate a Differentially Private Mean. Now, use Diffprivlib's mean function to compute the mean of the dataset with differential privacy:

```
# Compute the differentially private mean
dp_mean = mean(data, epsilon=1.0)
print("Differentially Private Mean:", dp_mean)
```

- Generate a Differentially Private Histogram. We can also compute a histogram with differential privacy:

```
# Compute a differentially private histogram
dp_hist, dp_bins = histogram(data, range=(20, 70), bins=5,
    epsilon=1.0)

print("Differentially Private Histogram:", dp_hist)
print("Histogram Bins:", dp_bins)
```

- Compare Results with Non-Private Outputs. For comparison, calculate the non-private mean and histogram using NumPy:

```
# Non-private mean
np_mean = np.mean(data)
print("Non-Private Mean:", np_mean)

# Non-private histogram
np_hist, np_bins = np.histogram(data, bins=5, range=(20, 70))
print("Non-Private Histogram:", np_hist)
```

Example 2: Using TensorFlow Privacy to Train a Linear Regression Model with Differential Privacy

- Install the Required Libraries. Ensure you have TensorFlow and TensorFlow Privacy installed. You can install them using pip:

```
pip install tensorflow tensorflow-privacy
```

- Import Libraries. Import necessary libraries.

```
import numpy as np
import tensorflow as tf
from tensorflow_privacy import DPKerasSGDOptimizer
from sklearn.datasets import make_regression
from sklearn.model_selection import train_test_split
```

- Generate Synthetic Data. We'll use synthetic data for simplicity. You can replace this with your own dataset.

```
# Generate synthetic data for linear regression
X, y = make_regression(n_samples=1000, n_features=5, noise
    =0.1, random_state=42)
y = y.reshape(-1, 1)  # Reshape target variable

# Split into train and test sets
X_train, X_test, y_train, y_test = train_test_split(X, y,
    test_size=0.2, random_state=42)
```

- Build a Linear Regression Model. Define the Linear Regression model using TensorFlow's Keras API.

```
# Define the linear regression model
model = tf.keras.Sequential([
    tf.keras.layers.Dense(1, input_shape=(X_train.shape[1],),
        activation=None)
])
```

- Use a Differentially Private Optimizer. To incorporate DP, use TensorFlow Privacy's DPKerasSGDOptimizer. This optimizer clips gradients and adds noise to ensure DP.

```
# Define the DP optimizer
dp_optimizer = DPKerasSGDOptimizer(
    l2_norm_clip=1.0,  # Maximum gradient norm (clipping
        threshold)
```

```
4        noise_multiplier=1.1,  # Noise multiplier for DP
5        num_microbatches=1,    # Number of microbatches
6        learning_rate=0.01     # Learning rate
7    )
8
9    # Compile the model with the DP optimizer
10   model.compile(
11       optimizer=dp_optimizer,
12       loss='mean_squared_error',
13       metrics=['mean_squared_error']
14   )
```

- Train the Model. Train the model with DP by fitting it to the training data.

```
1    # Train the model
2    history = model.fit(
3        X_train,
4        y_train,
5        epochs=10,      # Number of training epochs
6        batch_size=256, # Batch size
7        verbose=1
8    )
```

- Evaluate the Model. Evaluate the model on the test set to check its performance.

```
1    # Evaluate the model on test data
2    test_loss, test_mse = model.evaluate(X_test, y_test, verbose
        =0)
3    print(f"Test Loss: {test_loss:.4f}, Test MSE: {test_mse:.4f}")
```

References

1. Dwork, C., Roth, A.: The algorithmic foundations of differential privacy. Found. Trends Theor. Comput. Sci. **9**(3–4), 211–407 (2014)
2. Hellman, M., Raviv, J.: Probability of error, equivocation, and the Chernoff bound. IEEE Trans. Inf. Theory **16**(4), 368–372 (1970)
3. Bentkus, V.: On Hoeffding's inequalities. Ann. Probab., 1650–1673 (2004)
4. Abadi, M., Chu, A., Goodfellow, I., McMahan, H.B., Mironov, I., Talwar, K., Zhang, L.: Deep learning with differential privacy. In: Proc. 2016 ACM SIGSAC Conf. Comput. Commun. Secur., pp. 308–318 (2016)
5. Mironov, I.: Réyi differential privacy. In: 2017 IEEE 30th Computer Security Foundations Symp. (CSF), pp. 263–275 (2017)
6. Kairouz, P., Oh, S., Viswanath, P.: The composition theorem for differential privacy. In: Int. Conf. Mach. Learn., pp. 1376–1385 (2015)
7. Wilhelmsen, D.R.: A Markov inequality in several dimensions. J. Approx. Theory **11**(3), 216–220 (1974)
8. Bun, M., Steinke, T.: Concentrated differential privacy: Simplifications, extensions, and lower bounds. In: Theory of Cryptography Conf., pp. 635–658 (2016)
9. Bassily, R., Smith, A.: Local, private, efficient protocols for succinct histograms. In: Proc. 47th Annu. ACM Symp. Theory Comput., pp. 127–135 (2015)
10. Holohan, N., Leith, D.J., Mason, O.: Optimal differentially private mechanisms for randomised response. IEEE Trans. Inf. Forensics Secur. **12**(11), 2726–2735 (2017)

11. Kairouz, P., Bonawitz, K., Ramage, D.: Discrete distribution estimation under local privacy. In: Proc. 33rd Int. Conf. Mach. Learn. (ICML), pp. 2436–2444 (2016)
12. Wang, T., Blocki, J., Li, N., Jha, S.: Locally differentially private protocols for frequency estimation. In: 26th USENIX Secur. Symp. (USENIX Security 17), pp. 729–745 (2017)
13. Erlingsson, Ú., Pihur, V., Korolova, A.: RAPPOR: Randomized aggregatable privacy-preserving ordinal response. In: Proc. 2014 ACM SIGSAC Conf. Comput. Commun. Secur., pp. 1054–1067 (2014)
14. Zhao, D., Zhao, S., Chen, H., Liu, R., Li, C., Liang, W.: Efficient protocols for heavy hitter identification with local differential privacy. Front. Comput. Sci. **16**(5), 165825 (2022)
15. Shen, Z., Xia, Z., Yu, P.: PLDP: Personalized local differential privacy for multidimensional data aggregation. Secur. Commun. Netw. **2021**(1), 6684179 (2021)
16. Ju, Z., Li, Y.: Local differential privacy-based privacy-preserving data range query scheme for electric vehicle charging. IEEE Trans. Netw. Sci. Eng. (2023)
17. Duchi, J., Wainwright, M.J., Jordan, M.I.: Local privacy and minimax bounds: Sharp rates for probability estimation. Adv. Neural Inf. Process. Syst. **26** (2013)
18. Wang, N., Xiao, X., Yang, Y., Zhao, J., Hui, S.C., Shin, H., Shin, J., Yu, G.: Collecting and analyzing multidimensional data with local differential privacy. In: 2019 IEEE 35th Int. Conf. Data Eng. (ICDE), pp. 638–649 (2019)

Chapter 5
Privacy-Preserving Association Rule Mining in Cloud Computing

Abstract The paradigm of data mining-as-a-service in cloud computing environments has garnered significant attention in recent years. In this paradigm, a company (data owner) that lacks the necessary data storage, computational resources, and expertise stores its data in the cloud and outsources its mining tasks to a cloud service provider (server). To safeguard the privacy of the outsourced database and the association rules derived, techniques such as k-anonymity, k-support, and k-privacy have been proposed to perturb the data before it is uploaded to the server. However, these techniques are computationally intensive. If the data owner possesses the resources to employ these techniques, it is often capable of executing association rule mining locally. In this chapter, we explore a scenario in which the data owner encrypts its data and stores it in the cloud. To mine association rules from this encrypted data, the data owner outsources the task to n (≥ 2) "semi-honest" servers, which collaborate to perform the association rule mining on the encrypted data in the cloud and return the encrypted association rules to the data owner. We introduce three solutions aimed at ensuring data privacy during association rule mining. These solutions, based on the distributed ElGamal cryptosystem, protect item privacy, transaction privacy, and database privacy, provided that at least one of the n servers remains honest. To minimize the risk of all servers being compromised, the data owner can choose servers from different cloud providers.

5.1 Introduction

Cloud computing represents the evolution and widespread adoption of existing technologies and paradigms. The primary goal of cloud computing is to enable users to leverage these technologies without needing in-depth knowledge or expertise in each one. The cloud aims to reduce costs and allow users to focus on their core business, alleviating the challenges associated with IT management.

The discovery of frequent patterns, association rules, and correlations in large datasets is essential for business intelligence. A classic example of frequent itemset mining is market basket analysis, where customer purchasing behaviors are analyzed to uncover associations between the items placed in their "shopping baskets". Such discoveries can assist retailers in developing marketing strategies by understanding which items are frequently bought together. For example, if a customer buys milk, how likely are they to also purchase bread (and which type of bread) during the same supermarket visit? This type of information can drive targeted marketing efforts and optimize shelf placement to increase sales [4].

In recent years, there has been growing interest in the paradigm of data mining-as-a-service within cloud computing environments [3, 8, 10]. In this model, a company (the data owner), which lacks the necessary data storage, computational resources, and expertise, stores its data in the cloud and outsources its mining tasks to a cloud service provider (the server). While the data mining-as-a-service model offers significant benefits to business intelligence, it also raises critical privacy concerns. Specifically, the cloud service provider has access to the company's data and could potentially uncover sensitive business information.

In the context of outsourcing data mining tasks to the cloud, the data owner is assumed to lack sufficient data storage, computational resources, and expertise. However, converting a database to k-support anonymity or k-privacy is computationally expensive. If the data owner possesses the necessary resources to employ these techniques, it could often run association rule mining algorithms, such as the Apriori algorithm [1], locally.

This chapter considers a cloud computing environment where the data owner lacks substantial local data storage and opts to store its entire transaction database in the cloud. The data owner may need to continuously add new transactions to the cloud-based database. Additionally, the data owner does not possess the computational resources to run data mining algorithms locally and may not be familiar with techniques like k-anonymity or association rule mining algorithms.

To safeguard data privacy, the data owner uses the ElGamal cryptosystem [2] to encrypt all items in a transaction. Specifically, the data owner generates a public/private key pair and uses the public key to encrypt the transaction items before uploading them to the cloud. Notably, the ElGamal encryptions of identical items appear distinct, enhancing privacy.

When outsourcing the data mining task, the data owner selects n ($n \geq 2$) different servers in the cloud. The private key is split into n pieces and distributed to the n servers. The private key remains secure as long as not all of the n servers collude. To mitigate the risk of all servers being compromised, the data owner can select servers from different cloud providers. We assume that at least one of the n servers is honest. Additionally, the data owner sends an encryption of the minimum support threshold to the n servers. Each server is assumed to be "semi-honest," meaning that while it follows the data mining algorithm as prescribed, it may attempt to learn information about the privacy of the data, such as the supports of itemsets or association rules.

Under this model, we formally define three privacy requirements for association rule mining over the encrypted transaction database [12]. Based on these privacy requirements, we propose three solutions, primarily built on the distributed ElGamal cryptosystem, that allow the n servers to cooperate in mining association rules from the encrypted data in the cloud and return the data owner the encrypted association rules with encrypted supports.

The first solution utilizes the Plaintext Equality Test (PET) technique [6], which enables the n servers to collaboratively determine the equality of two plaintexts based on their encryptions, without the need for decryption. The core idea is to use PET to identify encryptions of the same item in the encrypted transactions and replace them with identical encryptions. The Apriori algorithm is then applied to this modified transaction database. This first solution does not conceal the support of each itemset and may be vulnerable to background knowledge-based attacks.

To conceal the support of each itemset, we propose a second solution where each server adds fake transactions (encrypted using the data owner's public key) to the encrypted database. The original and fake transactions are then mixed through ElGamal re-encryptions and random permutations performed by all servers. Like the first solution, PET is used to identify and replace the encryptions of identical items. When applying the Apriori algorithm to the modified database, the effect of the fake transactions is neutralized by all servers, and the real support of the itemset is compared with the encrypted minimum support threshold using the conditional gate [9]. This second solution hides the real support of each itemset but does not completely obscure the original transactions.

To fully conceal the original transactions, we present a third solution in which each server counts the support of an itemset in a subset of the database. The n servers then jointly compute the support of the itemset across the entire database. Before counting the support, all transactions are shuffled to hide their structure. As in the second solution, the encrypted support of the itemset and the encrypted minimum support threshold are compared using the conditional gate.

Compared to previous work on outsourcing association rule mining, our approach reduces the requirements for data storage, computational resources, and expertise for the data owner. The only actions required from the data owner are encrypting the data before uploading it to the cloud and decrypting the mined association rules received from the cloud.

The rest of the chapter is organized as the following: Sect. 5.2 reviews the works that have already been done in the literature; Sect. 5.3 introduces some techniques for developing our solutions; Sect. 5.4 describes our model for privacy-preserving association rule mining in cloud computing environments; Sects. 5.5 to 5.7 give our solutions on the basis of the model. The last section concludes our work.

5.2 Related Work

To protect the data privacy of a company while still enabling the server to perform association rule mining on data stored in the cloud, a straightforward solution is for the data owner to obscure the meanings of items in its transaction database by replacing them with integers (where the same item is replaced with the same integer and different items are replaced with different integers). This one-to-one substitution, however, does not conceal the frequencies of items. If the server has precise knowledge of the frequencies of certain items, it could re-identify some items, particularly the most frequent ones. For instance, if bread is the most frequent item in retail transaction databases and an integer a appears most frequently in the transformed database, the server can deduce that integer a corresponds to bread.

The issue of background knowledge-based attacks was first addressed by Wong et al. [11] in 2007. To counter this, they proposed a one-to-n item mapping that transforms transactions in a non-deterministic manner. The fundamental idea is to add fake items to the transaction database. Suppose I represents the set of items in the original database and m is a one-to-one mapping from items to integers. The data owner adds a set of fake items F to the dictionary J (where $|J| = |I| + |F|$) and maps each item x onto $M(x) = \{m(x)\} \cup f$, where f is a random subset of F. For example, if $I = \{a, b, c\}$, $m(a) = 1$, $m(b) = 2$, $m(c) = 3$, and $F = \{4, 5\}$, a possible one-to-n item mapping M could be defined as $M(a) = \{1, 4, 5\}$, $M(b) = \{2\}$, $M(c) = \{3, 5\}$, and $M(a, b) = M(a) \cup M(b) = \{1, 2, 4, 5\}$, and so on. In this manner, even if the server has rough estimates of itemset frequencies, it becomes difficult to discern the distribution of the actual itemsets in the transformed database.

However, the approach of adding fake items has two limitations. The first limitation is that each fake item has the same probability of being added to any transaction, and thus they appear with similar frequencies when the number of transactions is large. The second limitation is that fake items are added to transactions independently of the items already present. As a result, each fake item in f is independent of all other items x. This second limitation holds even if the frequencies of the fake items differ. Based on these observations, in 2009, Molloy et al. [7] introduced a frequency analysis-based attack on Wong et al.'s algorithm. This attack removes independently added fake items by detecting low correlations between items, thereby enabling the successful re-identification of some of the most frequent items.

In 2010, Tai et al. [10] proposed a novel approach based on k-support anonymity, where each sensitive item is protected by $k - 1$ other items with similar support. For example, consider a retail transaction database with five transactions: $t_1 = \{wine\}$, $t_2 = \{cigar, wine\}$, $t_3 = \{cigar, tea\}$, $t_4 = \{beer, cigar, wine\}$, and $t_5 = \{beer, tea, wine\}$. Suppose that the items $beer$, $cigar$, $wine$, and tea are replaced with a, b, c, and h, respectively. A transformed transaction database with 3-support anonymity would be: $t_1 = \{c, d, g\}$, $t_2 = \{b, d, g\}$, $t_3 = \{b, h\}$, $t_4 = \{a, b, c\}$, and $t_5 = \{a, c, d, h\}$, where d and g are fake items. In this case, each of the items a, g, and h has 2 supports, while each of the items b, c, and d has 3 supports. Tai et al. transformed frequent itemset mining into generalized/multi-level frequent itemset mining and limited the occurrence of additional items with the help of a pseudo-taxonomy tree. This approach can protect sensitive information through k-support anonymity.

More recently, Giannotti et al. [3] extended the concept of k-support anonymity to k-privacy. While k-support anonymity ensures that each transformed item cannot be distinguished from at least $k - 1$ other items, k-privacy requires that each transformed itemset (i.e., a set of items) cannot be distinguished from at least $k - 1$ other itemsets of the same size. They proposed a method to transform a database to achieve k-privacy. Their approach involves three main steps: (1) using a 1-to-1 mapping to replace each plain item, resulting in a database D; (2) grouping items for k-privacy; and (3) adding fake transactions to achieve k-privacy. Fake transactions are added to D to form a new database D^*, which is then submitted to the server with a minimum support threshold. After the server returns the computed frequent patterns from D^*, the data owner removes the effect of the fake transactions. For example, if the support of an itemset E in D^* is x and the support of E in $D^* \backslash D$ is y, the real support of E in D is $x - y$.

The key difference between the privacy-preserving approaches proposed by Tai et al. [10] and Giannotti et al. [3] is that Tai et al. added fake items to the data to achieve k-support anonymity, while Giannotti et al. added fake transactions to the data to achieve k-privacy.

5.3 Preliminaries

5.3.1 ElGamal Cryptosystem

The ElGamal cryptosystem [2] consists of key generation, encryption, and decryption algorithms as follows.

- Key Generation: On input a security parameter k, it publishes a multiplicative cyclic group $\mathbb{G}$ of prime order q with a generator g, such that discrete logarithm problem over the group $\mathbb{G}$ is hard. Then it chooses a private key x randomly from $\mathbb{Z}_q^* = \{1, 2, \cdots, q-1\}$ and computes a public key $y = g^x$. The private key x is kept secret while the public key y can be known to everyone.
- Encryption: On inputs a message $m \in \mathbb{G}$ and the public key y, it chooses an integer r randomly from $\mathbb{Z}_q^*$ and outputs a ciphertext $C = \mathsf{E}(m) = (A, B)$, where $A = g^r$ and $B = m \cdot y^r$.
- Decryption: On inputs a ciphertext (A, B), and the private key x, it outputs the plaintext $m = \mathsf{D}(C) = B/A^x$.

The ElGamal decryption algorithm is correct because

$$B/A^x = m \cdot y^r/(g^r)^x = m \cdot (g^x)^r/g^{rx} = m.$$

The ElGamal cryptosystem has two important properties for privacy protection as follows.

Homomorphic Property—Given the encryptions of two messages m_1 and m_2, denoted as $\mathsf{E}(m_1) = (A_1, B_1) = (g^{r_1}, m_1 y^{r_1})$ and $\mathsf{E}(m_2) = (A_2, B_2) = (g^{r_2}, m_2 y^{r_2})$, one can compute the encryption of the multiplication $m_1 m_2$ of the two messages, that is, $\mathsf{E}(m_1 m_2) = (A_1 A_2, B_1 B_2)$. This property is correct because

$$B_1 B_2/(A_1 A_2)^x = m_1 y^{r_1} m_2 y^{r_2}/(g^{r_1} g^{r_2})^x = m_1 m_2.$$

Based on this property, we can see that $\mathsf{E}(m^n) = (A^n, B^n)$ for any positive integer n. We denote $\mathsf{E}(m_1)\mathsf{E}(m_2) = (A_1 A_2, B_1 B_2)$ and $\mathsf{E}(m)^n = (A^n, B^n)$.

Re-encryption Property—Given an encryption of a message m, denoted as $\mathsf{E}(m) = (A, B)$, one can compute another encryption of the message, denoted as $\mathsf{RE}(\mathsf{E}(m))$, that is, $(A', B') = (Ag^{r'}, By^{r'})$, where r' is randomly chosen from Z_q^*. This property is correct because

$$By^{r'}/(Ag^{r'})^x = Bg^{xr'}/(A^x g^{r'x}) = B/A^x = m.$$

5.3.2 Distributed ElGamal Cryptosystem

The ElGamal cryptosystem has been used in a distributed environment in a threshold manner. A simplified distributed ElGamal cryptosystem for n servers $S_1, S_2, \cdots, S_n$ can be described as follows.

- Key Generation: On input a security parameter k, it publishes a multiplicative cyclic group $\mathbb{G}$ of prime order q with a generator g, such that discrete logarithm problem over the group $\mathbb{G}$ is hard. Then each server S_i chooses a private key x_i randomly from $\mathbb{Z}_q^*$ and computes a public key $y_i = g^{x_i}$. The private key x_i is known to S_i only while the public key y_i is known to everyone. Let $y = \prod_{i=1}^{n} y_i = g^{\sum_{i=1}^{n} x_i}$, which is the common public key for the n servers.
- Encryption: On inputs a message $m \in \mathbb{G}$ and the common public key y, it chooses an integer r randomly from $\mathbb{Z}_q^*$ and outputs a ciphertext $C = \mathsf{E}(m) = (A, B)$, where $A = g^r$ and $B = m \cdot y^r$.
- Decryption: On inputs a ciphertext (A, B), each server S_i outputs $A_i = A^{x_i}$. Then the algorithm outputs the plaintext $m = \mathsf{D}(C) = B/\prod_{i=1}^{n} A^{x_i}$.

The decryption algorithm of the distributed ElGamal scheme is correct because

$$B/\prod_{i=1}^{n} A^{x_i} = m \cdot y^r/g^{r \sum_{i=1}^{n} x_i} = m \cdot y^r/y^r = m.$$

5.3.3 Distributed Plaintext Equality Test

Distributed plaintext equality test (PET) was introduced by Jakobsson and Juels [6] to test if two plaintexts are equal, given their ElGamal encryptions, without disclosing the two plaintexts.

Assume the same setting as distributed ElGamal cryptosystem, a simplified PET can be described as follows:

- Blindness: Given two ElGamal encryptions $C_1 = (A_1, B_1)$ and $C_2 = (A_2, B_2)$, to test if they are encryptions of the same plaintext, each server S_i chooses an integer r_i randomly from $\mathbb{Z}_q^*$ and outputs $A'_i = (A_1/A_2)^{r_i}$ and $B'_i = (B_1/B_2)^{r_i}$.

5.3 Preliminaries

- Test: Let $C = (A', B') = (\prod_{i=1}^n A'_i, \prod_{i=1}^n B'_i)$. The n servers decrypt C as described in the decryption algorithm in Sect. 5.3.2. If $\mathsf{D}(C) = 1$, then the two ciphertexts represent the encryption of the same plaintext. This is denoted by $\mathsf{PET}(C_1, C_2) = 1$. Otherwise, the plaintexts are not equal. This is denoted by $\mathsf{PET}(C_1, C_2) = 0$.

When (A_1, B_1) and (A_2, B_2) are encryptions of the same plaintext, (A'_i, B'_i) is an encryption of 1 and therefore (A', B') is an encryption of 1 as well. Otherwise, (A'_i, B'_i) is an encryption of a random number and therefore (A', B') is an encryption of a random number as well.

5.3.4 Conditional Gate

Conditional gate [9] allows $n\ (\geq 2)$ parties to multiply two encrypted values a and b without disclosing a and b, as long as a is restricted to a two-valued domain. It can be realized by the distributed ElGamal cryptosystem.

Assume the same setting as distributed ElGamal cryptosystem. Let $\mathsf{E}(g^a)$, $\mathsf{E}(g^b)$ denote ElGamal encryptions, with $a \in \{-1, 1\}$ and $b \in \mathbb{Z}_q$, the n server jointly compute $W = \mathsf{E}(g^{ab})$ using the conditional gate as follows.

- Let $U_0 = \mathsf{E}(g^a)$ and $V_0 = \mathsf{E}(g^b)$
- For $i = 1, 2 \cdots, n$, the server S_i takes U_{i-1} and V_{i-1} as input and outputs $U_i = \mathsf{RE}(U_{i-1}^{r_i})$ and $V_i = \mathsf{RE}(V_{i-1}^{r_i})$, where r_i is randomly chosen from $\{-1, 1\}$.
- The n servers jointly decrypt U_n to obtain $g^{a \prod_{i=1}^n r_i}$. Because $a, r_i \in \{-1, 1\}$, it is easy to know $a' = a \prod_{i=1}^n r_i \in \{-1, 1\}$.
- Let $W = V_n^{a'}$.

The conditional gate correctly computes $\mathsf{E}(g^{ab})$ because

$$V_n^{a'} = \mathsf{E}(g^{b \prod_{i=1}^n r_i})^{a'} = \mathsf{E}(g^{ab(\prod_{i=1}^n r_i)^2}) = \mathsf{E}(g^{ab}).$$

Based on the conditional gate, the n servers can jointly compute $\mathsf{E}(g^{ab})$ given $\mathsf{E}(g^a)$ and $\mathsf{E}(g^b)$ for $a, b \in \{0, 1\}$ as follows.

$$\mathsf{E}(g^{ab}) = \mathsf{E}(g^{((2a-1)b+b)2^{-1}}) = (\mathsf{E}(g^{(2a-1)b})\mathsf{E}(g^b))^{2^{-1}}.$$

Note that $2a - 1 \in \{-1, 1\}$ when $a \in \{0, 1\}$ and $\mathsf{E}(g^{(2a-1)b})$ can be computed with the conditional gate as above.

Furthermore, given encrypted bit representations $\mathsf{E}(x) = (\mathsf{E}(g^{x_{m-1}}), \cdots, \mathsf{E}(g^{x_1}), \mathsf{E}(g^{x_0}))$ and $\mathsf{E}(y) = (\mathsf{E}(g^{y_{m-1}}), \cdots, \mathsf{E}(g^{y_1}), \mathsf{E}(g^{y_0}))$ of two numbers x and y, the n servers can jointly compute the bits of $x + y$, given by $\mathsf{E}(z) = \mathsf{E}(x + y) = (\mathsf{E}(g^{z_{m-1}}), \mathsf{E}(g^{z_{m-1}}), \cdots, \mathsf{E}(g^{z_1}), \mathsf{E}(g^{z_0}))$ as follows:

$$E(g^{c_i}) = E(g^{x_i y_i})E(g^{x_i c_{i-1}})E(g^{y_i c_{i-1}})E(g^{-2x_i y_i c_{i-1}}),$$

$$E(g^{z_i}) = E(g^{x_i})E(g^{y_i})E(g^{c_{i-1}})E(g^{-2c_i}),$$

for $i = 0, 1, \cdots, m-1$, where $c_{-1} = 0$. We denote $E(x + y) = E(x) \boxplus E(y)$.

5.3.5 Two's Complement

Two's complement system is the most common method of representing signed integers on computers [5]. A positive integer in two's complement always has a 0 in the leftmost bit (sign bit) and is represented the same way as an unsigned binary integer. To negate a number, we invert all the bits and add one.

Comparison of two integers is often implemented with two's complement. Given the m-bit representations $(x_{m-1} \cdots x_1 x_0)$ and $(y_{m-1} \cdots y_1 y_0)$ of two numbers x and y, where the sign bits $x_{m-1} = 0$ and $y_{m-1} = 0$, the two's complement of $-y$, denoted as $\overline{-y}$, is $(\overline{y_{m-1}} \cdots \overline{y_1}\, \overline{y_0}) + 1$ where $\overline{y_i} = 1 - y_i$, let $z = (z_{m-1} \cdots z_1 z_0) = x + \overline{-y}$. If the sign bit $z_{m-1} = 0$, $x \geq y$ and otherwise $x < y$.

5.4 Model for Association Rule Mining with Data Privacy

Our model is designed for a cloud computing environment involving clients and servers. Clients are assumed to lack data storage, computational resources, and expertise, while servers are capable of providing services such as data storage and data mining. The model involves one client, one database (DB) server, and n data mining (DM) servers, as illustrated in Fig. 5.1.

As shown in Fig. 5.1, the client stores its data in the DB server in the cloud. To protect its data privacy, the ElGamal cryptosystem is used. First, the client generates its ElGamal public/private key pair, and then encrypts the data with its public key before uploading it to the DB server. To perform association rule mining on the encrypted data, the client selects n DM servers in the cloud, where $n \geq 2$, and splits its private key into n pieces, distributing them to the n DM servers.

The n DM servers cooperate to decrypt the encrypted data on behalf of the client and can be viewed as decryption servers as well. To minimize the risk of all servers being compromised, the client can choose servers from different cloud providers.

All servers (whether DB or DM servers) are assumed to be "semi-honest", meaning they follow the protocols or algorithms precisely, but may be curious about the client's data privacy. Moreover, **at least one of the n DM servers is assumed to be trusted not to collude with the other DM servers**. This ensures that the n servers can cooperate to decrypt the encrypted data only when required by data mining algorithms.

5.4 Model for Association Rule Mining with Data Privacy

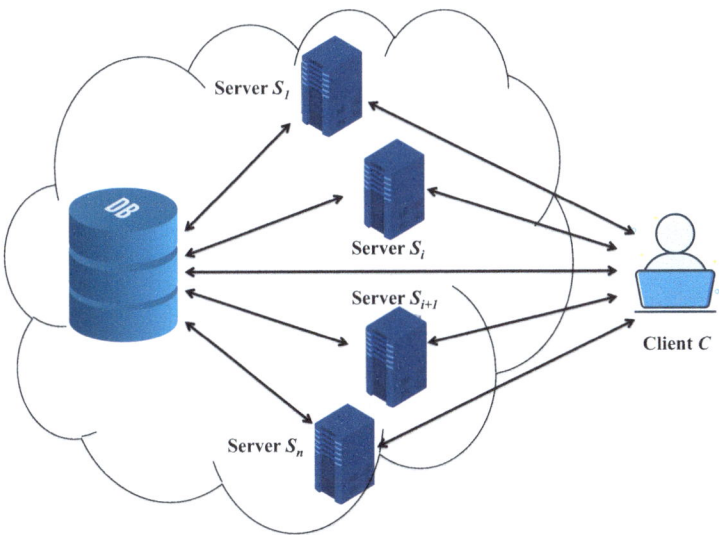

Fig. 5.1 Our model for privacy-preserving data mining in the cloud

Our model is designed to protect the client's data privacy while outsourcing the data mining task to the n DM servers and the DB server. Since this chapter focuses on association rule mining, we specifically consider a database composed of transactions. Each transaction contains a transaction identity (ID) and a set of items. Each item in the transaction is encrypted and stored in the DB server. The DB server does not know the client's private key, and therefore cannot decrypt any transaction.

When conducting association rule mining on the encrypted transaction database, the primary privacy concern is ensuring the confidentiality of items. We formally define item privacy in association rule mining over the encrypted transaction database using a game-theoretic approach as follows.

Definition 5.1 An association rule mining algorithm over the encrypted transaction database ensures *item privacy* if any adversary does not have a non-negligible advantage, greater than 1/2, of winning the following game:

- The adversary is provided with the public key of the client, selects two distinct items a_0 and a_1 from the transaction database, and challenges the system to return the encrypted form of either a_0 or a_1.
- The system encrypts a_0 and a_1 and returns their encrypted values to the adversary.
- The adversary then outputs a guess $\hat{a}$ of which item was encrypted and challenges the system to return the encrypted form of the other item.
- The system responds by encrypting and returning the encrypted form of the remaining item.

- The adversary wins the game if they can correctly guess which item was encrypted.

An association rule mining algorithm with item privacy cannot fully guarantee the privacy of the support of itemsets. For instance, in a transaction database, items are replaced with their encrypted versions, and the modified database is then sent to the DB server for association rule mining. According to Definition 5.1, the association rule mining algorithm provides item privacy. However, the adversary can still infer the supports of the encrypted items and carry out a background knowledge-based attack.

To prevent such attacks, one approach involves adding noisy transactions (encrypted with the client's public key) to the encrypted transaction database, as proposed by Giannotti et al. [3]. These noisy transactions are mixed with the original transactions through re-encryption and random shuffling, and their effects are removed during the association rule mining process. The noisy transactions obscure the actual supports of the itemsets. In this setup, any potential attacker, including insiders such as a DM server or a coalition of DM servers, must be unable to distinguish between the original and noisy transactions. Otherwise, the noisy transactions could be easily discarded.

We formally define the *transaction privacy* of association rule mining over the encrypted transaction database via a game, as follows.

Definition 5.2 An association rule mining algorithm over the encrypted transaction database, which includes both original transactions (encrypted) and noisy transactions (encrypted), has *transaction privacy* if no adversary can gain a non-negligible advantage over a random guess (i.e., a probability greater than 1/2) in the following game:

- The adversary is given the public key of the client, the set of encrypted transactions from the original transaction database, denoted as T, and a set of noisy transactions, denoted as T'. The adversary selects two distinct transactions, t_0 and t_1, from T and T', respectively, ensuring both transactions contain the same number of encrypted items. We assume that t_1 is not a noisy transaction generated by the adversary or its coalition if the adversary is a DM server.
- The challenger randomly chooses a bit b.
- The n DM servers and the DB server execute the association rule mining algorithm over the encrypted transaction database.
- The challenger retains all the information related to transaction t_b during the mining process and sends it back to the adversary.
- The adversary guesses the bit selected by the challenger, denoted as b'. If $b' = b$, the adversary wins the game.

Despite the inclusion of noisy transactions, an association rule mining algorithm with transaction privacy might still expose certain information about the original transactions, such as the probability that a transaction is an original one.

An ideal association rule mining process that ensures the privacy of the transaction database must not reveal any information about the original data, except for whether an encrypted itemset is frequent. We formally define the *database privacy* of association rule mining in the following game:

Definition 5.3 An association rule mining algorithm over the encrypted transaction database has *database privacy* if no adversary can gain a non-negligible advantage over a random guess (i.e., a probability greater than 1/2) in the following game:

- The adversary is given the public key of the client, selects two distinct transaction databases D_0 and D_1 that have the same frequent itemsets and the same size, and sends both D_0 and D_1 to the challenger.
- The challenger randomly chooses a bit b.
- The n DM servers and the DB server execute the association rule mining algorithm on the transaction database D_b and return all intermediate and final outcomes to the adversary.
- The adversary guesses the bit selected by the challenger, denoted as b'. If $b' = b$, the adversary wins the game.

In the three privacy requirements for association rule mining described above, item privacy is the basic level of privacy, transaction privacy is a higher level, and database privacy is the highest level of privacy.

5.5 Association Rule Mining with Item Privacy

In this section, we present our solution for association rule mining with item privacy.

Setup
Initially, upon receiving a security parameter k, the cloud publishes a multiplicative cyclic group $\mathbb{G}$ of prime order q with a generator g, where the discrete logarithm over $\mathbb{G}$ is computationally hard.

Additionally, the cloud publishes a standard item table that maps each item to a group element in $\mathbb{G}$, such that an item a is mapped to the code $g^{h(a)}$, where h is a collision-resistant hash function that maps strings $\{0, 1\}^*$ to elements in $\mathbb{Z}_q^*$.

A client chooses a private key x randomly from $\mathbb{Z}_q^* = \{1, 2, \ldots, q-1\}$ and computes the corresponding public key $y = g^x$. For a transaction $t =$ (TID, $a_1, a_2, \ldots, a_\lambda$), where TID is the transaction identity and a_i is an item, the client looks up the code for each item in the transaction from the standard item table and encrypts it with the client's public key y by ElGamal encryption. The client then uploads the encrypted transaction, consisting of encrypted items, to the DB server.

To outsource the association rule mining task to the cloud, the client selects $n \geq 2$ DM servers from different cloud providers. The client splits its private key into n pieces, $x_1, x_2, \ldots, x_n$, such that $\sum_{i=1}^{n} x_i = x \pmod{q}$. The client securely distributes these pieces to the n DM servers, for example, by encrypting each x_i with the public key of the i-th DM server S_i and sending it to the respective server.

For association rule mining, the client sends the minimum support and confidence values to the DB server.

Same Item Identification

In the encrypted transaction database, the same item is encrypted multiple times due to the probabilistic nature of the ElGamal encryption scheme. To mine frequent itemsets from the encrypted transaction database, denoted as $T = \{t_1, t_2, \ldots, t_\ell\}$, the n DM servers and the DB server cooperate to identify the encryptions of the same item and consolidate them into a single encryption, as described in Algorithm 1.

Algorithm 1 Same item identification (DB server and n DM servers)

Input: $T = \{t_1, t_2, \ldots, t_\ell\}$, y (public), $x_1, x_2, \ldots, x_n$ (private)
Output: $T' = \{t'_1, t'_2, \ldots, t'_\ell\}$
1: $t'_1 \leftarrow t_1, T' \leftarrow \{t'_1\}, t'_2 = t'_3 = \cdots = t'_\ell \leftarrow \emptyset$.
2: for $i = 1$ to $\ell - 1$ {
3:   if $(t_i \neq \emptyset)$ {
4:     for any $c \in t_i$ {
5:       for $j = i + 1$ to ℓ {
6:         if $(t_j \neq \emptyset)$ {
7:           for any $c' \in t_j$ {
8:             n DM servers cooperate to run PET(c, c')
9:             if (PET$(c, c') = 1$) $\{t'_j \leftarrow t'_j \cup \{c\}; t_j \leftarrow t_j - \{c'\}$;
                $j \leftarrow j + 1$; go to Step 5; }
10:           }
11:     }
12:   }
13:   $t'_{i+1} \leftarrow t'_{i+1} \cup t_{i+1}$.
14: }

In Algorithm 1, the DB server executes all lines except for the Plaintext Equality Test (PET) at Step 8, which requires those n DM servers to cooperate as described in Sects. 5.3.2 and 5.3.3. For efficiency, we assume that each DM server in PET sends (A'_i, B'_i) to the DB server which computes and broadcasts (A', B') to all DM servers, and each DM server in the distributed ElGamal decryption sends A_i to the DB server which computes and broadcasts the decryption result.

Theorem 5.1 (Correctness) *Algorithm 1 does not alter the content of the transaction database, i.e., $D(T') = D(T)$, where D denotes the ElGamal decryption.*

Proof We need to prove that for any i, $D(t'_i) = D(t_i)$. It is evident that $D(t'_1) = D(t_1)$.

Now, consider t_i and t'_i for $2 \leq i \leq \ell$. For the first $i - 1$ iterations at Step 2, the transaction t'_i is populated with items from t_i that are identical to items in transactions $t_1, t_2, \ldots, t_{i-1}$ at Step 9. At the conclusion of the $i - 1$ iterations, at

5.6 Association Rule Mining with Transaction Privacy

Step 15, the transaction t'_i is updated with items from t_i that do not match any items in transactions $t_1, t_2, \ldots, t_{i-1}$. Thus, the two transactions, t'_i and t_i, contain the same items, i.e., $\mathsf{D}(t'_i) = \mathsf{D}(t_i)$. □

Association Rule Mining on Encrypted Data

Once the different encryptions of the same item in the encrypted transaction database have been consolidated into a single encryption for each item, the DB server can proceed to count the support of any encrypted itemset. As a result, the DB server can apply the Apriori algorithm [1] on the modified encrypted database T', using the minimum support specified by the client.

By executing the Apriori algorithm, the DB server identifies all frequent encrypted itemsets that meet the minimum support criteria, and subsequently constructs the association rules. The DB server then returns the strong encrypted association rules that satisfy both the minimum support and confidence thresholds to the client.

Finally, the client decrypts the encrypted association rules and consults the standard item table to determine the specific items in the association rules.

Throughout the association rule mining process, all items in the transaction database remain encrypted. Therefore, this solution ensures item privacy. However, it does not protect against background knowledge-based attacks, as the DB server in the cloud can still observe the supports of encrypted itemsets in the encrypted transaction database.

5.6 Association Rule Mining with Transaction Privacy

In this section, we present our solution for association rule mining while preserving transaction privacy. This approach obfuscates the actual supports of the encrypted itemsets in the encrypted transaction database by introducing noisy transactions.

Setup

The setup of this solution is consistent with that described in Sect. 5.5, except that the client transmits an encrypted version of the minimum support δ to the DB server instead of the plaintext minimum support.

Suppose there are ℓ transactions in the database, with $m = \lceil \log_2 \ell \rceil$, and the binary representation of δ is given by $(w_{m-1} \cdots w_1 w_0)$. The encryption of δ is expressed as

$$\mathsf{E}(\delta) = \{\mathsf{E}(g^{w_{m-1}}), \cdots, \mathsf{E}(g^{w_1}), \mathsf{E}(g^{w_0})\},$$

where $w_i \in \{0, 1\}$.

Minimum confidence is not considered here. We assume the DB server and the n DM servers collaborate to mine all frequent encrypted itemsets and return them, along with their encrypted supports, to the client. After decryption, the client can locally construct strong association rules that meet the minimum support and confidence thresholds.

Item Dictionary Anonymization

To construct noisy transactions, the DM servers refer to a dictionary of possible items. Let the set of all possible items in the original transaction database be $\{a_1, a_2, \cdots, a_\gamma\}$. The DM servers may directly select and encrypt certain items from the dictionary and include them in a noisy transaction. This method complicates the removal of the noisy transactions without decryption during the mining of frequent itemsets.

To simplify the removal of noisy transactions, the DB server and the n DM servers cooperate to anonymize the dictionary of all possible items, producing a list of encrypted items $\{c_1, c_2, \cdots, c_\gamma\}$ using Algorithm 2. This ensures that no one can determine the mapping between the original dictionary $\{a_1, a_2, \cdots, a_\gamma\}$ and the anonymized dictionary $\{c_1, c_2, \cdots, c_\gamma\}$ as long as at least one of the n DM servers remains honest.

Algorithm 2 Item dictionary anonymization (DB server and n DM servers)

Input: $\{a_1, a_2, \cdots, a_\gamma\}$, y (public)
Output: $\{c_1, c_2, \cdots, c_\gamma\}$
1: for $i = 1$ to γ {
2:    DB server looks up the code a'_i of a_i from the standard item table;
3:    DB server encrypts a'_i to form $c_i = (g^r, a'_i y^r)$, where r is randomly chosen from $\mathbb{Z}_q^*$.
4: } //end for at Step 1
5: for $j = 1$ to n {
6:    Server S_j mixes $c_1, c_2, \cdots, c_\gamma$ by re-encryption and randomly shuffling;
7:    Server S_j forwards the results to Server S_{j+1} if $j < n$.
8: } // end for at Step 5
9: **return** $(c_1, c_2, \cdots, c_\gamma)$

In Algorithm 2, the DB server executes Steps 1 through 4 and forwards $c_1, c_2, \cdots, c_\gamma$ to the first DM server at Step 5. Subsequently, the n DM servers anonymize the dictionary alternately at Steps 6 and 7. Finally, the last DM server returns the anonymized dictionary to the DB server, which publishes it in the cloud.

Transaction Anonymization

Based on the anonymized dictionary, the DM servers can select and re-encrypt items from the dictionary and include them in noisy transactions, which are then forwarded to the DB server.

To generate noisy transactions that align with the original encrypted transaction database, the DB server groups the original transactions based on the number of items each transaction contains. Let the original transaction database $T =$

$\{t_1, t_2, \cdots, t_\ell\}$ be grouped into $\{T_1, T_2, \cdots, T_\lambda\}$, where T_i represents the set of all transactions with i items, and λ is the length of the longest transaction.

Each DM server appends $\lceil \phi \cdot |T_i| \rceil$ noisy transactions with i items to T_i, where $|T_i|$ denotes the total number of transactions in T_i, and ϕ is a positive parameter determined by the desired level of transaction privacy. Naturally, a larger ϕ enhances transaction privacy but reduces protocol efficiency.

Let T_i^* denote the set of transactions with i items after all n DM servers add noisy transactions. Then,

$$|T_i^*| = |T_i| + n \cdot \lceil \phi \cdot |T_i| \rceil$$

for $i = 1, 2, \cdots, \lambda$, and the combined transaction database, composed of original and noisy transactions, is represented as $T^* = \{T_1^*, T_2^*, \cdots, T_\lambda^*\}$. Evidently,

$$|T^*| = |T| + n \cdot \lceil \phi \cdot |T| \rceil.$$

If $\phi \approx 1/n$ and $|T_i| \geq n$ for all i, the size of the resulting transaction database T^* is approximately double that of the original transaction database T.

Next, the DB server and the n DM servers collaborate to execute Algorithm 3 to anonymize $T_\eta^* = \{t_1^*, t_2^*, \cdots, t_{\ell_\eta}^*\}$, where each transaction t_i^* contains η items (ℓ_η is the total number of transactions in T_η^*). Consequently, the two transactions t_i' and t_i possess identical items, i.e., $\mathsf{D}(t_i') = \mathsf{D}(t_i)$.

Algorithm 3 Transaction anonymization (DB server and n DM servers)

Input: $T_\eta^* = \{t_1^*, t_2^*, \cdots, t_{\ell_\eta}^*\}$, y (public)
Output: $T_\eta' = \{t_1', t_2', \cdots, t_{\ell_\eta}'\}$
1: $T_\eta' \leftarrow T_\eta^*$
2: for $i = 1$ to n {
3:    for $j = 1$ to ℓ_η {
4:      for any $c \in t_j'$ {
5:      Server S_i re-encrypts c to $c' = \mathsf{RE}(c)$ using
      the public key y of the client, as described in Sect. 5.3.1,
      and replaces c with c' in t_j'.
6:      } //end for at Step 4
7:    Server S_i randomly shuffles items in t_j'.
8:    } //end for at Step 3
9:    Server S_i randomly shuffles transactions in T_η' and forwards T_η' to Server S_{i+1} if $i < n$.
10: } //end for at Step 2
11: **return** T_η'

In Algorithm 3, the DB server initially forwards the transactions T_η' to the first DM server in Step 2. At the end of the process, the last DM server sends the anonymized transactions back to the DB server. Similarly, the DB server and the n DM servers can collaborate to anonymize $T_1^*, T_2^*, \cdots, T_\lambda^*$.

Same Item Identification Based on Dictionary

Algorithm 1, described in Sect. 5.5, identifies the same items in transactions. This process can be simplified into Algorithm 4 by utilizing the anonymized dictionary of potential items, i.e., $(c_1, c_2, \cdots, c_\gamma)$.

Algorithm 4 Same item identification based on dictionary (DB server and n DM servers)

Input: $(c_1, c_2, \cdots, c_\gamma), T' = \{t'_1, t'_2, \cdots, t'_L\}$, y (public), $x_1, x_2, \cdots, x_n$ (private)
Output: $T'' = \{t''_1, t''_2, \cdots, t''_L\}$
1: $T'' \leftarrow T'$
2: for $i = 1$ to γ {
3:   for $j = 1$ to L {
4:     for any $c \in t''_j$ {
5:     n DM servers cooperate to run $\mathsf{PET}(c_i, c)$
6:     If $\mathsf{PET}(c_i, c) = 1$ {
7:       Replace c in t''_j with c_i;
8:       $j \leftarrow j + 1$; go to Step 3
9:     } //end if at Step 6
10:    } //end for at Step 4
11:   } //end for at Step 3
12: } //end for at Step 2
13: **return** T''

Using Algorithm 4, all encrypted items in the transaction database are replaced with the encrypted counterparts from the dictionary.

Similar to Algorithm 1, Algorithm 4 does not alter the content of the transaction database, as it replaces the encryption of each item with an equivalent encryption of the same item.

After executing Algorithm 4, a DM server may identify the noisy transactions it added in T'' but remains unable to differentiate between the original transactions and noisy transactions introduced by other DM servers.

Frequent Itemset Mining

After identifying the same item, the DB server can calculate the support of any encrypted itemsets within the transformed transaction database. However, due to the presence of noisy transactions, the calculated support does not reflect the true support of the itemset. To address this, each DM server is required to retain the noisy transactions it introduces into the transaction database.

Given the encrypted minimum support threshold $\mathsf{E}(\delta) = \{\mathsf{E}(g^{w_{m-1}}), \cdots, \mathsf{E}(g^{w_1}), \mathsf{E}(g^{w_0})\}$ provided by the client, consider the critical step in the Apriori algorithm applied to the encrypted transaction database: determining whether an encrypted k-itemset $\alpha = \{\mathsf{E}(a_1), \mathsf{E}(a_2), \cdots, \mathsf{E}(a_k)\}$–constructed using the anonymized dictionary—is frequent (i.e., if its support exceeds the minimum support threshold).

5.6 Association Rule Mining with Transaction Privacy

Assume the support of the encrypted itemset α in the transformed transaction database is f, represented in binary form as $(u_{m'-1} \cdots u_1 u_0)$, known to the DB server and the n DM servers. The support of α in the noisy transactions added by Server S_i is f_i, with binary representation $(v^i_{m'-1} \cdots v^i_1 v^i_0)$, known only to S_i. Here, $m' = \lceil \log_2 |T''| \rceil + 1$, and the most significant bits $u_{m'-1} = v_{m'-1} = 0$. Since $|T''| > |T|$, it follows that $m' > m$.

Because the encrypted itemset α consists of encrypted items from the dictionary, each DM server can calculate the support of α within its noisy transactions.

Based on $\mathsf{E}(\delta)$, f, and f_i ($i = 1, 2, \cdots, n$), the DB server and the n DM servers collaborate to execute Algorithm 5 to determine if α is frequent.

Algorithm 5 Private comparison 1 (DB Server and n DM servers)

Input: $\mathsf{E}(\delta)$, f, y (public); (f_i, x_i) ($i = 1, 2, \cdots, n$) (private)
Output: Returns 1 if α is frequent, 0 otherwise.
1: $\mathsf{E}(z) \leftarrow \mathsf{E}(\delta)$.
2: **for** $i = 1$ to n **do**
3:     Server S_i computes
        $\mathsf{E}(f_i) = (\mathsf{E}(g^{v^i_{m'-1}}), \cdots, \mathsf{E}(g^{v^i_1}), \mathsf{E}(g^{v^i_0}))$
4:     n DM servers collaborate to compute
        $\mathsf{E}(z) \leftarrow \mathsf{E}(z) \boxplus \mathsf{E}(f_i)$ as described in Sect. 5.3.4.
5: **end for**
6: n DM servers compute
   $\mathsf{E}(z) = \mathsf{E}(z) \boxplus \mathsf{E}(\overline{-1})$ as outlined in Sects. 5.3.4 and 5.3.5.
7: The DB server computes $\overline{-f} \leftarrow (\overline{u_{m'-1}} \cdots \overline{u_1 u_0}) + 1$, where $\overline{-f}$ represents the two's complement of $-f$, and $\overline{u_j} = 1 - u_j$.
8: The DB server and n DM servers compute
   $\mathsf{E}(z) = \mathsf{E}(z) \boxplus \mathsf{E}(\overline{-f})$ as described in Sect. 5.3.4.
9: n DM servers decrypt $\mathsf{E}(g^{z_{m'-1}})$ as outlined in Sect. 5.3.2.
10: If $z_{m'-1} = 1$, then $\beta = 1$; otherwise, $\beta = 0$.
11: **return** β

Theorem 5.2 (Correctness) *Algorithm 5 outputs 1 if the encrypted itemset α is frequent and 0 otherwise.*

Proof Since f represents the support of α in the transformed transaction database with noisy transactions, and f_i is the support of α in the noisy transactions added by Server S_i, the actual support F of α in the original transaction database is given by

$$F = f - \sum_{i=1}^{n} f_i$$

Thus,

$$(\delta - 1) - F = (\sum_{i=1}^{n} f_i + \delta - 1) - f$$

Since $m' = \lceil \log_2 |T''| \rceil + 1$, $|T''| = |T'| = |T^*| = |T| + n\lceil \phi|T|\rceil$, $\delta \leq |T|$, and $\sum_{i=1}^{n} f_i \leq \lceil \phi|T|\rceil$, the binary representation $\sum_{i=1}^{n} f_i + \delta - 1$ contains fewer than $m' - 1$ bits. Therefore, after Step 6 in Algorithm 5, the most significant bit of $z = \sum_{i=1}^{n} f_i + \delta - 1$ is 0, i.e., $z_{m'-1} = 0$.

In Step 8 of Algorithm 5, $z = (\sum_{i=1}^{n} f_i + \delta - 1) + \overline{-f}$. According to Sect. 5.3.5, if the sign bit $z_{m'-1}$ is 1, then $\delta - 1 < F$, implying $F \geq \delta$ and $\beta = 1$. Otherwise, $\delta - 1 \geq F$, implying $F < \delta$ and $\beta = 0$. □

Using Algorithm 5, the DB server and n DM servers can collaboratively identify all frequent itemsets in T without decrypting any encrypted items. This ensures item privacy during association rule mining. Moreover, the original and noisy transactions are entirely mixed through re-encryption and random shuffling, making them indistinguishable to adversaries. Thus, association rule mining also ensures transaction privacy. However, Algorithm 4 may partially reveal information about the original transactions.

5.7 Association Rule Mining with Database Privacy

In this section, we present our solution for association rule mining while ensuring database privacy. The setup of this solution aligns with the description in Sect. 5.6.

Item Dictionary Anonymization
Initially, the DB server and the n DM servers execute Algorithm 2 to generate an anonymized dictionary of encrypted items $(c_1, c_2, \cdots, c_\gamma)$.

Transaction Anonymization
Utilizing the anonymized dictionary, the DM servers select and re-encrypt specific encrypted items from the dictionary, incorporating them into noisy transactions, which are then transmitted to the DB server, as outlined in Sect. 5.6. Subsequently, the DB server and DM servers employ Algorithm 3 to mix the original and noisy transactions, producing a new database D'.

Same Item Identification Based on Dictionary
The DB server partitions D' into n subsets $D'_1, D'_2, \cdots, D'_n$ such that $D'_i \cap D'_j = \emptyset$ and assigns D'_i to Server S_i. The n DM servers then collaboratively execute Algorithm 6 to identify items in the database D'. Following Algorithm 6, all items in the subset D'_i are replaced with their corresponding entries from the anonymized dictionary, which are only known to the DM Server S_i.

Frequent Itemset Mining
For a k-itemset $\alpha = \{\mathsf{E}(a_1), \mathsf{E}(a_2), \cdots, \mathsf{E}(a_k)\}$ constructed based on the anonymized dictionary, we examine whether α is frequent.

The DM Server S_i calculates the support of the k-item α in D''_i, denoted as F_i, and the support of α in the noisy transactions it added, denoted as f_i. Both F_i and

5.7 Association Rule Mining with Database Privacy

Algorithm 6 Same item identification based on dictionary (n DM servers)

Input: $(c_1, c_2, \cdots, c_\gamma), D'_1, D'_2, \cdots, D'_n, y$ (public), $x_1, x_2, \cdots, x_n$ (private)
Output: $D''_1, D''_2, \cdots, D''_n$
1: for $i = 1$ to n {
2:   $D''_i \leftarrow D'_i$
3:   for $j = 1$ to γ {
4:     for any $c \in D''_i$ {
5:       n DM servers cooperate to run $\mathsf{PET}(c_j, c)$ as described in Sects. 5.3.2 and 5.3.3, where **all DM servers send their computation results to** S_i. S_i checks if $\mathsf{PET}(c_j, c) = 1$ and keeps it confidential.
6:       If $\mathsf{PET}(c_j, c) = 1$ {
7:         S_i replaces c in D''_i with c_j.}
8:     } //end for at Step 4
9:   } //end for at Step 3
10: } //end for at Step 1
11: **return** $D''_1, D''_2, \cdots, D''_n$

---

f_i are only known to S_i. The real support of α in the original transaction database is given by:

$$F = \sum_{i=1}^{n}(F_i - f_i).$$

Using the encrypted binary representation of the minimum support $\mathsf{E}(\delta) = \{\mathsf{E}(g^{w_{m-1}}), \cdots, \mathsf{E}(g^{w_1}), \mathsf{E}(g^{w_0})\}$ provided by the client, and $F_i - f_i = (v^i_{m'-1} \cdots v^i_1 v^i_0)$ (where $i = 1, 2, \cdots, n$) for the encrypted itemset α, with $m' = \lceil \log_2 |T''| \rceil + 1$ and T'' as the anonymized database, the n DM servers run Algorithm 7 (similar to Algorithm 5) to determine if α is frequent.

Algorithm 7 Private comparison 2 (n DM servers)

Input: $\mathsf{E}(\delta), y$ (public), $(F_i - f_i, x_i)$ ($i = 1, 2, \cdots, n$) (private)
Output: 1 if α is frequent and 0 otherwise.
1: n DM servers cooperate to compute
   $\mathsf{E}(\overline{-\delta}) = (\mathsf{E}(g^{1-w_{m'}}), \cdots, \mathsf{E}(g^{1-w_1}), \mathsf{E}(g^{1-w_0})) \boxplus \mathsf{E}(1)$
   as in Sect. 5.3.4, where $\mathsf{E}(g^{1-w_i}) = \mathsf{E}(g)/\mathsf{E}(g^{w_i})$ and $w_{m'} = \cdots = w_{m+1} = w_m = 0$. $\overline{-\delta}$ is the two's complement of $-\delta$ with $m' + 1$ bits, as detailed in Sect. 5.3.5.
2: $\mathsf{E}(z) \leftarrow \mathsf{E}(\overline{-\delta})$
3: for $i = 1$ to n {
4:   Server S_i sets $v^i_{m'} = 0$ if $F_i - f_i \geq 0$ and 1 otherwise.
   $\mathsf{E}(F_i - f_i) = (\mathsf{E}(g^{v^i_{m'}}), \mathsf{E}(g^{v^i_{m'-1}}), \cdots, \mathsf{E}(g^{v^i_1}), \mathsf{E}(g^{v^i_0}))$
5:   n DM servers collaborate to compute
   $\mathsf{E}(z) = \mathsf{E}(z) \boxplus \mathsf{E}(F_i - f_i)$ as described in Sect. 5.3.4.
6: } //end for at Step 3
7: n DM servers cooperate to decrypt $\mathsf{E}(g^{z_{m'}})$ as explained in Sect. 5.3.2.
8: if $z'_m = 0$, then $\beta = 1$ else $\beta = 0$
9: **return** β

Theorem 5.3 (Correctness) *Algorithm 7 returns 1 if the encrypted itemset α is frequent and 0 otherwise.*

Proof After Step 6 in Algorithm 7, the value of z is computed as $z = \sum_{i=1}^{n}(F_i - f_i) + \overline{-\delta}$. Based on the explanation in Sect. 5.3.5, if the sign bit $z_{m'}$ equals 0, it implies $F \geq \delta$, resulting in $\beta = 1$. Conversely, if $z_{m'} = 1$, it implies $\delta > F$, and hence $\beta = 0$. □

5.8 Conclusion

In this chapter, we have introduced three solutions for privacy-preserving association rule mining within a cloud computing environment. These solutions ensure the privacy of items, transactions, and databases, respectively. Our implementation and experimental results have demonstrated that these solutions are practical for real-world scenarios.

Additionally, we observed that performance can be significantly enhanced through parallel computation. Specifically, the time efficiency of our protocol could be improved by dividing the encrypted transactions into disjoint subsets, where each subset is processed by an independent group of n DM servers. However, further modifications to our protocols would be required to ensure correctness under such an arrangement.

We plan to explore this optimization in future work and aim to evaluate its performance in real-world deployments, particularly within cloud environments provided by commercial cloud service providers.

References

1. Rakesh Agrawal and Ramakrishnan Srikant. Fast algorithms for mining association rules in large databases. In *VLDB'94, Proceedings of 20th International Conference on Very Large Data Bases, Santiago de Chile, Chile, September 12-15, 1994*, pages 487–499, 1994. Available at: http://www.vldb.org/conf/1994/P487.PDF.
2. Taher El Gamal. A public key cryptosystem and a signature scheme based on discrete logarithms. *IEEE Transactions on Information Theory*, 31(4):469–472, 1985. http://dx.doi.org/10.1109/TIT.1985.1057074.
3. Fosca Giannotti, Laks V. S. Lakshmanan, Anna Monreale, Dino Pedreschi, and Wendy Hui Wang. Privacy-preserving mining of association rules from outsourced transaction databases. *IEEE Systems Journal*, 7(3):385–395, 2013. http://dx.doi.org/10.1109/JSYST.2012.2221854.
4. Jiawei Han and Micheline Kamber. *Data Mining: Concepts and Techniques*. Morgan Kaufmann, 2000. ISBN: 1-55860-489-8.
5. David Money Harris and Sarah L. Harris. *Digital Design and Computer Architecture*. Morgan Kaufmann, 2007. ISBN: 978-0-12-370497-9. Available at: http://opac.inria.fr/record=b1133648.
6. Markus Jakobsson and Ari Juels. Mix and match: Secure function evaluation via ciphertexts. In *Advances in Cryptology—ASIACRYPT 2000, Proceedings*, pages 162–177, Kyoto, Japan, December 3-7, 2000. http://dx.doi.org/10.1007/3-540-44448-3_13.

7. Ian Molloy, Ninghui Li, and Tiancheng Li. On the (in)security and (im)practicality of outsourcing precise association rule mining. In *ICDM 2009, The Ninth IEEE International Conference on Data Mining, Miami, Florida, USA, December 6-9, 2009*, pages 872–877, 2009. http://dx.doi.org/10.1109/ICDM.2009.122.
8. Sunita Sarawagi and Sree Hari Nagaralu. Data mining models as services on the Internet. *SIGKDD Explorations*, 2(1):24–28, 2000. http://doi.acm.org/10.1145/360402.360412.
9. Berry Schoenmakers and Pim Tuyls. Practical two-party computation based on the conditional gate. In *Advances in Cryptology—ASIACRYPT 2004, Proceedings*, pages 119–136, Jeju Island, Korea, December 5-9, 2004. http://dx.doi.org/10.1007/978-3-540-30539-2_10.
10. Chih-Hua Tai, Philip S. Yu, and Ming-Syan Chen. k-Support anonymity based on pseudo taxonomy for outsourcing of frequent itemset mining. In *Proceedings of the 16th ACM SIGKDD International Conference on Knowledge Discovery and Data Mining*, pages 473–482, Washington, DC, USA, July 25–28, 2010. http://doi.acm.org/10.1145/1835804.1835866.
11. Wai Kit Wong, David W. Cheung, Edward Hung, Ben Kao, and Nikos Mamoulis. Security in outsourcing of association rule mining. In *Proceedings of the 33rd International Conference on Very Large Data Bases*, pages 111–122, University of Vienna, Austria, September 23-27, 2007. Available at: http://www.vldb.org/conf/2007/papers/research/p111-wong.pdf.
12. Xun Yi, Fang-Yu Rao, Elisa Bertino, Athman Bouguettaya: Privacy-Preserving Association Rule Mining in Cloud Computing. AsiaCCS 2015: 439-450

Chapter 6
Single-Database Private Information Retrieval from FHE

Abstract Private Information Retrieval (PIR) allows a user to retrieve the i-th bit of an n-bit database without revealing to the database server the value of i. In this chapter, we introduce a PIR protocol with the communication complexity of $O(\gamma \log n)$ bits, where γ is the ciphertext size. Furthermore, we extend the PIR protocol to a private block retrieval (PBR) protocol, a natural and more practical extension of PIR in which the user retrieves a block of bits, instead of retrieving single bit. Our protocols are built on the state-of-the-art fully homomorphic encryption techniques and provide privacy for the user if the underlying fully homomorphic encryption scheme is semantically secure. The total communication complexity of our PBR is $O(\gamma \log m + \gamma n/m)$ bits, where m is the number of blocks. The total computation complexity of our PBR is $O(m \log m)$ modular multiplications plus $O(n/2)$ modular additions. In terms of total protocol execution time, our PBR protocol is more efficient than existing PBR protocols which usually require to compute $O(n/2)$ modular multiplications when the size of a block in the database is large and a high speed network is available.

6.1 Introduction

Private information retrieval (PIR) protocol allows a user to retrieval the i-th bit of an n-bit database, without revealing to the database server the value of i. A trivial solution is for the user to retrieve the entire database, but this approach may incur enormous communication cost. A good PIR protocol is expected to have considerably lower communication complexity. Private Block Retrieval (PBR) is a natural and more practical extension of PIR in which, instead of retrieving only a single bit, the user retrieves a block of bits from the database.

PIR was first introduced by Chor et al. [9] in 1995 in a multi-server setting, where the user retrieves information from multiple database servers, each has a copy of the same database. To ensure user privacy in the multi-server setting, the servers must be trusted not to collude. In [9], Chor et al. have shown that if only single database is used, n bits must be communicated in the information-theoretic sense, that is, the user's query gives absolutely no information about i. They have also shown that any PIR protocol can be converted to a PBR protocol.

In 1997, using the quadratic residuosity computational assumption, Kushilevitz and Ostrovsky [21] constructed a single-database PIR with communication complexity of $O(2^{\sqrt{\log n \log \log N}})$ which is less then $O(n^\epsilon)$ for any $\epsilon > 0$, where N is the composite modulus. Their basic idea is viewing the database as a matrix $M = (x_{ij})_{s \times t}$ of bits. To retrieve the (a, b) entry of the matrix, the user sends to the database server a composite (hard-to-factor) modulus N and t randomly chosen integers $y_1, y_2, \cdots, y_t$ such that only y_a is not a quadratic residuosity modulo N, that is, $y_i \neq \alpha^2 (mod\ N)$ for any integer α. The server sends back $z_i = \prod_{j=1}^{t} y_j^{2-x_{ij}} (mod\ N)$ for $1 \leq i \leq s$. The user concludes that $x_{ij} = 0$ if z_a is a quadratic residuosity modulo N, and $x_{ij} = 1$ otherwise.

In 1999, Cachin et al. [7] constructed the first single-database PIR with polylogarithmic communication complexity $O(\log^8 n)$. The security of their protocol is based on the Φ-hiding number-theoretic assumption, that is, it is hard to distinguish which of two primes divides $\phi(N)$ for the composite modulus N. Their basic idea is mapping each index i to a distinct prime p_i. To retrieve bit b_i from a database $B = b_1 b_2 \cdots b_n$, the user sends to the database server a composite (hard-to-factor) modulus N such that p_i divides $\phi(N)$ and a generator g with order divisible by p_i. The server sends back $r = g^P (mod\ N)$ where $P = \prod_j p_j^{b_j}$. The user concludes that $b_i = 1$ if r is a p_i-residue modulo N, otherwise, $b_i = 0$.

In 2000, Kushilevitz and Ostrovsky [22] constructed a PIR protocol with total communication complexity $n - \frac{cn}{2k} + O(k^2)$, where k is a security parameter and c is a constant. Their protocol is built on the Naor-Yung one-way 2-to-1 trapdoor permutations [25] and the Goldreich-Levin hard-core predicates [19]. Their basic idea is dividing an n-bit database into k-bit blocks and organizing the database into pairs of blocks, denoted by $z_{i,L}$ and $z_{i,R}$, where $i = 1, 2, \cdots, \frac{n}{2k}$. Suppose that the user wishes to retrieve $z_{s,L}$. The user sends to the database server the descriptions of one-way trapdoor permutations f_L and f_R, to which the user has the trapdoors. The server computes $f_L(z_{i,L})$ and $f_R(z_{i,R})$ for all i and returns these values to the user. With trapdoors, the user computes two possible pre-images $(z_{s,L}, z'_{s,L})$. Next, the user sends to the server two hardcore predicates r_L and r_R such that $r_L(z_{s,L}) \neq r_L(z'_{s,L})$, but $r_R(z_{s,R}) = r_R(z'_{s,R})$. The server responds with $r_L(z_{i,L}) \oplus r_R(z_{i,R})$ for all i. At the end, the user learns which pre-image is $z_{s,L}$ from $r_L(z_{s,L})$.

In 2005, Gentry and Ramzan [14] extended the single-database PIR of Cachin et al. [7] to a PBR with communication complexity $O(\log^2 n)$, the current best bound for communication complexity. The security of their protocol is also based on the Φ-hiding assumption. Assume that an n-bit database B is partitioned into m blocks, each has ℓ bits, denoted as $B = C_1 \| C_2 \| \cdots \| C_m$. Their basic idea is associating C_i with a distinct small prime p_i, rather than associating a (largish) prime with each bit. The database server uses the Chinese Reminder Theorem to determine an integer e such that $e = C_i (mod\ p_i^{c_i})$ for $1 \leq i \leq m$, where c_i is the smallest integer such that $p_i^{c_i} \geq 2^\ell$. To retrieve block C_i, the user sends to the database server a composite (hard-to-factor) modulus N such that $p_i^{c_i}$ divides $\phi(N)$ and a

6.1 Introduction

generator g with order divisible by $p_i^{C_i}$. The server sends back $r = g^e \pmod{N}$. Let $q = order(g)/p_i^{C_i}$, then $order(g^q) = p_i^{C_i}$. Since p_i is a small prime, the user can compute the discrete logarithm $\log_{(g^q)}(r^q) = e \pmod{p_i^{C_i}} = C_i$ using the Pohlig-Hellman algorithm [30].

Like the original single-database PIR of Kushilevitz and Ostrovsky [21], Chang [8] and Lipmaa [23, 24] also constructed PIR protocols with communication complexity of $O(\log^2 n)$. The difference is that the former is based on the Goldwasser-Micali homomorphic encryption [20], but the later is built on the Damgard-Jurik homomorphic encryption [11], a variant of the Paillier homomorphic encryption [29].

These single-database PIR protocols provide almost optimal communication cost, but require the database to use an enormous amount of computational power. In 2007, Aguilar-Melchor and Gaborit [1, 2] presented a lattice-based computationally-efficient PIR protocol, in which the computational cost is a few thousand bit-operations per bit in the database. In this protocol, the user, who wants to retrieve an element of index i from a database composed of n elements, generates a query formed of n matrices $B_1, B_2, \cdots, B_n$, one for each database element. All matrices are soft disturbed matrices except B_i, which is a hard disturbed matrix. The user sends the query to the database server which encodes the n elements to n matrices $A_1, A_2, \cdots, A_n$ and computes $R = (A_1, A_2, \cdots, A_n)(B_1, B_2, \cdots, B_N)^T$, and responds to the user with R. At last, the user retrieves B_i from R. Using analytical and experimental techniques, Olumofin and Goldberg [27] analyzed the performance of the lattice-based PIR protocol in 2010 and reported that the end-to-end response time of the protocol is one to three orders of magnitude less than the trivial protocol for realistic computation power and network bandwidth.

In 2008, Aguilar-Melchor et al. [3] provided a solution for securely evaluating multivariate polynomials of degree d with additively homomorphic encryption scheme. This scheme, further improved in 2010, can be used in private information retrieval which requires secure evaluation of low-degree multivariate polynomials with a large number of monomials. The basic idea is to build a compound ciphertext $\alpha \otimes \beta = (\alpha^{(1)}\beta, \alpha^{(2)}\beta, \cdots, \alpha^{(t)}\beta)$, given the encryptions $\alpha = \mathsf{E}(a, pk) = (\alpha^{(1)}, \alpha^{(2)}, \cdots, \alpha^{(t)})$ and $\beta = \mathsf{E}(b, pk)$, where $a, b \in \{0, 1\}$. To decrypt the compound ciphertext, one decrypts each coordinate at first and then reconstructs the inner ciphertext $\alpha = \sum_i 2^{i-1}\alpha^{(i)}$ and decrypts it again to ab. This allows to evaluate degree 2 polynomials over encrypted data. The idea can be generalised by iterating the construction to evaluate polynomials of degree d securely, at the price of an expansion factor for the length of the ciphertext which is exponential in d.

Single-database PIR has a close connection to the notion of Oblivious Transfer (OT), introduced by Rabin [31] in 1981. A different variant of OT, called 1-out-of-2 OT, was introduced by Even et al. [13] in 1985 and, more generally, 1-out-of-n OT was considered in Brassard et al. [6] in 1987. Informally, OT is a two-party protocol, where a sender with n messages $M_1, M_2, \cdots, M_n$ and a receiver with an index i ($1 \leq i \leq n$) interact, and at the end of the protocol the receiver obtains M_i without learning anything about other messages, while the sender does not learn anything about the index i. OT is different from PIR in that there is no communication complexity requirement (beyond being polynomially bounded) but, on the other hand, secrecy is required for both players, while for PIR it is required only for the user. Naor and Pinkas [26] have shown how to turn any PIR protocol into 1-out-of-n OT protocol with one invocation of a single-database PIR protocol and logarithmic number of invocations of 1-out-of-2 OT. DiCrescenzo et al. [10] showed that any single-database PIR protocol implies 1-out-of-n OT.

Contributions Homomorphic encryption techniques are often very natural ways to construct a variety of privacy-preserving protocols. For example, the PIR protocol of Kushilevitz and Ostrovsky [21] is based on the Goldwasser-Micali homomorphic encryption E [20], in which $\mathsf{E}(b_1)\mathsf{E}(b_2) = \mathsf{E}(b_1 \oplus b_2)$ for any $b_1, b_2 \in \{0, 1\}$, while the PIR protocol of Chang [8] and Lipmaa [23] are based on the Damgard-Jurik homomorphic encryption E' [11], in which $\mathsf{E}'(m_1)\mathsf{E}'(m_2) = \mathsf{E}'(m_1 + m_2)$ for any $m_1, m_2 \in \mathbb{Z}_N$. A generic method to construct a single-database PIR from a homomorphic encryption scheme was given by Ostrovsky and Skeith in [28]. These underlying encryption schemes allows homomorphic computation of only one operation (either addition or multiplication) on plaintexts.

In 2009, Gentry [15–18] constructed the first fully homomorphic encryption (FHE) scheme using lattice-based cryptography. FHE allows homomorphic computation of two operations (both addition and multiplication) of plaintexts. In the same year, Dijk et al. [12] presented the second FHE scheme, which uses many of the tools of Gentry's construction, but which does not require ideal lattices. The scheme is therefore conceptually simpler than Gentry's ideal lattice scheme, but has similar properties with regards to homomorphic operations and efficiency. In 2010, Smart and Vercauteren [32] presented a refinement of Gentry's scheme giving smaller key and ciphertext sizes. In 2011, Brakerski et al. [4, 5] presented fully homomorphic encryption schemes based on the learning with error assumption.

Motivated by recent breakthrough in FHE, we study single-database PIR and PBR protocols from FHE in this chapter [33]. In [15], Gentry briefly described a single-database PIR protocol with communication complexity $O(\gamma \log n)$. The basic idea is that the user, who wishes to retrieve the i-th bit from a database with n bits, sends to the database server the encryption of the index i, and the server sends back the encryption of the i-th bit computed by fully homomorphic properties.

We extend Gentry's basic idea to a PBR protocol, where the user, who wishes to retrieve the i-th block from a database with m blocks, sends to the server the encryption of the index i, and the server sends back the encryption of the i-th block computed by fully homomorphic properties, as if the user sends the index i and the server replies the i-th block. It can be seen that this solution is conceptually simpler than any existing PBR protocols without FHE.

Based on the formal model for security of single-database PIR [7, 14], we show that such PIR and PBR protocols from FHE provide privacy of the user as long as the underlying FHE scheme is semantically secure.

The performance of our protocols depends on the underlying FHE scheme. The user and the server need to exchange $O(\gamma \log n)$ bits in our PIR protocol, and exchange $O(\gamma \log m + \gamma n/m)$ in our PBR protocol, where n is the size of the database, m is the number of blocks in the database, γ is the ciphertext size, and the base of the logarithm is 2.

So far, existing FHE schemes have not been practical. We give a variant of Dijk et al.'s somewhat homomorphic encryption scheme, from which we construct a practical PBR protocol. In addition, we have implemented the practical PBR protocol for a database composed of 10,000 elements of size of 200 kbits. Our experiment shows that our PBR protocol is practical.

The rest of the chapter is as follows. Section 6.2 introduces the definitions of FHE and PIR; Sects. 6.3 and 6.4 describes generic and practical PIR and PBR protocols from FHE; Conclusions are in the last section.

6.2 Prelimaries

In this section, we introduce the concepts of fully homomorphic encryption (FHE) scheme and single-database private information retrieval (PIR) protocol.

6.2.1 Fully Homomorphic Encryption

Formally, a FHE scheme consists of five algorithms as follows.

(1) Key Generation (KG): The algorithm takes as an input a security parameter k and outputs a public and private key pair (pk, sk), where pk is public, while sk is kept secret.
(2) Encryption (E): The algorithm takes as input a plaintext $m \in \{0, 1\}$ and the public key pk, and output a ciphertext c, denoted as $c = \mathsf{E}(m, pk)$.
(3) Decryption (D): The algorithm takes as input a ciphertext c and the private key sk, and outputs a plaintext $m \in \{0, 1\}$, denoted as $m = \mathsf{D}(c, sk)$.
(4) Homomorphic Addition (Add): The algorithm takes as input two ciphertexts $c_1 = \mathsf{E}(m_1, pk)$, $c_2 = \mathsf{E}(m_2, pk)$ and the public key pk, and outputs a ciphertext c, denoted as $c = \mathsf{Add}(c_1, c_2, pk) = c_1 \boxplus c_2$, such that

$$\mathsf{D}(c, sk) = m_1 \oplus m_2.$$

(5) Homomorphic Multiplication (Mult): The algorithm takes as input two ciphertexts $c_1 = \mathsf{E}(m_1, pk)$, $c_2 = \mathsf{E}(m_2, pk)$ and the public key, and outputs a ciphertext $c = \mathsf{Mult}(c_1, c_2, pk) = c_1 \boxtimes c_2$, such that

$$\mathsf{D}(c, sk) = m_1 \cdot m_2.$$

A FHE scheme (KG, E, D, Add, Mult) is semantically secure if, given any public key pk, no probabilistic polynomial-time (PPT) adversary has success probability greater than ϵ to distinguish $\mathsf{E}(0, pk)$ and $\mathsf{E}(1, pk)$, where ϵ is negligible in k.

6.2.2 DGHV Somewhat Homomorphic Encryption

In this section, we instantiate "somewhat" homomorphic encryption scheme proposed by Dijk, Gentry, Halevi and Vaikuntanathan [12], called DGHV somewhat scheme for brevity. The scheme does not require ideal lattices.

In [12], a symmetric homomorphic encryption scheme was proposed as follows.

1. KeyGen: The key is an odd integer, chosen from some interval $p \in [2^{\eta-1}, 2^{\eta})$.
2. Encrypt(p, M): To encrypt a bit $M \in \{0, 1\}$, set the ciphertext as an integer $c = M + 2r + qp$, where the integers q, r are chosen at random in some other prescribed intervals, such that $2r$ is smaller than $p/2$ in absolute value.
3. Decrypt(p, c): Output $(c \bmod p) \bmod 2$.

Given two ciphertexts $c_1 = m_1 + 2r_1 + q_1 p$ and $c_2 = m_2 + 2r_2 + q_2 p$, we have

$$c_1 + c_2 = (m_1 + m_2) + 2(r_1 + r_2)$$
$$+ (q_1 + q_2)p, \qquad (6.1)$$
$$c_1 c_2 = m_1 m_2 + 2(2r_1 r_2 + r_1 m_2 + m_1 r_2)$$
$$+ (c_1 q_2 + q_1 c_2 + q_1 q_2 p)p \qquad (6.2)$$

Furthermore, $(c_1 + c_2 \bmod p) \bmod 2 = m_1 \oplus m_2$ and $(c_1 c_2 \bmod p) \bmod 2) = m_1 m_2$. Therefore, the symmetric encryption scheme supports both additional and multiplicative homomorphisms.

In [12], an asymmetric homomorphic encryption scheme was also proposed as follows.

(1) KeyGen(k): Takes a security parameter λ and determines a (convenient) parameter set $\rho = \lambda, \rho' = 2\lambda, \eta = \tilde{O}(\lambda^2), \gamma = \tilde{O}(\lambda^5), \tau = \gamma + \lambda$, where γ is the bit-length of the ciphertext, η is the bit-length of the secret key, ρ is the bit-length of the noise, τ is the number of integers in the public key. Chooses a random odd η-bit integer p from $(2\mathbb{Z} + 1) \cap (2^{\eta-1}, 2^{\eta})$ as the secret key sk. Randomly chooses $q_0, q_1, \cdots, q_\tau$ from $[1, 2^\gamma/p)$ subject to the condition that the largest q_i is odd and relabels $q_0, q_1, \cdots, q_\tau$ so that q_0 is the largest. Randomly chooses $r_1, \cdots, r_\tau$ from $\mathbb{Z} \cap (-2^\rho, 2^\rho)$ and sets $x_0 = q_0 p$ and $x_i = q_i p + r_i$. The public key is $pk = <x_0, x_1, \cdots, x_\tau>$.
(2) Encrypt(pk, M): To encrypt $M \in \{0, 1\}$, chooses a random subset $S \subset \{1, 2, \cdots, \tau\}$ and a random integer r from $(-2^{\rho'}, 2^{\rho'})$ and outputs the ciphertext

$$c = \mathsf{E}(M, pk) = [M + 2r + 2\sum_{i \in S} x_i]_{x_0},$$

where $[z]_{x_0}$ stands for $z(\bmod\ x_0)$.
(3) Decrypt(sk, c): To decrypt c, outputs

$$M' = \mathsf{D}(c, sk) = (c\ mod\ p) mod\ 2.$$

(4) Homomorphic Addition (Add): Given two ciphertext $c_1 = \mathsf{E}(m_1, pk)$, $c_2 = \mathsf{E}(m_2, pk)$ and the public key pk, outputs a ciphertext

$$c = \mathsf{Add}(c_1, c_2, pk) = [\mathsf{E}(m_1) + \mathsf{E}(m_2)]_{x_0},$$

(5) Homomorphic Multiplication (Mult): Given two ciphertext $c_1 = \mathsf{E}(m_1, pk)$, $c_2 = \mathsf{E}(m_2, pk)$ and the public key pk, and outputs a ciphertext

$$c = \mathsf{Mult}(c_1, c_2, pk) = [\mathsf{E}(m_1)\mathsf{E}(m_2)]_{x_0}.$$

This asymmetric encryption scheme supports both additional and multiplicative homomorphisms, too.

The security of the DGHV somewhat scheme is based on the approximate-gcd problem, that is, for a randomly chosen η-bit odd integer p, given polynomially many samples in the form $qp + r$, where q is randomly chosen from $[1, 2^\gamma/p)$ and r is randomly chosen from $\mathbb{Z} \cap (2^{-\rho}, 2^\rho)$, determine p.

The DGHV somewhat scheme is semantically secure if the approximate-gcd problem is hard. The choice of parameters in the DGHV somewhat scheme achieves at least 2^λ security against all of known attacks.

6.2.3 Single-Database Private Information Retrieval

Informally, a single-database PIR protocol is a two-party protocol, where a user retrieves the i-th bit from an n-bit database $DB = b_1 b_2 \cdots b_n$, without revealing to the database server the value of i. Formally, a single-database PIR protocol consists of three algorithms as in [7, 14].

(1) Query Generation (QG): Takes as input a security parameter k, the size n of the database, and the index i of a bit in the database, outputs a query Q and a secret s, denoted as $(Q, s) = \mathsf{QG}(n, i, 1^k)$.
(2) Response Generation (RG): Takes as input the security parameter k, the query Q and the database DB, outputs a response R, denoted as $R = \mathsf{RG}(DB, Q, 1^k)$.
(3) Response Retrieval (RR): Takes as input the security parameter k, the response R, the index i of the bit, the size n of the database, the query Q, and the secret s, output a bit b', denoted as $b' = \mathsf{RR}(n, i, (Q, s), R, 1^k)$.

A single-database PIR protocol is correct if, for any security parameter k, any database DB with any size n, and any index i for $1 \leq i \leq n$, $b_i = \mathsf{RR}(n, i, (Q, s), R, 1^k)$ holds, where $(Q, s) = \mathsf{QG}(n, i, 1^k)$ and $R = \mathsf{RG}(DB, Q, 1^k)$.

The security of single-database PIR protocol can be defined with a game as follows.

Give an n-bit database $DB = b_1 b_2 \cdots b_n$. Consider the following game between an adversary (the database server) $\mathcal{A}$, and a challenger C. The game consists of the following steps:

(1) The adversary $\mathcal{A}$ chooses two different indices $1 \le i, j \le n$ and sends them to C.
(2) Let $\lambda_0 = i$ and $\lambda_1 = j$. The challenger C chooses a random bit $b \in \{0, 1\}$, and executes $\mathsf{QG}(n, \lambda_b, 1^k)$ to obtain (Q_b, s), and then sends Q_b back to $\mathcal{A}$.
(3) The adversary $\mathcal{A}$ can experiment with the code of Q_b in an arbitrary non-black-box way, and finally outputs $b' \in \{0, 1\}$.

The adversary wins the game if $b' = b$ and loses otherwise. We define the adversary $\mathcal{A}$'s advantage in this game to be

$$\mathsf{Adv}_{\mathcal{A}}(k) = |\Pr(b' = b) - 1/2|.$$

Definition 6.1 (Security Definition) A single-database PIR protocol is semantically secure if for any probabilistic polynomial time (PPT) adversary $\mathcal{A}$, we have that $\mathsf{Adv}_{\mathcal{A}}(k)$ is a negligible function, where the probability is taken over coin-tosses of the challenger and the adversary.

Similarly, single-database private block retrieval (PBR) protocol can be defined by viewing the n-bit database as $DB = B_1 \| B_2 \| \cdots \| B_m$, where B_i is a block with n/m bits.

6.3 Generic Single-Database PIR from FHE

In this section, we present a response generation circuit, from which we construct a generic single-database PIR protocol based on FHE and then extend it to a generic single-database PBR protocol.

6.3.1 Response Generation Circuit

Without taking security into account, a response generation circuit can be described as follows.

Inputs: An index $i \in [1, n]$ and an n-bit database $DB = b_1 b_2 \cdots b_n$

Output: b_i

Response Generation Circuit:

(1) Write the index i in the binary representation, denoted as $i = \alpha_1 \alpha_2 \cdots \alpha_\ell$, where $\ell = \lceil \log n \rceil$.

(2) For each index $j \in [1, n]$, write j in the binary representation, denoted as $j = \beta_{j,1}\beta_{j,2}\cdots\beta_{j,\ell}$. Compute

$$\gamma_j = \prod_{t=1}^{\ell}(\alpha_t \oplus \beta_{j,t} \oplus 1), \tag{6.3}$$

where $\oplus$ stands for XOR operation. If $j = i$, $\gamma_j = 1$ and 0 otherwise. This means only $\gamma_i = 1$.

(3) Output

$$R = \bigoplus_{b_j=1} \gamma_j. \tag{6.4}$$

If $b_i = 1$, then $\bigoplus_{b_j=1} \gamma_j = \gamma_i = 1$. If $b_i = 0$, then $\bigoplus_{b_j=1} \gamma_j = 0$. Therefore, $R = \bigoplus_{b_j=1} \gamma_j = b_i$.

The response generation circuit is implemented with two simple operations, $\oplus$ and $\cdot$, where $\cdot$ is equivalent to AND. It does not need IF-THEN statement.

The circuit requires the plain index i to generate the response R and thus cannot preserve user privacy. To do so, we make use of the FHE technique to evaluate the response generation circuit with the encrypted index as input in next section.

6.3.2 Generic Single-Database PIR from FHE

The generic single-database PIR protocol is built on a FHE scheme (KG, E, D, Add, Mult) and consists of three algorithms (Query Generation QG, Response Generation RG, and Response Retrieval RR).

At a high level, the user generates a public and private key pair (pk, sk) for the FHE scheme, sends the public key pk to the database server, but keeps the private key sk secret. Then the user chooses an index i, where $1 \leq i \leq n$, and encrypts i with the public key pk, and sends the ciphertext as a query to the database server. Based on the response generation circuit and homomorphic properties, the server computes an encryption of the i-th bit as a response based on the database, the query and the public key pk, and sends the response back. At the end, the user decrypts the response to obtain the i-th bit.

Assume that the user and the database server have agreed upon a FHE scheme (KG, E, D, Add, Mult) in advance, our single-database PIR can be described as follows.

Query Generation $\mathsf{QG}(n, i, 1^k)$

Inputs: The size n of the database DB, an index $i \in [1, n]$, the key generation algorithm KG, the encryption algorithm E, and a security parameter k.

Outputs: A query $Q = (pk, \mathsf{E}(i, pk))$ and a secret $s = sk$, where (pk, sk) is a public and private key pair for the FHE scheme and $\mathsf{E}(i, pk)$ is the encryption of i with the public key pk.

Algorithm $\mathsf{QG}(n, i, 1^k)$:

(1) (The user) generates a public and private key pair (pk, sk) with the key generation algorithm (KG) and the security parameter k, i.e., $(pk, sk) = \mathsf{KG}(1^k)$.
(2) Assume that the binary representation of i is $\alpha_1 \alpha_2 \cdots \alpha_\ell$, where $\alpha_i \in \{0, 1\}$ and $\ell = \lceil \log n \rceil$. (The user) encrypts each a_j with the public key pk, denoted as $\hat{\alpha}_j = \mathsf{E}(\alpha_i, pk)$. Let $\mathsf{E}(i, pk) = (\hat{\alpha}_1, \hat{\alpha}_2, \cdots, \hat{\alpha}_\ell)$.
(3) Output the query $Q = (pk, \mathsf{E}(i, pk))$ and a secret $s = sk$.

Response Generation $\mathsf{RG}(DB, Q, 1^k)$

Inputs: An n-bit database $DB = b_1 b_2 \cdots b_n$, a query $Q = (pk, \mathsf{E}(i, pk))$, E, Add, Mult, and a security parameter k.

Output: A response R.

Algorithm $\mathsf{RG}(DB, Q, 1^k)$:

(1) For each index $j \in [1, n]$, (the database server) writes j in the binary representation $\beta_{j,1} \beta_{j,2} \cdots \beta_{j,\ell}$. (The database server) encrypts each bit $\beta_{j,t}$ with the public key pk, denoted as $\hat{\beta}_{j,t} = \mathsf{E}(\beta_{j,t}, pk)$ for $1 \leq t \leq \ell$, and computes

$$\hat{\gamma}_j = \boxplus_{t=1}^{\ell} (\hat{\alpha}_t \boxplus \hat{\beta}_{j,t} \boxplus \hat{1}), \tag{6.5}$$

where $\hat{1}$ is an encryption of 1.
(2) (The database server) computes

$$R = \boxplus_{b_j=1} \hat{\gamma}_j. \tag{6.6}$$

(3) Output the response R.

Response Retrieval $\mathsf{RR}((Q, s), R, 1^k)$

Inputs: $s = sk$, an output of $\mathsf{QG}(n, i, 1^k)$; R, an output of $\mathsf{RG}(DB, Q, 1^k)$, and the decryption algorithm D.

Output: A bit $b' = \mathsf{D}(R, sk)$.

Theorem 6.1 (Correctness) *The generic single-database PIR from FHE is correct for any security parameter k, any database DB with any size n, and any index $1 \leq i \leq n$.*

Proof By comparing our response generation circuit and our response generation algorithm (RG), we can see that $\hat{\gamma}_j$ is an encryption of 1 when $j = i$ and an encryption of 0 otherwise, on the basis of fully homomorphic properties. Therefore,

6.3 Generic Single-Database PIR from FHE

if $b_i = 1$, $R = \boxplus_{b_j=1} \hat{\gamma}_j = \hat{\gamma}_i = \hat{1}$, if $b_i = 0$, $R = \boxplus_{b_j=1} \hat{\gamma}_j = \hat{0}$. This means R is an encryption of b_i and thus $b' = \mathsf{D}(R, sk) = b_i$. △

6.3.3 Generic Single-Database PBR from FHE

Now, we extend the single-database PIR from FHE to a single-database PBR from FHE, which also consists of three algorithms (QG, RG, RR).

Assume that an n-bit database DB is equally partitioned into m blocks, denoted as $DB = B_1 \| B_2 \cdots \| B_m$, our single-database PBR is described as follows.

Query Generation $\mathsf{QG}(m, i, 1^k)$

Inputs: The number m of blocks in the database DB, an index $i \in [1, m]$, KG, E, and a security parameter k.

Outputs: A query $Q = (pk, \mathsf{E}(i, pk))$ and a secret $s = sk$, where (pk, sk) is a public and private key pair for the FHE scheme.

Algorithm $\mathsf{QG}(m, i, 1^k)$:

(1) (The user) generates a public and private key pair (pk, sk) with the key generation algorithm (KG) and the security parameter k, i.e., $(pk, sk) = \mathsf{KG}(1^k)$.
(2) Assume that the binary representation of i is $\alpha_1\alpha_2\cdots\alpha_\ell$, where $\alpha_i \in \{0, 1\}$ and $\ell = \lceil \log m \rceil$. (The user) encrypts each α_j with the public key pk, denoted as $\hat{\alpha}_j = \mathsf{E}(\alpha_j, pk)$. Let $\mathsf{E}(i, pk) = (\hat{\alpha}_1, \hat{\alpha}_2, \cdots, \hat{\alpha}_\ell)$.
(3) Output the query $Q = (pk, \mathsf{E}(i, pk))$ and a secret $s = sk$.

Response Generation $\mathsf{RG}(DB, Q, 1^k)$

Inputs: An n-bit database $DB = B_1 \| B_2 \cdots \| B_m$, where $B_j = (b_{j,1}, b_{j,2}, \cdots, b_{j,L})$ and $L = n/m$, a query $Q = (pk, \mathsf{E}(i, pk))$, E, Add, Mult, and a security parameter k.

Output: A response R.

Algorithm $\mathsf{RG}(DB, Q, 1^k)$:

(1) For each index $j \in [1, m]$, (the database server) writes j in the binary representation $\beta_{j,1}\beta_{j,2}\cdots\beta_{j,\ell}$. (The database server) encrypts each bit $\beta_{j,t}$ with the public key pk, denoted as $\hat{\beta}_{j,t} = \mathsf{E}(\beta_{j,t}, pk)$ for $1 \leq t \leq \ell$, and computes

$$\hat{\gamma}_j = \boxtimes_{t=1}^{\ell} (\hat{\alpha}_t \boxplus \hat{\beta}_{j,t} \boxplus \hat{1}). \tag{6.7}$$

(2) For each $c \in [1, L]$, (the database server) computes

$$R_c = \boxplus_{b_{j,c}=1} \hat{\gamma}_j. \tag{6.8}$$

(3) Output the response

$$R = (R_1, R_2, \cdots, R_L).$$

Response Retrieval $\mathsf{RR}((Q, s), R)$

Inputs: $s = sk$, an output of $\mathsf{QG}(n, i, 1^k)$; R, an output of $\mathsf{RG}(DB, Q, 1^k)$, and D.
Output: A block $B' = (\mathsf{D}(R_1, sk), \mathsf{D}(R_2, sk), \cdots, \mathsf{D}(R_L, sk))$.

Theorem 6.2 (Correctness) *The generic single-database PBR from FHE is correct for any security parameter k, any database DB with any size n and any number m of blocks, and any index $1 \leq i \leq m$.*

Proof The generic single-database PBR can be viewed as running L generic single-database PIR protocols in parallel. In each single-database PIR, the user retrieves the i-th bit from an m-bit database $DB_c = b_{1,c} b_{2,c} \cdots b_{m,c}$ for $1 \leq c \leq L$.

Based on Theorem 6.1, we know each of L single-database PIR protocol is correct, that is, $\mathsf{D}(R_c, sk) = b_{i,c}$ for $1 \leq c \leq L$. Therefore, we have $B' = (\mathsf{D}(R_1, sk), \mathsf{D}(R_2, sk), \cdots, \mathsf{D}(R_L, sk)) = B_i$. △.

6.4 Practical Single-Database PIR from FHE

So far, existing fully homomorphic encryption schemes have not been practical for application in private information retrieval. In this section, we present a variant of DGHV somewhat scheme for PBR and then construct a practical PBR protocol from it.

6.4.1 A Variant of DGHV Somewhat Scheme

An obstruct to make the DGHV somewhat scheme practical is the size of the public keys. For the purpose of PIR and PBR, the database server does not need to perform any encryption operation except from some addition and multiplication operations of ciphertexts which require x_0 only. In addition, the user, who generates the private key p, is able to encrypt the index of the interested bit or block directly by using the secret key p and x_0. Therefore, the public keys $x_1, x_2, \cdots, x_\tau$ become redundant in PIR and PBR. In view of this, we introduce a variant of the DGHV somewhat scheme (V-DGHV) for PBR with a database of m blocks as follows.

(1) $\mathsf{KeyGen}(\lambda)$: The user takes a security parameter λ and determines a parameter set $\rho = \lambda, \eta = (\lambda + 3)\lceil \log m \rceil, \gamma = 5(\lambda + 3)\lceil \log m \rceil/2$. Chooses a random odd η-bit integer p from $(2\mathbb{Z} + 1) \cap (2^{\eta-1}, 2^\eta)$ as the secret key sk. Randomly chooses q_0 from $(2\mathbb{Z} + 1) \cap [1, 2^\gamma/p)$ and sets $x_0 = q_0 p$. The public key is $pk = x_0$.

6.4 Practical Single-Database PIR from FHE

(2) **Encrypt**(pk, M): To encrypt $M \in \{0, 1\}$, the user, who knows the secret key $sk = p$, randomly chooses q from $[1, 2^\gamma/p)$ and an integer r from $(-2^\rho, 2^\rho)$ and outputs the ciphertext

$$c = \mathsf{E}(M, pk) = (M + 2 \cdot r + q \cdot p) \bmod x_0.$$

(3) **Decrypt**(sk, c): With the secret key p, the user decrypts a ciphertext as the DGHV somewhat scheme, that is,

$$M = \mathsf{D}(c, sk) = (c \bmod p) \bmod 2.$$

(4) **Homomorphic Addition** (**Add**): Using the public key x_0, the database server adds two ciphertexts c_1 and c_2 as the DGHV somewhat scheme, that is,

$$\mathsf{Add}(c_1, c_2) = (c_1 + c_2) \bmod x_0.$$

(5) **Homomorphic Multiplication** (**Mult**): Using the public key x_0, the database server multiplies two ciphertexts c_1 and c_2 as the DGHV somewhat scheme, that is,

$$\mathsf{Mult}(c_1, c_2) = (c_1 \cdot c_2) \bmod x_0.$$

Like the DGHV somewhat scheme, the choice of parameters in the variant scheme achieves at least 2^λ security against all of known attacks.

6.4.2 Practical Single-Database PBR from V-DGHV Scheme

Assume that an n-bit database DB is equally partitioned into m blocks, denoted as $DB = B_1 \| B_2 \cdots \| B_m$, the practical single-database PBR from the variant of DGHV scheme is described as follows.

Query Generation $\mathsf{QG}(m, i, 1^k)$

Inputs: The number m of blocks in the database DB, an index $i \in [1, m]$, KG, E, and a security parameter k.

Outputs: A query $Q = (x_0, \mathsf{E}(i, pk))$ and a secret $sk = p$, where (x_0, p) is a public and private key pair for the V-DGHV scheme.

Algorithm $\mathsf{QG}(m, i, 1^k)$:

(1) (The user) generates a public and private key pair (x_0, p) with the key generation algorithm (KG) of the V-DGHV scheme and the security parameter k.
(2) Assume that the binary representation of i is $\alpha_1 \alpha_2 \cdots \alpha_\ell$, where $\alpha_i \in \{0, 1\}$ and $\ell = \lceil \log m \rceil$. (The user) encrypts each a_j with the public key x_0, denoted as

$\hat{\alpha}_j = \mathsf{E}(\alpha_j, x_0) = (\alpha_j + 2 \cdot r_j + q_j \cdot p) \bmod x_0$, where r_j and q_j are randomly chosen on the basis of V-DGHV scheme. Let $\mathsf{E}(i, x_0) = (\hat{\alpha}_1, \hat{\alpha}_2, \cdots, \hat{\alpha}_\ell)$.

(3) Output the query $Q = (x_0, \mathsf{E}(i, x_0))$ and a secret $sk = p$.

Response Generation $\mathsf{RG}(DB, Q, 1^k)$

Inputs: An n-bit database $DB = B_1 \| B_2 \cdots \| B_m$, where $B_j = (b_{j,1}, b_{j,2}, \cdots, b_{j,L})$ and $L = n/m$, a query $Q = (x_0, \mathsf{E}(i, x_0))$, E, Add, Mult, and a security parameter k.

Output: A response R.

Algorithm $\mathsf{RG}(DB, Q, 1^k)$:

(1) For each index $j \in [1, m]$, (the database server) writes j in the binary representation $\beta_{j,1}\beta_{j,2}\cdots\beta_{j,\ell}$. (The database server) computes

$$\hat{\gamma}_j = \prod_{t=1}^{\ell} (\hat{\alpha}_t + (\beta_{j,t} \oplus 1)) \bmod x_0 \tag{6.9}$$

(2) For each $c \in [1, L]$, (the database server) computes

$$R_c = \sum_{b_{j,c}=1} \hat{\gamma}_j \bmod x_0 vspace * -3pt \tag{6.10}$$

(3) Output the response

$$R = (R_1, R_2, \cdots, R_L).$$

Response Retrieval $\mathsf{RR}((Q, p), R)$

Inputs: $sk = p$, an output of $\mathsf{QG}(n, i, 1^k)$; R, an output of $\mathsf{RG}(DB, Q, 1^k)$, and D.
Output: A block $B' = ((R_1 \bmod p) \bmod 2, (R_2 \bmod p) \bmod 2, \cdots, (R_L \bmod p) \bmod 2)$.

Theorem 6.3 *The V-DGHV scheme can correctly evaluate the response generation circuit of our practical PBR protocol.*

Proof Suppose that the size of the noise in $\prod_{t=1}^{s} c_i$ is $\mathcal{N}(s)$, where $c_t = (m_t + 2r_t + q_t p) \bmod x_0$ is a fresh ciphertext and $r_t \in (-2^\lambda, 2^\lambda)$. According to Eq. (2), the part of the noise in $c_1 c_2$ is $2r_1 r_2 + r_1 m_2 + r_2 m_1$ and

$$\mathcal{N}(2) \leq 2 \cdot 2^\lambda \cdot 2^\lambda + 2^\lambda + 2^\lambda < 2^{2\lambda+2}. \tag{6.11}$$

For any $s > 2$, we have

$$\mathcal{N}(s) \leq 2 \cdot \mathcal{N}(s-1) \cdot 2^\lambda + \mathcal{N}(s-1) + 2^\lambda$$

$$< 2^{s\lambda + 2(s-1)} = 2^{(\lambda+2)s-2}. \tag{6.12}$$

Therefore, for each $1 \le j \le m$, the size of the noise in $\hat{\gamma}_j = \prod_{t=1}^{\ell}(\hat{\alpha}_t + (\beta_{j,t} \oplus 1))$ mod x_0 ($\ell = \lceil \log m \rceil$) is less than $2^{(\lambda+2)\lceil \log m \rceil - 2}$ and thus the size of the noise in $R_c = \boxplus_{b_{j,c}=1} \hat{\gamma}_j$ for each c is less than $2^{(\lambda+2)\lceil \log m \rceil - 2} m \le 2^{(\lambda+3)\lceil \log m \rceil}/4$, which is less than $p/2$.

In view of it, the V-DGHV scheme can correctly decrypt R_c for any $1 \le c \le n/m$. △

Based on the correctness of the generic PBR from FHE (Theorem 2), we can see that our practical PBR protocol is correct as well.

6.5 Conclusion

In this chapter, we have presented generic single-database PIR and PBR protocols from fully homomorphic encryption, and a practical single-database PBR protocol from a variant of DGHV scheme. Our protocols work as if the user sends the index of a bit or a block to the database server and then receives the bit or the block from the database server. Security analysis has shown that the generic single-database PIR is semantically secure if the underlying fully homomorphic encryption scheme is semantically secure.

We have implemented our practical PBR protocol based on a variant of the DGHV somewhat homomorphic encryption scheme for a database composed of 10,000 elements of size of 200 kbits. On an Intel(R) Core(TM)2 Duo CPU E4600 with clock speed of 2.40 GHz, our experiment has shown that our PBR protocol is practical.

Compared with existing PIR and PBR protocols, our PIR and PBR protocols are conceptually simpler. Our practical PBR protocol has lower computation complexity but higher communication complexity than existing PBR protocols. Overall, our practical PBR protocol is more efficient than existing PBR protocols in terms of total protocol execution time when a high speed network is available.

Our future work will further improve efficiency of PIR and PBR protocols from variants of existing fully homomorphic encryption schemes, such as [3–5].

References

1. C. Aguilar-Melchor and P. Gaborit. A lattice-based computationally-efficient private information retrieval protocol. In *Proc. WEWORC'07*, 2007.
2. C. Aguilar-Melchor and P. Gaborit. A fast private information retrieval protocol. In *Proc. ISIT'08*, 2008.
3. C. Aguilar-Melchor, P. Gaborit and J. Herranz. Additively homomorphic encryption with d-operand multiplications. http://eprint.iacr.org/2008/378.
4. Z. Brakerski, C. Gentry and V. Vaikuntanathan. Fully homomorphic encryption without bootstrapping. http://eprint.iacr.org/2011/277.

5. Z. Brakerski and V. Vaikuntanathan. Efficient fully homomorphic encryption from (standard) LWE. http://eprint.iacr.org/2011/344.
6. G. Brassard, C. Crepeau and J.M. Robert. All-or-nothing disclosure of secrets. In *Proc. CRYPTO'86*, pages 234–238, 1986.
7. C. Cachin, S. Micali, and M. Stadler. Computationally private information retrieval with polylogarithmic communication. In *Proc. EUROCRYPT'99*, pages 402–414, 1999.
8. Y. C. Chang. Single Database Private Information Retrieval with Logarithmic Communication. In *Proc. ACISP'04*, pages 50–61, 2004
9. B. Chor, O. Goldreich, E. Kushilevitz, and M. Sudan. Private information retrieval. In *Proc. 36th Annual IEEE Conference on Foundations of Computer Science*, pages 41–50, 1998.
10. G. Di Crescenzo, T. Malkin, and R. Ostrovsky. Single-database private information retrieval implies oblivious transfer. In *Proc. EUROCRYPT'00*, pages 122–138, 2000.
11. I. Damgard and M. Jurik. A Generalisation, a simplification and some applications of Paillier's probabilistic public-key system. In *Proc. PKC'01*, pages 119–136, 2001.
12. M. Dijk, C. Gentry, S. Halevi and V. Vaikuntanathan. Fully homomorphic encryption over the integers. In *Proc. EUROCRYPT'10*, pages 24–43, 2010.
13. S. Even, O. Goldreich and A. Lempel. A randomized protocol for signing contracts. *Communications of the ACM*, 28(6): 637–647, 1985.
14. C. Gentry and Z. Ramzan. Single database private information retrieval with constant communication rate. In *Proc. ICALP'05*, pages 803–815, 2005.
15. C. Gentry. *Fully Homomorphic Encryption Scheme*. PhD thesis, Stanford University, 2009. Manuscript available at http://crypto.stanford.edu/craig.
16. C. Gentry. Fully homomorphic encryption using ideal lattices. In *Proc. STOC'09*, pages 169–178, 2009.
17. C. Gentry. Computing arbitrary functions of encrypted data. *Communications of the ACM*, 53(3): 97–105, 2010.
18. C. Gentry. Toward basing fully homomorphic encryption on worst-case hardness. In *Proc. CRYPTO'10*, pages 116–137, 2010.
19. O. Goldreich and L. Levin. A hard predicate for all one-way functions. In *Proc. STOC'89*, pages 23–32, 1989.
20. S. Goldwasser and S. Micali. Probabilistic encryption. *Journal of Computer and Systems Sciences*, 28(2):270–299, 1984.
21. E. Kushilevitz and R. Ostrovsky. Replication is not needed: Single database, computationally-private information retrieval. In *Proc. 38th Annual IEEE Symposium on the Foundations of Computer Science*, pages 364–373, 1997.
22. E. Kushilevitz and R. Ostrovsky. One-way trapdoor permutations are sufficient for non-trivial single-server private information retrieval. In *Proc. EUROCRYPT'00*, pages 104–121, 2000.
23. H. Lipmaa. An oblivious transfer protocol with log-squared communication. In *Proc. 8th Information Security Conference*, pages 314–328, 2005.
24. H. Lipmaa. First CPIR protocol with data-dependent computation. In *Proc. ICISC'09*, pages 193–210, 2009.
25. M. Naor and M. Yung. Universal one-way hash functions and their cryptographic applications. In *Proc. STOC'89*, pages 33–43, 1989.
26. M. Naor and B. Pinkas. Oblivious transfer and polynomial evaluation, In *Proc. STOC'99*, pages 245–254, 1999.
27. F. Olumofin and I. Goldberg. Revisiting the computational practicality of private information retrieval. Technical Report CACR 2010-17, University of Waterloo, 2010.
28. R. Ostovsky and W. E. Skeith III. A survey of single-database PIR: techniques and applications. In *Proc. PKC'07*, pages 393–411, 2007.
29. P. Paillier. Public key cryptosystems based on composite degree residue classes. In *Proc. EUROCRYPT'99*, pages 223–238, 1999.
30. S. Pohlig and M. Hellman. An improved algorithm for computing logarithms over GF(p) and its cryptographic significance. *IEEE Transactions on Information Theory*, 24(1): 106–110, 1978.

References

31. M. O. Rabin. How to exchange secrets by oblivious transfer. *Technical Report TR-81*, Aiken Computation Laboratory, Harvard University, 1981.
32. N. Smart and F. Vercauteren. Fully homomorphic encryption with relatively small key and ciphertext sizes. In *Proc. PKC'10*, pages 420–443, 2010.
33. X Yi, MG Kaosar, R Paulet, E Bertino. Single-database private information retrieval from fully homomorphic encryption IEEE Transactions on Knowledge and Data Engineering 25 (5), 1125–1134, 2012.

Chapter 7
Privacy-Preserving Lightweight Deep Learning in Medical Diagnostic Service

Abstract In this chapter, we present a lightweight and secure neural network (NN) inference service framework, MediSC. MediSC is specifically designed for medical diagnostic services, enabling enterprises to provide secure medical diagnostic capabilities to their customers by executing NN inference in the ciphertext domain. MediSC guarantees the privacy of both parties through robust cryptographic techniques. At its core, we introduce an efficient and communication-optimized secure inference protocol. Our protocol relies exclusively on lightweight secret-sharing techniques and is capable of handling commonly-used linear and non-linear NN layers effectively. In comparison to solutions based on garbled circuits, MediSC achieves significantly better performance, with $24\times$ lower latency and $868\times$ less communication for secure ReLU, and $20\times$ lower latency and $314\times$ less communication for secure Max-pool. We evaluate MediSC using two benchmark datasets and four real-world medical datasets, performing a comprehensive comparison with prior works. The results highlight the exceptional performance of MediSC, demonstrating that it is substantially more bandwidth-efficient than existing approaches.

7.1 Introduction

Recent advancements in deep learning techniques have fueled a wide range of medical applications, spanning radiotherapy [8], clinical trials and research [10], and medical imaging diagnostics [9]. Enterprises leverage neural networks (NNs) to provide medical diagnostic services, enabling hospitals and researchers to make faster and more accurate decisions based on their medical data. However, with the proliferation of such services comes heightened awareness of significant privacy concerns. Medical data is inherently sensitive and must remain confidential at all times [11, 16, 23, 24]. Simultaneously, the NN models employed in these services are considered valuable intellectual property and encapsulate private training data [17].

The typical setup for such NN-powered service scenarios aligns with the field of secure multi-party computation (MPC). By designing specialized MPC protocols, recent studies [2, 3, 20, 21, 26] have facilitated secure joint execution of NN inference over encrypted customer data and/or service provider models. Nonetheless, these systems face significant performance challenges, making them less practical for real-world medical scenarios. During inference, customers are required to perform computationally intensive cryptographic operations, such as homomorphic encryption (HE) and garbled circuits (GC), resulting in high computational and communication overheads. These challenges are particularly pronounced when the services are deployed in resource-constrained environments, such as hospitals using portable medical imaging devices [7]. Additionally, some protocols [2, 3] lack compatibility with widely-used non-linear functions, such as ReLU, limiting their applicability in modern NN architectures.

To address these challenges, we propose, implement, and evaluate MediSC, a lightweight and secure NN inference system designed specifically for medical diagnostic services [1]. MediSC facilitates secure inference between the hospital and the medical service provider over encrypted inputs, ensuring that only the hospital receives the diagnostic result while preserving the privacy of both the medical data and the model. Leveraging insights from cryptography and digital circuit design, MediSC offers an efficient, low-interaction solution suitable for realistic medical scenarios. Our contributions are summarized as follows:

- We introduce a secure NN inference system framework, MediSC, which exclusively employs lightweight secret-sharing techniques. This approach eliminates the need for heavy cryptographic computations and large ciphertext transmissions.
- We present a hybrid protocol design comprising a preprocessing phase and an online phase. The preprocessing phase performs as much computation as possible to alleviate the workload during the online phase, utilizing only lightweight operations in the secret-sharing domain.
- We design an efficient and communication-optimized secure comparison function to handle the most challenging and widely-used non-linear functions, including ReLU and Max-pool. By incorporating cryptographic insights and digital circuit design principles, MediSC's secure ReLU achieves 24× faster performance with 868× less communication, while the secure Max-pool is 20× faster with 314× less communication compared to conventional GC solutions.
- We perform a formal security analysis, implement a prototype of MediSC, and conduct extensive evaluations using two benchmarking datasets and four real-world medical datasets. Our experimental results demonstrate that MediSC achieves up to 413×, 19×, and 10× bandwidth savings for MNIST, CIFAR-10, and medical applications, respectively, compared to prior works.
- MediSC outperforms the state-of-the-art (SOTA) [26] in terms of bandwidth cost, achieving 10× savings with comparable latency.[1]

---

[1] Performance comparisons are based on results reported in SOTA, which rely on highly optimized implementations with GPU acceleration. Our results are derived without such optimizations.

7.2 Related Work

Recent years have witnessed a growing interest in secure neural network inference. Numerous prior studies [2, 3, 15, 20, 21, 26, 27, 29] have focused on scenarios where an *interactive* protocol is executed between the service provider and the customer. Other works [4, 5, 22] operate under the assumption of *non-colluding cloud servers*, where two independent servers collaboratively perform secure inference on outsourced models and data. Despite variations in their system models, these approaches generally rely on computationally intensive cryptographic tools, such as homomorphic encryption (HE) and garbled circuits (GC), during the latency-sensitive online inference process. Additionally, certain methods lack comprehensive support for modern neural network models [2, 3]. These approaches often approximate non-linear functions using cryptographically convenient polynomials, sacrificing accuracy and applicability for computational efficiency [19, 25]. Such compromises could lead to critical consequences, particularly in medical scenarios.

The state-of-the-art (SOTA) approach [26] introduces a hybrid and interactive inference protocol, where cryptographic operations are preprocessed to expedite online inference execution. However, this method still imposes significant computational burdens on the customer, requiring intensive cryptographic computations during the preprocessing phase. Moreover, it relies on costly GC-based methods to evaluate ReLU functions. In contrast, MediSC adopts a similar hybrid framework but exclusively utilizes lightweight secret-sharing techniques throughout the entire secure inference process. This design offers a significant advantage of *simplified implementation*, facilitating real-world deployment compared to SOTA approaches, which necessitate extensive optimization of GC and homomorphic encryption components.

7.3 Preliminaries

In this section, we provide an overview of the cryptographic primitive employed in our design: *additive secret sharing*. This high-level description is sufficient for understanding our approach. Additive secret sharing [12] protects an ℓ-bit value $x \in \mathbb{R}_{2^\ell}$ by splitting it into two secret shares: $\langle x \rangle_0 = r \pmod{2^\ell}$ and $\langle x \rangle_1 = x - r \pmod{2^\ell}$, such that $\langle x \rangle_0^A + \langle x \rangle_1^A \equiv x \pmod{2^\ell}$. Here, $\mathbb{Z}_{2^\ell}$ denotes the ring, and r is a random value chosen from the ring ($r \in_R \mathbb{Z}_{2^\ell}$). This method perfectly hides the value x, as each share is random and reveals no information about x. Given two parties, P_0 and P_1, each party holds the corresponding shares of two secret values, x and y. Additive secret sharing supports efficient local operations such as addition and subtraction over shares, i.e., $\langle z \rangle_i = \langle x \rangle_i \pm \langle y \rangle_i$, and scalar multiplication, i.e., $\langle z \rangle_i = \eta \cdot \langle x \rangle_i$ (η is a public value). These operations are performed by each party P_i ($i \in \{0, 1\}$) independently, without requiring interaction between the parties. Multiplication of two shares, $\langle z \rangle = \langle x \rangle \cdot \langle y \rangle$, is supported by Beaver's triple [14].

Specifically, P_i holds $(\langle t_1 \rangle_i, \langle t_2 \rangle_i, \langle t_3 \rangle_i)$ such that $t_3 = t_1 \cdot t_2$. This multiplication protocol is standard for secure computation, enabling P_i to obtain the shares $\langle z \rangle_i$ of the product xy. Beaver's triples are data-independent and can be efficiently generated through a one-time computation by a third party [4, 30]. Additive secret shares can also support boolean operations over binary values. For instance, when the bit length is $\ell = 1$ and the ring is $\mathbb{Z}_2$, a secret binary value x is shared as $[\![x]\!]_0 = r \in \mathbb{Z}_2$ and $[\![x]\!]_1 = r \oplus [\![x]\!]_0$. The bitwise XOR ($\oplus$) and AND ($\wedge$) operations over shares are computed in the same manner as the addition and multiplication operations described above.

7.4 Model for Privacy-Preserving Machine Learning

7.4.1 System Architecture

MediSC targets a typical scenario of secure neural network (NN) inference in a medical diagnostic service. As illustrated in Fig. 7.1, MediSC operates between two parties: the *hospital* (the customer) and the *medical service* provider. The *medical service* provider possesses a proprietary NN model, pre-trained on medical datasets. The *hospital* holds confidential medical records (e.g., brain MRI images) and intends to use the deep learning service to assist in making a medical diagnosis. In practice, the role of the hospital in MediSC can be filled by any healthcare institution, medical research laboratory, or life-science organization. To initiate a secure medical diagnostic service, the two parties execute MediSC's secure NN inference protocol over their encrypted model and encrypted medical records. At the conclusion of the process, an encrypted inference result is returned to the hospital, which can then decrypt it to obtain the plaintext inference result. MediSC ensures that the hospital learns only the inference result and no other information, while the medical service provider learns nothing about the hospital's medical records.

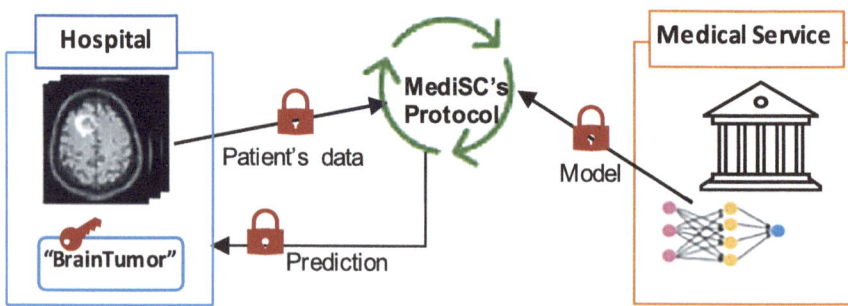

Fig. 7.1 System architecture for MediSC, a secure and lightweight inference system for medical diagnostic services

7.4.2 Threat Model

MediSC is designed for the *semi-honest* two-party model, where both the hospital and the medical service provider faithfully follow the protocol but may attempt to deduce information about the counterparty's private input from the messages exchanged during protocol execution. It is important to note that the behavior of the hospital is governed by ethics, law, and privacy regulations [11, 16]. The medical service is typically provided by reputable companies (e.g., Microsoft Project InnerEye [9], Google DeepMind Health [8]), which would not jeopardize their business model or reputation by acting maliciously [13]. This adversarial model is also adopted in prior secure NN inference works [21, 26]. MediSC is designed to ensure the privacy of the hospital's medical records and the NN model (i.e., the values of the trained weights). Similar to prior works [21, 22, 26], MediSC does not conceal the data-independent aspects of the model architecture, such as kernel size and the number of layers. Finally, MediSC assumes that adversarial machine learning attacks, which attempt to exploit the inference process as a black-box oracle to extract private information, are orthogonal to the goal of the system. Mitigation strategies for such attacks could include differentially private learning techniques [28].

7.5 Our Proposed Design

In this section, we present MediSC's secure neural network (NN) inference protocol, specifically designed for medical diagnostic applications. At a high level, our design consists of two types of secure layer evaluations: secure *linear* and *non-linear* layers, which are well-suited for typical NN layers, such as convolutional, fully-connected, batch normalization, and average pooling layers (linear), as well as ReLU activation and max pooling layers (non-linear). Each secure layer evaluation performs a specific function on the encrypted inputs (features) and generates encrypted outputs, which are then passed to the subsequent secure layer. Our overarching goal is to develop a lightweight protocol for secure NN inference, while minimizing interactions between the hospital and the medical service provider, thus ensuring a low-latency diagnosis. Based on this goal, we have identified two key design insights.

Eliminating Heavy Cryptography for Linear Layers We begin by dividing MediSC's protocol into a preprocessing phase and an online phase, shifting as much computation as possible to the preprocessing phase. Inspired by Mishra et al. [26], we preprocess the model as secret shares and deliver the corresponding shares to the hospital before the medical records are available. Thus, the online phase can operate directly on the secret shares without relying on heavy cryptographic techniques (such as homomorphic encryption, HE) or multi-round ciphertext transmissions. However, we note that the protocol in [26] involves substantial HE during preprocessing to generate and transmit model shares as ciphertexts, which may not be suitable for resource-limited hospitals, such as COVID-19 pandemic screening centers equipped with handheld medical imaging scanners [7]. Instead, our protocol effectively leverages the insights from Chameleon [4], enabling the preprocessing phase to rely purely on lightweight computations in the secret sharing domain. As a result, our entire protocol operates only with small shares, leading to a 20× improvement in preprocessing efficiency and a 10× reduction in overall communication costs compared to [26].

Eliminating the Use of Garbled Circuits (GC) for Non-linear Layers For the secure evaluation of non-linear layers, previous works either rely on heavy cryptographic techniques (e.g., garbled circuits) [21, 26] or bypass the non-linearities using polynomial approximations [2, 3]. Unfortunately, these methods may incur high communication overheads or cause instability in the NN when handling complex tasks [19, 25]. In MediSC, we draw inspiration from the field of digital circuit design [18] and introduce a secure comparison function that efficiently evaluates comparison-based non-linear layers, such as ReLU. At its core, this function is based entirely on lightweight secret sharing, with optimized interactions between the hospital and the medical service provider. With these design optimizations, our experiment demonstrates a 413× reduction in bandwidth compared to prior works.

7.5.1 Secure Linear Layers

In this section, we present MediSC's secure inference protocol, which is divided into two distinct phases: the *preprocessing* phase and the *online inference* phase. Before delving into the specifics, it is important to note that MediSC operates within the secret sharing domain. That is, any floating-point number v (e.g., model weights and medical record values) is mapped to a signed fixed-point integer $\bar{v} = \lfloor v \cdot 2^s \rfloor \mod 2^\ell$, where 2^s is the scaling factor. The most significant bit (MSB) indicates the sign ($1 \rightarrow$ negative; $0 \rightarrow$ non-negative). This conversion ensures that both the value and the sign information are securely hidden. Multiplication of two fixed-point integers can result in overflow within the ring $\mathbb{Z}_{2^\ell}$, as the fractional part of the product grows to $2s$ bits. To maintain correctness, all intermediate results after multiplying two shares are rescaled down by 2^s before any subsequent operation. We adopt a secure local truncation scheme, as proposed in [5], which discards the last s fractional bits to adjust the product to fit within ℓ bits, following the approach from prior works [6, 26].

Preprocessing Phase
The preprocessing phase is depicted in Fig. 7.2. In this phase, the hospital and the medical service pre-generate secret shares of the NN model in an appropriate form, which will be used during the online inference phase. This computation is performed only once and is independent of the hospital's medical records. Let L denote the

7.5 Our Proposed Design

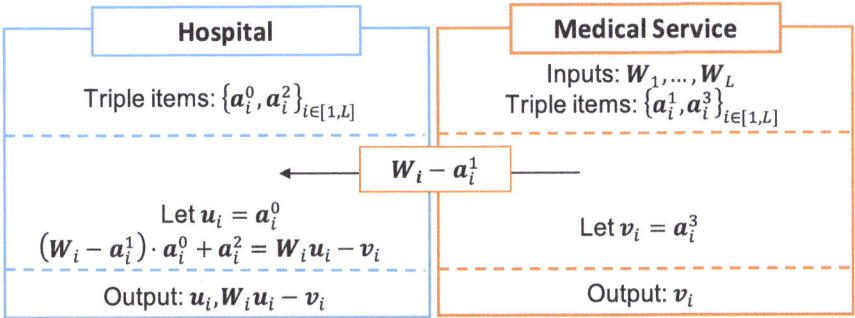

Fig. 7.2 MediSC's preprocessing phase

number of layers in the model. The hospital receives L sets of randomnesses (in tensor form) $\{\mathbf{a}_i^0, \mathbf{a}_i^2\}$, where $i \in [1, L]$.

Similarly, the medical service receives the weight tensors for each layer $\mathbf{W}_1, \ldots, \mathbf{W}_L$ and randomness tensors $\{\mathbf{a}_i^1, \mathbf{a}_i^3\}$. These randomness tensors $\{\mathbf{a}_i^0, \mathbf{a}_i^1, \mathbf{a}_i^2, \mathbf{a}_i^3\}$ are independent of any party's input and can be pre-distributed to the parties, satisfying the relationship $\mathbf{a}_i^3 = \mathbf{a}_i^0 \cdot \mathbf{a}_i^1 - \mathbf{a}_i^2$. It is important to note that the dimensions of each randomness tensor align with the dimensions of each layer's filter. Before MediSC's execution, a third party can prepare these items in a format where $\mathbf{a}_i^0, \mathbf{a}_i^1, \mathbf{a}_i^2$ are randomly distributed in ring $\mathbb{Z}_{2^\ell}$, and $\mathbf{a}_i^3 = \mathbf{a}_i^0 \cdot \mathbf{a}_i^1 - \mathbf{a}_i^2$. In MediSC, we assume these items are already distributed and available at both the hospital and medical service. With these inputs in place, the two parties perform the following steps:

1. For each $i \in [1, L]$, the medical service computes $\mathbf{W}_i - \mathbf{a}_i^1$ over the weight tensors and sends the result to the hospital.
2. The hospital computes $(\mathbf{W}_i - \mathbf{a}_i^1) \cdot \mathbf{a}_i^0 = \mathbf{W}_i \mathbf{a}_i^0 - \mathbf{a}_i^0 \mathbf{a}_i^1 + \mathbf{a}_i^2$ for each layer.
3. Let $\mathbf{u}_i$ denote $\mathbf{a}_i^0$, and $\mathbf{v}_i$ denote $\mathbf{a}_i^3$. The medical service holds $\mathbf{v}_i$, while the hospital holds $\mathbf{W}_i \mathbf{u}_i - \mathbf{v}_i$, which represents an additively secret-shared weight tensor $\mathbf{W}_i \mathbf{u}_i$.

Online Inference Phase

During the online inference phase, the hospital inputs the tensor of the medical record $\mathbf{X}_1$, the randomness tensors $\mathbf{u}_i$, and the weight shares $\mathbf{W}_i \mathbf{u}_i - \mathbf{v}_i$, as shown in Fig. 7.3. Meanwhile, the medical service inputs the weight tensors $\mathbf{W}_1, \ldots, \mathbf{W}_L$ and the randomness tensors $\mathbf{v}_i$.

The two parties then proceed to securely evaluate the layer functions in a pipeline as follows. *The first linear layer ($i = 1$):*

1. The hospital computes and sends $\mathbf{X}_1 - \mathbf{u}_1$ to the medical service, and uses $\langle \mathbf{X}_2 \rangle_0$ to denote $\mathbf{W}_1 \mathbf{u}_1 - \mathbf{v}_1$.
2. The medical service computes $\langle \mathbf{X}_2 \rangle_1 = \mathbf{W}_1(\mathbf{X}_1 - \mathbf{u}_1) + \mathbf{v}_1 = \mathbf{W}_1 \mathbf{X}_1 - \mathbf{W}_1 \mathbf{u}_1 + \mathbf{v}_1$.

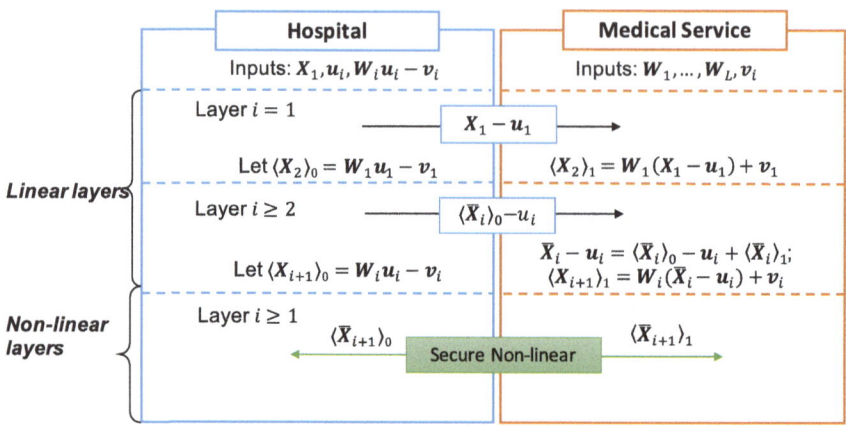

Fig. 7.3 MediSC's online phase

3. At this stage, the hospital and the medical service each hold additive secret shares (i.e., $\langle X_2\rangle_0, \langle X_2\rangle_1$) of the features[2] output from the first linear layer $W_1 X_1$.

Subsequent linear layers ($i \geq 2$):

1. Similar to the first layer, the hospital computes $\langle \bar{X}_i\rangle_0 - u_i$ over its share $\langle \bar{X}_i\rangle_0$ of the activation produced by the secure ReLU evaluation (which will be detailed later) and sends it to the medical service. This procedure perfectly hides the hospital's share and protects the activation $\bar{X}_i$ from the medical service. Next, the hospital sets $\langle X_{i+1}\rangle_0 = W_i u_i - v_i$.
2. The medical service computes $\bar{X}_i - u_i = \langle \bar{X}_i\rangle_0 - u_i + \langle \bar{X}_i\rangle_1$. Then, the medical service computes $\langle X_{i+1}\rangle_1 = W_i(\bar{X}_i - u_i) + v_i$, ensuring that both parties hold additive secret shares (i.e., $\langle X_{i+1}\rangle_0, \langle X_{i+1}\rangle_1$) of the layer result $W_i \bar{X}_i$.

Non-linear layers: The shares formed by the secure linear layer evaluation can be fed into the secure non-linear layer (e.g., ReLU), which then outputs shares $\langle \bar{X}_{i+1}\rangle_0$, $\langle \bar{X}_{i+1}\rangle_1$ of the activations to each party.

Output layer: The medical service sends $\langle X_L\rangle_1$ to the hospital, which then integrates $\langle X_L\rangle_0$ to reconstruct the final inference result X_L.

7.5.2 Secure Non-linear Layers

MediSC supports highly efficient evaluation of secure non-linear layers in the domain of secret sharing. As mentioned earlier, MediSC represents all values as signed fixed-point integers, with the most significant bit (MSB) indicating the sign.

---

[2] Biases can be locally added to the medical service's shares.

7.5 Our Proposed Design

Algorithm 1 Secure MSB($\cdot$) extraction function

Input: Arithmetic shared integer feature $\langle x \rangle \in \mathbb{Z}_{2^\ell}$.
Output: Boolean shared MSB $[\![x_\ell]\!] \in \mathbb{Z}_2$.
Decompose $\langle x \rangle$ into bit strings:
1: Let e denote $\langle x \rangle_0$ and f denote $\langle x \rangle_1$.
2: Decompose to bit strings $e \rightarrow e_\ell, ..., e_1$ and $f \rightarrow f_\ell, ..., f_1$.
3: **for** $k \in [1, \ell]$ **do**
4: Set $[\![e_k]\!]_0 = e_k$, $[\![e_k]\!]_1 = 0$ and $[\![f_k]\!]_0 = 0$, $[\![f_k]\!]_1 = f_k$.
Compute signal tuples (g, p) in Eq. 7.2:
5: $[\![g_k^0]\!] = [\![e_k]\!] \cdot [\![f_k]\!]$, $[\![p_k^0]\!] = [\![e_k]\!] + [\![f_k]\!]$.
6: **end for**
Compute PPA tree based on Eq. 7.3:
Round $\mathcal{R} = 1$:
7: **for** $k \in [2, \ell/2]$ **do**
8: Set $([\![g_1^1]\!], [\![p_1^1]\!]) = ([\![g_1^0]\!], [\![p_1^0]\!])$ as a dummy node.
9: Let $in_1 = 2k - 2$, $in_2 = 2k - 1$.
10: $([\![g_k^1]\!], [\![p_k^1]\!]) = ([\![g_{in_1}^0]\!], [\![p_{in_1}^0]\!]) \odot ([\![g_{in_2}^0]\!], [\![p_{in_2}^0]\!])$.
11: **end for**
Round $\mathcal{R} = 2, ..., \log \ell$:
12: **for** $k \in [1, \ell/2^\mathcal{R}]$ **do**
13: Let $in_1 = 2k - 1$, $in_2 = 2k$.
14: $([\![g_k^\mathcal{R}]\!], [\![p_k^\mathcal{R}]\!]) = ([\![g_{in_1}^{\mathcal{R}-1}]\!], [\![p_{in_1}^{\mathcal{R}-1}]\!]) \odot ([\![g_{in_2}^{\mathcal{R}-1}]\!], [\![p_{in_2}^{\mathcal{R}-1}]\!])$.
15: **end for**
Compute MSB:
16: Set $[\![c_\ell]\!] = [\![g_1^{\log \ell}]\!]$, $[\![x_\ell]\!] = [\![p_\ell^0]\!] + [\![c_\ell]\!]$.

Specifically, the MSB is '0' for non-negative values and '1' for negative values. This representation allows us to observe that non-linear layers, particularly those that rely on comparison operations, can be simplified to an MSB extraction problem coupled with some linear operations (such as addition and multiplication). For clarity, we focus on the most widely adopted ReLU function.

$$\text{ReLU}(x) = \max(x, 0) \rightarrow \neg \text{MSB}(x) \cdot x = \begin{cases} 1 \cdot x & \text{if } x \geq 0 \\ 0 \cdot x & \text{if } x < 0 \end{cases}, \quad (7.1)$$

where x represents the feature output from the previous linear layer. Through this conversion, we identify four atomic steps: the *secure* MSB($\cdot$) *extraction*, the *secure NOT*, the *secure* B2A (Boolean-to-Additive shares conversion), and the *secure multiplication*. The most challenging computation is the secure MSB($\cdot$) extraction. To address this, we propose an efficient and communication-optimized construction, with details presented in the following section. For the other steps, the secure B2A operation converts boolean shares $[\![x]\!]$ in the ring $\mathbb{Z}_2$ into additive shares in $\mathbb{Z}_{2^\ell}$, i.e., $\langle x \rangle \leftarrow \text{B2A}([\![x]\!])$. Meanwhile, the *secure NOT* and *secure multiplication* are linear operations that are efficiently handled by additive secret sharing.

Communication-Optimized Secure MSB Extraction

The secure MSB(·) extraction function is employed to securely extract the most significant bit (MSB) of an additive-shared data $\langle x \rangle$ and generate a boolean-shared MSB $[\![x_\ell]\!]$, where ℓ represents the bit length of the value. The process of extracting the MSB in the secret sharing domain is performed through binary addition over two secret shares' bit strings using an ℓ-bit full adder, as explained in the following. Suppose we are given an ℓ-bit value x with its decomposed bit string $x = \{x_\ell, ..., x_1\}$ and its secret shares $\langle x \rangle_0, \langle x \rangle_1$. Let $e = \{e_\ell, ..., e_1\}$ and $f = \{f_\ell, ..., f_1\}$ represent the bit strings of $\langle x \rangle_0$ and $\langle x \rangle_1$, respectively. Therefore, $x = e + f \pmod{2^\ell}$. An ℓ-bit full adder is then used to perform binary addition ($\{e_k\} + \{f_k\}$) in the secret sharing domain, producing the carry bits $c_\ell, ..., c_1$. Finally, the MSB is calculated as $x_\ell = c_\ell \oplus e_\ell \oplus f_\ell$, where $k \in [1, \ell]$. The key insight for extracting the MSB lies in producing the most significant carry bit c_ℓ through the full adder logic. For simplicity, unless otherwise specified, the operator '+' applied to two binary values (including boolean shares) refers to the bitwise-XOR operation in the following discussions.

We draw inspiration from the field of digital circuit design, specifically from the parallel prefix adder [18] (PPA), which offers an efficient implementation of full adder logic with logarithmic round complexity, $O(\log \ell)$.

To construct the PPA, we introduce a *signal tuple* (g_i, p_i): the carry generate signal g_i and the carry propagate signal p_i, both of which can be computed in parallel using the following equations:

$$g_i = e_i \cdot f_i; \quad p_i = e_i + f_i. \tag{7.2}$$

The full adder logic, $c_{i+1} = (e_i \cdot f_i) + c_i \cdot (e_i + f_i)$, can then be reformulated as $c_{i+1} = g_i + c_i \cdot p_i$. The ℓ-th carry bit is generated using the following expression:

$$c_\ell = g_{\ell-1} + (p_{\ell-1} \cdot g_{\ell-2}) + \cdots + (p_{\ell-1} \cdots p_2 \cdot g_1).$$

This reformulation allows the carry to be derived without waiting for the previous carry to propagate through all preceding adders. As a result, the PPA can extract the most significant bit (MSB) in $O(\log \ell)$ communication round latency, without increasing the computational overhead.

A concrete example of an 8-bit PPA is provided in Fig. 7.4. As illustrated, this construction forms a $\log \ell$-depth binary tree (with 3 levels in this example), where each node performs a binary operation $\odot$. Each layer of the tree corresponds to a round of communication.

The binary operator $\odot$ takes as input two adjacent signal tuples (g_{in_1}, p_{in_1}) and (g_{in_2}, p_{in_2}), performing the following operations:

$$(g_{out}, p_{out}) = (g_{in_1}, p_{in_1}) \odot (g_{in_2}, p_{in_2}); \tag{7.3}$$

$$g_{out} = g_{in_2} + g_{in_1} \cdot p_{in_2}; \quad p_{out} = p_{in_2} \cdot p_{in_1},$$

7.5 Our Proposed Design

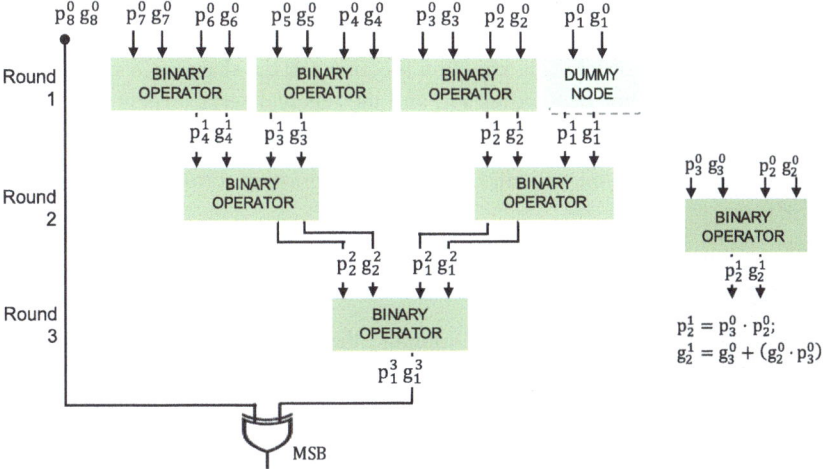

Fig. 7.4 An illustration of the 8-bit parallel prefix adder (PPA)

and outputs a new signal tuple (g_{out}, p_{out}).

PPA iteratively applies this binary operation to the input tuples associated with each leaf node and propagates the output signal tuples to the next layer of nodes as inputs. The process continues until the root node is reached, such as the node with (p_1^3, g_1^3) in Fig. 7.4. At this point, the carry bit c_ℓ is obtained, and the MSB is computed as $x_\ell = c_\ell + p_\ell = c_\ell + (e_\ell + f_\ell)$.

Following this approach, we present the details of the secure MSB extraction function in Algorithm 1.

Secure B2A Function Given a secret value x, the secure B2A function is used to convert its boolean shares $[\![x]\!]$ in the ring $\mathbb{Z}_2$ to the corresponding additive secret shares $\langle x \rangle$ in the ring $\mathbb{Z}_{2^\ell}$.

Recall that the secure ReLU function applied to each feature x is formulated as follows: $max(x, 0) \rightarrow \neg MSB(x) \cdot x$.

The secure B2A function is invoked after securely extracting the boolean shares of the NOT MSB, $[\![\neg x_\ell]\!]$.

However, the produced boolean shares cannot be directly multiplied with the additively-shared feature $\langle x \rangle$ as they are shared with different moduli. Specifically, $[\![\neg x_\ell]\!] = [\![\neg x_\ell]\!]_0 + [\![\neg x_\ell]\!]_1 \pmod 2$, and $\langle x \rangle = \langle x \rangle_0 + \langle x \rangle_1 \pmod{2^\ell}$.

Therefore, we need to convert $[\![\neg x_\ell]\!]$ into its additive form $\langle \neg x_\ell \rangle$.

Our secure B2A function follows the standard realization outlined in [30]. Given two parties, the hospital (denoted as P_0) and the medical service (denoted as P_1), the secure B2A($[\![x]\!]$) function is performed as follows:

1. P_0 sets $\langle e \rangle_0 = [\![x]\!]_0$, $\langle f \rangle_0 = 0$, and P_1 sets $\langle e \rangle_1 = 0$, $\langle f \rangle_1 = [\![x]\!]_1$;
2. P_0 and P_1 compute $\langle x \rangle_i = \langle e \rangle_i + \langle f \rangle_i - 2 \cdot \langle e \rangle \cdot \langle f \rangle$.

Secure ReLU Function For simplicity, we present the secure ReLU function applied to each neuron over a single feature element x.

Given the above secure MSB extraction function and the shares of a single input feature $\langle x \rangle$, the hospital (denoted as P_0) and the medical service (denoted as P_1) perform the secure ReLU function as follows:

1. *Secure MSB Extraction:* P_0 and P_1 invoke Algorithm 1 to obtain $[\![x_\ell]\!] \leftarrow$ MSB($\langle x \rangle$).
2. *Secure NOT:* P_i computes the NOT MSB $[\![\neg x_\ell]\!] = [\![x_\ell]\!] + i$.
3. *Secure B2A:* P_0 and P_1 run $\langle \neg x_\ell \rangle \leftarrow$ B2A($[\![\neg x_\ell]\!]$) to convert the boolean-shared NOT MSB into additive shares.
4. *Secure Multiplication:* P_0 and P_1 compute the activation for each neuron as $\langle \bar{x} \rangle = \langle \neg x_\ell \rangle \cdot \langle x \rangle$.

Secure Pooling Layer

Within an n-width pooling window, the max pooling layer $\max(x_1, \ldots, x_n)$ can be transformed into a pairwise maximum operation, which can be realized based on the secure MSB($\cdot$) extraction. Specifically, we use $\mathfrak{b} \leftarrow$ MSB($x_1 - x_2$) and $\max(x_1, x_2) = (1 - \mathfrak{b}) \cdot x_1 + \mathfrak{b} \cdot x_2$.

Given the secure MSB extraction function, the secure B2A function, and the shares of a set of activations $\langle x_1 \rangle, \ldots, \langle x_n \rangle$ within the n-width pooling window, the hospital (denoted as P_0) and the medical service (denoted as P_1) perform the secure MaxPool function as follows:

1. For $k \in [1, n-1]$:
2. *Secure MSB Extraction:* P_0 and P_1 invoke Algorithm 1 to obtain the boolean shares of the MSB: $[\![\mathfrak{b}]\!] \leftarrow$ MSB($\langle x_k \rangle - \langle x_{k+1} \rangle$).
3. *Secure B2A:* P_0 and P_1 run $\langle \mathfrak{b} \rangle \leftarrow$ B2A($[\![\mathfrak{b}]\!]$) to convert the boolean-shared MSB into additive shares.
4. *Secure Branching:* P_0 and P_1 use the MSB to choose the maximum value as follows: $\langle \mathfrak{b}' \rangle_i = i - \langle \mathfrak{b} \rangle_i$, where $i \in \{0, 1\}$ is the identifier of party P_i, and then compute $\langle z_k \rangle = \langle \mathfrak{b}' \rangle \cdot \langle x_k \rangle + \langle \mathfrak{b} \rangle \cdot \langle x_{k+1} \rangle$. P_i then updates $\langle x_{k+1} \rangle := \langle z_k \rangle$.
5. Finally, P_i outputs $\langle z_n \rangle_i$ as the shares of the MaxPool result.

The average pooling layer $\lfloor (x_1 + \cdots + x_n)/n \rfloor$ can be directly computed over additive secret shares via secure addition, where n is a cleartext hyperparameter.

7.6 Conclusion

In this chapter, we present MediSC, a secure and lightweight NN inference system towards secure intelligent medical diagnostic services. Our protocol fully resorts to the lightweight additive secret sharing techniques, free of heavy cryptographic operations as seen in prior art. The commonly-used non-linear ReLU and max pooling layer functions are well supported in a secure and efficient manner. With

MediSC, the privacy of the medical record of the hospital and the NN model of the medical service is provably ensured with practical performance.

References

1. Xiaoning Liu, Yifeng Zheng, Xingliang Yuan, Xun Yi: MediSC: Towards Secure and Lightweight Deep Learning as a Medical Diagnostic Service. ESORICS (1) 2021: 519–541
2. Chiraag Juvekar, Vinod Vaikuntanathan, and Anantha Chandrakasan. GAZELLE: A low latency framework for secure neural network inference. In *Proceedings of the 27th USENIX Security Symposium*, 2018.
3. Ran Gilad-Bachrach, Nathan Dowlin, Kim Laine, Kristin Lauter, Michael Naehrig, and John Wernsing. Cryptonets: Applying neural networks to encrypted data with high throughput and accuracy. In *Proceedings of the International Conference on Machine Learning (ICML)*, 2016.
4. M. Sadegh Riazi, Mohammad Samragh, Hao Chen, Kim Laine, Kristin Lauter, and Farinaz Koushanfar. XONN: XNOR-based Oblivious Deep Neural Network Inference. In *Proceedings of the 28th USENIX Security Symposium*, 2019.
5. Payman Mohassel and Peter Rindal. ABY3: A mixed protocol framework for machine learning. In *Proceedings of the ACM Conference on Computer and Communications Security (CCS)*, 2018.
6. Sameer Wagh, Divya Gupta, and Nishanth Chandran. SecureNN: 3-party secure computation for neural network training. In *Proceedings of the Privacy Enhancing Technologies Symposium (PETS)*, 2019.
7. Adam Jacobi, Michael Chung, Adam Bernheim, and Corey Eber. Portable chest X-ray in coronavirus disease-19 (COVID-19): A pictorial review. *Clinical Imaging*, 2020.
8. Google DeepMind Health. Online at https://deepmind.com/blog/announcements/deepmind-health-joins-google-health (2020).
9. Microsoft Project InnerEye. Online at https://www.microsoft.com/en-us/research/project/medical-image-analysis/ (2020).
10. PathAI. Online at https://www.pathai.com/ (2020).
11. 104th United States Congress: Health Insurance Portability and Accountability Act of 1996 (HIPAA). Online at https://www.hhs.gov/hipaa/index.html (1996).
12. Atallah, M., Bykova, M., Li, J., Frikken, K., Topkara, M.: Private collaborative forecasting and benchmarking. In: Proc. of WPES (2004).
13. Barni, M., Failla, P., Lazzeretti, R., Sadeghi, A.R., Schneider, T.: Privacy-preserving ECG classification with branching programs and neural networks. IEEE Trans. on Information Forensics and Security (2011).
14. Beaver, D.: Efficient multiparty protocols using circuit randomization. In: Proc. of Crypto (1991).
15. Brutzkus, A., Gilad-Bachrach, R., Elisha, O.: Low latency privacy-preserving inference. In: Proc. of ICML. pp. 812–821. PMLR (2019).
16. European Parliament and the Council: The General Data Protection Regulation (GDPR). Online at http://data.europa.eu/eli/reg/2016/679/2016-05-04 (2016).
17. Fredrikson, M., Jha, S., Ristenpart, T.: Model inversion attacks that exploit confidence information and basic countermeasures. In: Proc. of ACM CCS (2015).
18. Harris, D.: A taxonomy of parallel prefix networks. In: The Thirty-Seventh Asilomar Conference on Signals, Systems & Computers, 2003. vol. 2, pp. 2213–2217. IEEE (2003).

19. Leshno, M., Lin, V.Y., Pinkus, A., Schocken, S.: Multilayer feedforward networks with a nonpolynomial activation function can approximate any function. Neural Networks **6**(6), 861–867 (1993).
20. Li, S., Xue, K., Zhu, B., Ding, C., Gao, X., Wei, D., Wan, T.: Falcon: A Fourier transform-based approach for fast and secure convolutional neural network predictions. In: Proc. of IEEE/CVF CVPR (2020).
21. Liu, J., Juuti, M., Lu, Y., Asokan, N.: Oblivious neural network predictions via MiniONN transformations. In: Proc. of ACM CCS (2017).
22. Liu, X., Wu, B., Yuan, X., Yi, X.: Leia: A lightweight cryptographic neural network inference system at the edge. IACR Cryptol. ePrint Arch. **2020**, 463 (2020).
23. Liu, X., Yi, X.: Privacy-preserving collaborative medical time series analysis based on dynamic time warping. In: European Symposium on Research in Computer Security. pp. 439–460. Springer (2019).
24. Liu, X., Zheng, Y., Yi, X., Nepal, S.: Privacy-preserving collaborative analytics on medical time series data. IEEE Transactions on Dependable and Secure Computing (2020).
25. Lou, Q., Jiang, L.: She: A fast and accurate deep neural network for encrypted data. In: Proc. of NeurIPS. pp. 10035–10043 (2019).
26. Mishra, P., Lehmkuhl, R., Srinivasan, A., Zheng, W., Popa, R.A.: Delphi: A cryptographic inference service for neural networks. In: USENIX Security Symposium (2020).
27. Xie, P., Wu, B., Sun, G.: BayHENN: Combining Bayesian deep learning and homomorphic encryption for secure DNN inference. In: Proc. of IJCAI. pp. 4831–4837 (2019).
28. Yu, L., Liu, L., Pu, C., Gursoy, M.E., Truex, S.: Differentially private model publishing for deep learning. In: Proc. of S&P. IEEE (2019).
29. Zhang, Q., Wang, C., Wu, H., Xin, C., Phuong, T.V.: GELU-Net: A globally encrypted, locally unencrypted deep neural network for privacy-preserved learning. In: Proc. of IJCAI. pp. 3933–3939 (2018).
30. Zheng, Y., Duan, H., Wang, C.: Towards secure and efficient outsourcing of machine learning classification. In: Proc. of ESORICS. Springer (2019).

Chapter 8
Privacy-Preserving User Profile Matching in Social Networks

Abstract In this chapter, we address a scenario in which a user queries a user profile database, maintained by a social networking service provider, to identify users whose profiles are similar to the one specified by the querying user. A common example of this application is online dating. Recently, the online dating site Ashley Madison was hacked, leading to the exposure of a large number of dating user profiles. This data breach has prompted researchers to investigate practical privacy protection methods for user profiles in social networks. In this chapter, we introduce a privacy-preserving solution for profile matching in social networks using multiple servers. Our solution is based on homomorphic encryption and enables a user to find matching profiles with the assistance of multiple servers, without revealing the query or the queried user profiles in plaintext. The proposed solution ensures both user profile privacy and query privacy, as long as at least one of the multiple servers is honest.

8.1 Introduction

Matching two or more users with similar interests is a significant and widely applicable problem, relevant in various contexts such as job searches, finding friends, and dating services. Current online matching services require users to trust a third-party server with their preferences. As a result, the matching server has full access to the users' preferences, raising privacy concerns, as the server might unintentionally or intentionally disclose users' profiles.

When users sign up for an online matching service, they create a "profile" that others can view. This profile may include sensitive information such as age, gender, education, occupation, number of children, religion, geographic location, sexual preferences, drinking habits, hobbies, income, ethnicity, drug use, and both home and work addresses. In some cases, even after a user cancels their account, online matching sites may retain this data.

Users may disclose sensitive personal information with the hope of attracting potential matches. However, such information may also be re-disclosed to advertisers and data aggregators, who use it for purposes unrelated to the matching service, often without the user's consent. Additionally, risks such as scammers, sexual predators, and reputational damage are prevalent in online matching services.

Many online matching platforms fail to adequately safeguard the privacy and security of their users. Frequently, these services offer misleading "privacy" settings, and their data management systems often contain serious security vulnerabilities.

In July 2015, the hacker group "The Impact Team" stole user data from Ashley Madison, a website designed to facilitate extramarital affairs. The group threatened to release the stolen information unless the site was immediately shut down. In August 2015, more than 25 gigabytes of company data, including user details, were leaked. Ashley Madison's policy of retaining users' personal information, including real names, home addresses, search history, and credit card records, led to significant public concern. Many users feared public exposure and shaming. On August 24, 2015, Toronto police reported that two unconfirmed suicides were linked to this breach.

Such incidents have raised significant concerns about the dangers of disclosing personal information. Users of these platforms must also be aware of the risk of data theft. A primary challenge, therefore, is to protect the privacy of user profiles in social networks. While encryption–where users encrypt their profiles before uploading them to social networks–seems to be the most effective solution, it complicates the process of matching users with similar profiles.

In this chapter, we consider a scenario where a user queries a user profile database maintained by a social networking service provider to identify users whose profiles resemble the one specified by the querying user. A typical example of this application is online dating. We propose a privacy-preserving solution for user profile matching in social networks using multiple servers.

The basic idea of our solution is as follows [33]: before uploading their profile to a social network, each user encrypts their profile using a homomorphic encryption scheme with a common encryption key. Thus, even if the user profile database is compromised, the attacker only obtains encrypted data. When a user wants to find matching profiles, they encrypt their preferred profile and a dissimilarity threshold, then submit the query to the social network provider. Based on the query, multiple servers–who secretly share the decryption key–compare the preferred user profile with each record in the database. If the dissimilarity is below the threshold, the matching user's contact information is returned to the querying user.

Our main contributions are as follows:

1. We formally define the user profile matching model, user profile privacy, and user query privacy.
2. We propose a solution for privacy-preserving user profile matching for a single dissimilarity threshold, and then extend this solution to handle multiple dissimilarity thresholds.
3. We conduct a security analysis of our protocols. If at least one of the multiple servers is honest, our protocols ensure both user profile privacy and user query privacy.

4. We perform extensive experiments using a real dataset to evaluate the performance of our proposed protocols under various parameter settings. Our experiments demonstrate that the solutions are both practical and efficient.

This chapter extends our previous work [31] in the following ways:

1. In our previous work, we employed a variant of the ElGamal encryption scheme [32] as the underlying homomorphic encryption method, assuming the two prime factors of the modulus were public parameters. However, Rao [22] identified a security vulnerability in this scheme, where an attacker could decrypt the ciphertexts without possessing the decryption key. In this chapter, we address this security flaw by keeping the factorization of the modulus secret.
2. In our previous work, the user profile data was shared across all matching servers, requiring each matching server to maintain a user profile database. In this chapter, we change this approach by storing the user profile data solely within the social service provider, eliminating the need for each matching server to maintain a user profile database.
3. In our previous work, the query user could specify only a single dissimilarity threshold for matching users. In this chapter, we extend this functionality by allowing the query user to specify multiple dissimilarity thresholds for matching users.
4. In our previous work, user profile matching was limited to numerical attributes only. In this chapter, we broaden our approach to include user profile matching based on categorical attributes as well.
5. In this chapter, we enhance both the security and performance of the private sharing and multiplication algorithms introduced in our previous work. Additionally, we introduce two new collaborative algorithms for encryption and decryption.

The rest of the chapter is organized as: Sect. 8.2 surveys related work. Section 8.3 introduces preliminaries. Section 8.4 describes our model for privacy-preserving user profile matching. Sections 8.5 and 8.6 present our solutions. The last section concludes the paper.

8.2 Related Work

In 2003, Agrawal et al. [1] proposed a solution to the private two-party set-intersection problem. In this setup, one party, P_1, inputs $X = \{x_1, x_2, \ldots, x_n\}$, and the other party, P_2, inputs $Y = \{y_1, y_2, \ldots, y_n\}$. The goal is for one party to learn the intersection $X \cap Y$ and nothing else, while the other party learns nothing. Their solution is based on commutative encryption with the property:

$$E_{k_1}(E_{k_2}(x)) = E_{k_2}(E_{k_1}(x)),$$

where k_1 and k_2 are known to P_1 and P_2, respectively. The process works as follows: P_1 (or P_2) encrypts its input set X (or Y) with its key k_1 (or k_2), producing $E_{k_1}(X)$ (or $E_{k_2}(Y)$), and exchanges them. Next, P_1 sends a pair $(E_{k_2}(Y), E_{k_1}(E_{k_2}(Y)))$ to P_2, who computes $E_{k_2}(E_{k_1}(X))$, compares it with the pair $(E_{k_2}(Y), E_{k_1}(E_{k_2}(Y)))$, and decrypts $E_{k_2}(y)$ if $E_{k_2}(E_{k_1}(x)) = E_{k_1}(E_{k_2}(y))$. Later, Vaidya et al. [27] extended this solution to the n-party setting, and Arb et al. [2] applied the idea to detect friend-of-friend relationships in MSN.

In 2004, Freedman et al. [9] provided another solution to the private two-party set-intersection problem based on polynomial evaluation. Here, P_1 defines a polynomial $P(y) = (x_1 - y)(x_2 - y)\ldots(x_n - y)$ and sends homomorphic encryptions of the coefficients of this polynomial to P_2. P_2 uses the homomorphic properties of the encryption system to evaluate the polynomial at each of its own inputs, multiplies each result by a random number r, and then adds the encryption of his input value. Specifically, P_2 computes $\mathsf{E}(rP(y) + y)$. This approach ensures that for each element in the intersection of the two sets, the result corresponds to the element's value, while for all other values, the result is random. In 2005, Kissner and Song [19] improved the solution to enable set-intersection, cardinality set-intersection, and over-threshold set-union operations on multisets. This approach is built upon set operations using polynomial representations and the mathematical properties of polynomials. Subsequent work by Sang et al. [23], Ye et al. [30], and Dachman-Soled et al. [7] further improved and extended these solutions. In 2007, Li and Wu [17] introduced an unconditionally secure protocol for multi-party set intersection, which relies on secret sharing [25] for input distribution and computations on shared inputs. In 2010, Narayanan et al. [20] proposed an improved solution, all of which are based on polynomial evaluation.

In 2008, Hazay and Lindell [13] proposed another private two-party set-intersection solution, where P_1 with a set X inputs a key k to a pseudorandom function F, and P_2 with a set Y inputs the elements of its set. At the end of the process, P_2 holds the set $F_k(y)$ for each element $y \in Y$, while P_1 learns nothing. P_1 then computes the set $F_k(x)$ for each element $x \in X$ and sends it to P_2. By comparing the sets, P_2 learns the intersection, but nothing more. This solution requires a large number of exponentiations. In 2009, Jarecki and Liu [16] improved the efficiency of this approach. In 2010, Cristofaro and Tsudik [6] proposed a solution based on blind RSA signatures [5], where the server computes $(H(x_j))^d$ for the client without learning the inputs, where x_j are the client's inputs and d is the server's one-time RSA signing key. Their solution is more efficient than previous methods. All of these solutions are based on pseudorandom functions.

In 2010, Yang et al. [28] proposed the E-SmallTalker scheme for matching people's interests before initiating a small talk. E-SmallTalker considers n potential communication users $U_1, U_2, \ldots, U_n$, each with a set of interests $A_i = \{a_{i,1}, a_{i,2}, \ldots, a_{i,n_i}\}$. The scheme allows each user U_i to discover common interests between their set A_i and the sets A_j of other users, $1 \leq i, j \leq n, i \neq j$. E-SmallTalker is based on the Bloom filter [3], which utilizes a bit array of m bits and k hash functions to perform membership tests on set elements. In 2011, Li et al. [18] proposed the FindU scheme for profile matching in mobile social networks. FindU

8.2 Related Work

builds on private set-intersection techniques, but improves efficiency by using secure multi-party computation based on polynomial secret sharing [25].

In 2011, Huang et al. [15] proposed a privacy-preserving solution for biometric matching, where the server holds a database $\{v_i, p_i\}_{i=1}^{n}$ (v_i is the biometric data corresponding to the identity profile p_i), and the client with a biometric reading v' wants to identify the closest match in the database within some distance threshold δ. They used Yao's garbled circuit technique [29] for secure matching, where each gate in the circuit is associated with a set of ciphertexts, and the circuit is evaluated via oblivious transfer, ensuring that the parties learn only the result of the function applied to their inputs. This work is closely related to our own, with the main difference being that we assume the database is encrypted, while their method assumes it is in cleartext.

In 2012, Shahandashti et al. [24] proposed a private fingerprint matching protocol, which enables two parties to determine if their fingerprints belong to the same individual. The protocol uses homomorphic encryption to compute and compare the Euclidean distance and angular difference for each pair of fingerprint minutiae. The construction uses aided computation, where Bob can compute $E(P(x))$ given $E(x)$ and the polynomial coefficients $\{a_k\}_{k=1}^{n}$. Using this technique, Bob calculates $E(R(z_{ij}))$ where z_{ij} is zero if the minutiae match, and non-zero otherwise. The final step uses the homomorphic property of the encryption to calculate the total number of minutiae matches and determine whether it exceeds a given threshold τ.

This work is also closely related to our own. The key distinction is that while we assume the user data is encrypted, they assume that both parties hold private fingerprints in cleartext.

More recently, Sun et al. [26] proposed privacy-preserving, distance-aware encoding mechanisms for comparing numerical values in an anonymous space. This approach embeds numerical values into the Hamming space using a low-computation encoding algorithm with randomized bit vectors. To ensure robust privacy guarantees, the method incorporates random responses based on differential privacy to maintain global indistinguishability of the original data, while Laplace noise, applied via the pufferfish mechanism, ensures local indistinguishability. Although similar to our work, their approach focuses on embedding original data into an anonymous space, whereas we encrypt the original data.

Additionally, our work is closely related to homomorphic encryption, a cryptographic method that allows computations to be performed on ciphertexts. The result of these computations, when decrypted, corresponds to the result of operations as if they had been executed on the plaintext data. The primary objective of homomorphic encryption is to enable computation on encrypted data without revealing the data itself. Notable homomorphic encryption schemes include ElGamal [8], Goldwasser-Micali [11], and Paillier [21] schemes, which support either addition or multiplication on encrypted data. The Boneh-Goh-Nissim [4] scheme, on the other hand, supports arbitrary additions and one multiplication (followed by further additions) on encrypted data. Fully homomorphic encryption schemes, such as the Gentry scheme [10], are capable of supporting arbitrary computations on encrypted data, though they have not yet become practically feasible.

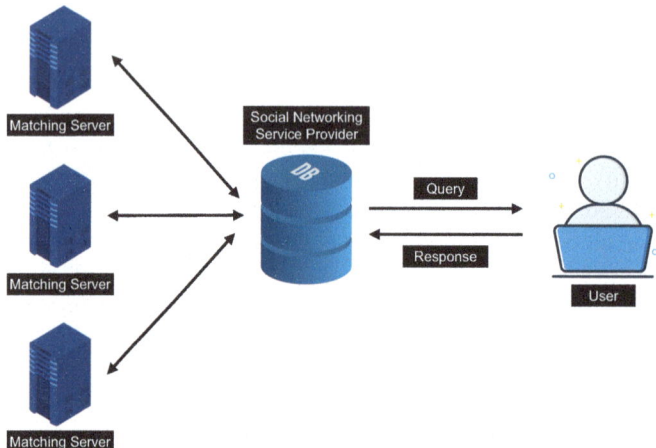

Fig. 8.1 Our model for privacy-preserving user profile matching

8.3 Model for Privacy-Preserving User Profile Matching

Our Model

Our model is designed for a social networking service environment consisting of users and servers. Specifically, it includes one user, one database (DB) server, and n matching servers, as illustrated in Fig. 8.1.

In this model, users store their profiles in the DB server of the service provider. User profile attributes can be either sensitive or non-sensitive. We focus on the protection of sensitive attributes only. Additionally, user profile attributes may be numeric (e.g., income) or categorical (e.g., address). For the purposes of this chapter, we consider only numeric attributes, with the exception of Sect. 8.6.

Dissimilarity Definitions

Definition 8.1 (Dissimilarity Definition for Single Attribute) For any numerical attribute A, the dissimilarity between two users, U and U', with $A = a$ and $A = a'$, respectively, is defined as $d_A(U, U') = (a - a')^2$, where a and a' are integers.

Definition 8.2 (Dissimilarity Definition for Multiple Attributes) Assume that $A_1, A_2, \cdots, A_m$ are m numerical attributes of a user profile. The dissimilarity between two users, U and U', is defined as $d(U, U') = \sum_{i=0}^{m} d_{A_i}(U, U')$.

For a user querying the user profile database, different attributes may have varying impacts on the overall dissimilarity. Some attributes may be more significant for determining dissimilarity than others. To account for this, we introduce the concept of weight to measure the importance of an attribute. The weight of an attribute is a non-negative integer specified by the user submitting the query.

8.3 Model for Privacy-Preserving User Profile Matching

Definition 8.3 (Weighted Dissimilarity Definition for Multiple Attributes)
Assume that $A_1, A_2, \cdots, A_m$ are m numerical attributes of a user profile, and that the weight for each attribute A_i is w_i. The weighted dissimilarity between two users, U and U', is defined as $d(U, U') = \sum_{i=0}^{m} w_i d_{A_i}(U, U')$.

Security Definitions To safeguard the privacy of user profiles, we use ElGamal encryption. Initially, each matching server S_i generates its own ElGamal public/private key pair, denoted as (pk_i, sk_i). The common public key for the social networking service is then defined as $PK = \prod_{i=1}^{n} pk_i$.

Using the common public key PK, a user encrypts each attribute of their profile along with their contact information and submits the encrypted profile to the DB server for storage in the user profile database. The encrypted profile can only be decrypted through the cooperation of the n matching servers.

To find users with similar profiles, the querying user specifies a profile, encrypts it, and submits it, along with a dissimilarity threshold, to the social networking service provider.

The n matching servers collaborate to identify users with matching profiles from the user profile database, based on the provided profile and dissimilarity threshold. The matching users' contact information, where the dissimilarity is below the threshold, is then returned to the querying user.

Considering m attributes $A_1, A_2, \cdots, A_m$ for the user profile, we formally define the above process through a user profile matching protocol, consisting of four algorithms as follows:

(1) Profile Generation (PG): This algorithm takes as input the common public key PK of the social networking service, the profile $a_1, a_2, \cdots, a_m$ of the user, and outputs an encrypted profile P, denoted as $P = \text{PG}(PK, a_1, a_2, \cdots, a_m)$. The encrypted profile P is then submitted to the social networking service provider and stored in the user profile database DB.

(2) Query Generation (QG): This algorithm takes as input the common public key PK of the social networking service, the profile $a_1, a_2, \cdots, a_m$ specified by the user, the corresponding weights $w_1, w_2, \cdots, w_m$, and the dissimilarity threshold δ. It outputs a query Q, denoted as $Q = \text{QG}(PK, a_1, w_1, a_2, w_2, \cdots, a_m, w_m, \delta)$, which is then submitted to the social networking service provider.

(3) Response Generation (RG): This algorithm takes as input the query Q, the user profile database DB, and the private keys $sk_1, sk_2, \cdots, sk_n$. The n matching servers cooperate to generate a response R, denoted as $R = \text{RG}(Q, DB, sk_1, sk_2, \cdots, sk_n)$. The response R contains the contact information of the first t matching users from DB and is returned to the user, where t is either fixed or specified by the user in Q.

(4) Response Retrieval (RR): This algorithm takes as input the response R and outputs the first t matching users in DB.

A user profile matching protocol is considered correct if the response R lists the contact information of the first t' matching users whose dissimilarity is below the threshold δ, where $t' = \min(t, T)$ and T denotes the total number of matching users in DB.

Now, we formally define user profile privacy within the context of a game, as described below.

Given the common public key PK, consider the following game between an adversary $\mathcal{A}$ and a challenger C. The game consists of the following steps:

(1) The adversary $\mathcal{A}$ selects two user profiles: $(a_{01}, a_{02}, \cdots, a_{0m})$ and $(a_{11}, a_{12}, \cdots, a_{1m})$, and sends them to the challenger C.
(2) The challenger C randomly selects a bit $b \in \{0, 1\}$, and executes the Profile Generation (PG) algorithm to generate an encrypted profile $P_b = \mathsf{PG}(PK, a_{b1}, a_{b2}, \cdots, a_{bm})$. The encrypted profile P_b is then sent back to the adversary $\mathcal{A}$.
(3) The adversary $\mathcal{A}$ may experiment with the code of P_b in any non-black-box manner, and eventually outputs a bit $b' \in \{0, 1\}$.

The adversary wins the game if $b' = b$, and loses otherwise. We define the advantage of the adversary $\mathcal{A}$ in this game as $\mathsf{Adv}_{\mathcal{A}}(k) = |\Pr(b' = b) - 1/2|$, where k is the security parameter.

Definition 8.4 (User Profile Privacy Definition) In a user profile matching protocol, a user's profile is considered private if, for any probabilistic polynomial-time (PPT) adversary $\mathcal{A}$, the advantage $\mathsf{Adv}_{\mathcal{A}}(k)$ is a negligible function. This probability is taken over the coin tosses of both the challenger and the adversary.

User profile privacy ensures that an attacker cannot determine a user's profile, even if some of the matching servers are compromised by the attacker.

Next, we formally define user query privacy within a similar game framework:

Given the common public key PK, consider the following game between an adversary $\mathcal{A}$ and a challenger C. The game consists of the following steps:

(1) The adversary $\mathcal{A}$ selects two user profiles: $(a_{01}, a_{02}, \cdots, a_{0m})$ with corresponding weights $(w_{01}, w_{02}, \cdots, w_{0m})$ and a dissimilarity threshold δ_0, and $(a_{11}, a_{12}, \cdots, a_{1m})$ with corresponding weights $(w_{11}, w_{12}, \cdots, w_{1m})$ and a different dissimilarity threshold δ_1. These are then sent to the challenger C.
(2) The challenger C randomly selects a bit $b \in \{0, 1\}$, and executes the Query Generation (QG) algorithm to generate a query $Q_b = \mathsf{QG}(PK, a_{b1}, w_{b1}, a_{b2}, w_{b2}, \cdots, a_{bm}, w_{bm}, \delta_b)$. The query Q_b is then sent back to the adversary $\mathcal{A}$.
(3) The adversary $\mathcal{A}$ may experiment with the code of Q_b in any non-black-box manner, and eventually outputs a bit $b' \in \{0, 1\}$.

The adversary wins the game if $b' = b$, and loses otherwise. We define the advantage of the adversary $\mathcal{A}$ in this game as $\mathsf{Adv}_{\mathcal{A}}(k) = |\Pr(b' = b) - 1/2|$, where k is the security parameter.

Definition 8.5 (User Query Privacy Definition) In a user profile matching protocol, the user's query privacy is preserved if, for any probabilistic polynomial-time

(PPT) adversary $\mathcal{A}$, the advantage $\mathsf{Adv}_{\mathcal{A}}(k)$ is a negligible function. The probability is taken over the coin tosses of both the challenger and the adversary.

User query privacy ensures that an attacker cannot determine the user's query, even if some of the matching servers are compromised by the attacker.

8.4 User Profile Matching Protocol

Initialization
In this section, we describe our solution for privacy-preserving user profile matching.

We utilize the ElGamal encryption scheme [8] and set the public parameters based on the Paillier encryption scheme [21].

Given a security parameter k, the system generates two large distinct primes p and q, and computes $N = pq$ and $g = (1+N)^p r^N \pmod{N^2}$ for a randomly chosen integer $r \in \{2, 3, \cdots, N^2 - 1\}$, ensuring that $(g-1)/N$ is not an integer. The system then publishes (N, g) and erases p and q from memory. It is important to note that the pair (N, g) can be privately generated by multiple parties, as described in [14], such that no one knows p and q if at least one party can be trusted. Alternatively, the generation can be performed using Intel Software Guard Extensions (Intel SGX). Anyone can verify that $(g-1)/N$ is not an integer.

Based on the public parameters (N, g), each matching server S_i selects a private key SK_i randomly from $\mathbb{Z}_\rho^* = \{1, 2, \cdots, \rho - 1\}$, where ρ is a large integer smaller than N. The server computes and publishes its public key $PK_i = g^{SK_i} \pmod{N^2}$ and provides a zero-knowledge proof of knowledge of SK_i. The common public key for the social networking service is then defined as $PK = \prod_{i=1}^n PK_i \pmod{N^2}$.

Given the common public key PK and the public parameters (N, g), each user U_i encrypts their profile $(a_{i1}, a_{i2}, \cdots, a_{im})$ as follows:

$$\mathsf{E}(g_1^{a_{ij}}) = (g^{r_{ij}} \pmod{N^2}, g_1^{a_{ij}} PK^{r_{ij}} \pmod{N^2})$$

where r_{ij} is randomly chosen from $\mathbb{Z}_\rho^*$, and $g_1 = N+1$. We assume that $|a_{ij}| \ll N$.

Similarly, the user U_i also encrypts $(a_{i1}^2, a_{i2}^2, \cdots, a_{im}^2)$. Additionally, the user encrypts their contact information CI_i (e.g., email address or mobile phone number), where we assume the binary representation of CI_i is smaller than N, i.e., $CI_i < N$.

The user U_i then submits the following to the social networking service provider:

$$\mathsf{E}(g_1^{CI_i}), \mathsf{E}(g_1^{a_{i1}}), \mathsf{E}(g_1^{a_{i1}^2}), \cdots, \mathsf{E}(g_1^{a_{im}}), \mathsf{E}(g_1^{a_{im}^2})$$

The service provider stores these in the user profile database DB.

The database DB containing l users' profiles is represented as:

$$DB = \{\mathsf{E}(g_1^{CI_i}), [\mathsf{E}(g_1^{a_{ij}}), \mathsf{E}(g_1^{a_{ij}^2})]_{1 \leq j \leq m} \mid 1 \leq i \leq l\}.$$

Since the user profiles are encrypted using the common public key PK, no one knows the user profiles unless the n matching servers in the social network collude. However, we assume that at least one of the n servers is trusted not to collude with others to attack the user profiles. Therefore, the privacy of the user profiles remains protected.

Private User Profile Matching

When a user U wishes to identify matching users within a social network, they specify their preferred profile attributes $A_1 = a_1, A_2 = a_2, \ldots, A_m = a_m$, along with the corresponding weights $w_1, w_2, \ldots, w_m$, and a dissimilarity threshold δ. We assume that all w_j values are integers ranging from 0 to $2^{\ell_w} - 1$, and δ is an integer between 0 and $2^{\ell_\delta} - 1$, where ℓ_w and ℓ_δ represent the bit-lengths of w_j and δ, respectively. Furthermore, based on the public parameters (N, g), the user U randomly selects the private key SK_U from $\mathbb{Z}_p^*$ and computes the corresponding public key $PK_U = g^{SK} \pmod{N^2}$.

Subsequently, the user U generates a query in the following form:

$$Q = \{\mathsf{E}(g_1^{a_1}), \mathsf{E}(g_1^{w_1}), \ldots, \mathsf{E}(g_1^{a_m}), \mathsf{E}(g_1^{w_m}), \mathsf{E}(g_1^\delta), PK_U\}$$

and submits it to the social networking service provider.

Upon receiving the query Q from the user U, the n matching servers collaborate to identify matching users from the database DB through four distinct steps: as follows.

Private Sharing of Query Given an encryption of g_1^x, i.e., $\mathsf{E}(g_1^x)$, where we assume that the absolute value of x is less than L, with $L = 2^{\ell_x + \ell} \ll N$, and ℓ_x being the bit-length of x and ℓ as a sufficiently large security parameter, the n matching servers can cooperate to privately share x by executing Algorithm 1.

Theorem 8.1 (Private Sharing Correctness) *In Algorithm 1, we have $\sum_{k=1}^n x_k = x$.*

Proof Assume $A = g^r$ and $B = g_1^x PK^r$ for a random integer r, we have:

$$B \prod_{k=1}^{n-1}(g_1^{x_k} A^{SK_k})^{-1}/A^{SK_n}$$

$$= g_1^{x - \sum_{k=1}^{n-1} x_k} PK^r / \prod_{k=1}^n g^{SK_k r}$$

$$= g_1^{x - \sum_{k=1}^{n-1} x_k} = (1+N)^{x - \sum_{k=1}^{n-1} x_k}$$

8.4 User Profile Matching Protocol

Algorithm 1 Private sharing (PS) (n servers)

Input: $\mathsf{E}(g_1^x)$, N, g, PK (public), $S_1 : SK_1, S_2 : SK_2, \ldots, S_n : SK_n$ (private)
Output: $S_1 : x_1, \ldots, S_n : x_n$ such that $\sum_{k=1}^{n} x_k = x$.
1: let $(A, B) = \mathsf{E}(g_1^x)$
2: for $k = 1$ to $n - 1$ {
3:    The server S_k randomly selects an integer x_k from $(-L, L)$, and computes $(g_1^{x_k} A^{SK_k})^{-1} (mod\ N^2)$, then sends it to server S_n.
4: }
5: S_n computes

$$x_n = [B \prod_{k=1}^{n-1} (g_1^{x_k} A^{SK_k})^{-1} / A^{SK_n} (mod\ N^2) - 1]/N$$

6: **return** $S_1 : x_1, S_2 : x_2, \ldots, S_n : x_n$

$$= 1 + (x - \sum_{k=1}^{n-1} x_k) N\ (mod\ N^2)$$

$$[B \prod_{k=1}^{n-1} (g_1^{x_k} A^{SK_k})^{-1} / A^{SK_n} (mod\ N^2) - 1]/N = x - \sum_{k=1}^{n-1} x_k$$

Thus, $x_n = x - \sum_{k=1}^{n-1} x_k$. The theorem is thus proven. $\triangle$

After executing Algorithm 1 for each attribute value a_j and weight w_j ($j = 1, 2, \ldots, m$), the server S_k ($k = 1, 2, \ldots, n$) will privately hold $a_j^{(k)}$ such that $\sum_{k=1}^{n} a_j^{(k)} = a_j$, and $w_j^{(k)}$ such that $\sum_{k=1}^{n} w_j^{(k)} = w_j$.

Private Computation of Dissimilarity Assuming that server S_k ($k = 1, 2, \ldots, n$) privately holds x_j such that $\sum_{k=1}^{n} x_j = x$, given the encryption of g_1^y, i.e., $\mathsf{E}(g_1^y)$, the n matching servers can cooperate to privately compute $\mathsf{E}(g_1^{xy})$ by executing Algorithm 2.

Algorithm 2 Private multiplication (PM) (n servers)

Input: $S_1 : x_1, S_2 : x_2, \ldots, S_n : x_n$ (private), $\mathsf{E}(g_1^y)$, N, g, PK (public)
Output: $\mathsf{E}(g^{xy})$.
1: for $k = 1$ to n {
2:    S_k computes and re-encrypts $\mathsf{E}(g_1^y)^{x_k} = \mathsf{E}(g_1^{x_k y})$
3:    S_k broadcasts $\mathsf{E}(g_1^{x_k y})$ to the other servers.
4:    S_k computes $(A, B) = \mathsf{E}(g_1^{x_1 y}) \mathsf{E}(g_1^{x_2 y}) \cdots \mathsf{E}(g_1^{x_n y})$
5: }
6: **return** (A, B)

Theorem 8.2 (Private Multiplication Correctness) *In Algorithm 2, we have* $(A, B) = \mathsf{E}(g_1^{xy})$.

Proof Since

$$(A, B) = \mathsf{E}(g_1^{x_1 y})\mathsf{E}(g_1^{x_2 y}) \cdots \mathsf{E}(g^{x_n y})$$
$$= \mathsf{E}(g_1^{(x_1 + x_2 + \cdots + x_n)y}) = \mathsf{E}(g_1^{xy})$$

The theorem is proven. △

Next, the n servers collaborate to calculate the dissimilarity between the user profile and a given record $(\mathsf{E}(g_1^{a_{i1}}), \mathsf{E}(g^{a_{i1}^2}), \mathsf{E}(g^{a_{i2}}), \mathsf{E}(g^{a_{i2}^2}) \cdots, \mathsf{E}(g_1^{a_{im}}), \mathsf{E}(g_1^{a_{im}^2}))$ of a user U_i from the database DB.

Since server S_k ($k = 1, 2, \ldots, n$) holds $a_j^{(k)}$ privately, such that $\sum_{k=1}^n a_j^{(k)} = a_j$, the n servers can privately collaborate to compute $\mathsf{E}(g_1^{(a_j - 2a_{ij})a_j})$ using Algorithm 2. Then, they compute

$$\mathsf{E}(g_1^{(a_j - 2a_{ij})a_j})\mathsf{E}(g_1^{a_{ij}^2}) = \mathsf{E}(g_1^{(a_j - a_{ij})^2}),$$

where $j = 1, 2, \ldots, m$.

Subsequently, because each server S_k ($k = 1, 2, \ldots, n$) holds $w_j^{(k)}$ privately, such that $\sum_{k=1}^n w_j^{(k)} = w_j$ ($j = 1, 2, \ldots, m$), each server S_k can compute and broadcast the following:

$$\prod_{j=1}^m \mathsf{E}(g_1^{w_j^{(k)}(a_j - a_{ij})^2}) = \mathsf{E}(g_1^{\sum_{j=1}^m w_j^{(k)}(a_j - a_{ij})^2}).$$

Finally, the n servers compute

$$\prod_{k=1}^n \mathsf{E}(g_1^{\sum_{j=1}^m w_j^{(k)}(a_j - a_{ij})^2}) = \mathsf{E}(g_1^{\sum_{k=1}^n \sum_{j=1}^m w_j^{(k)}(a_j - a_{ij})^2})$$

$$= \mathsf{E}(g_1^{\sum_{j=1}^m w_j(a_j - a_{ij})^2}) = \mathsf{E}(g_1^{d(U, U_i)}).$$

Given the encryption of the threshold, i.e., $\mathsf{E}(g_1^\delta)$, the n servers can compute

$$\mathsf{E}(g_1^\delta)/\mathsf{E}(g_1^{d(U, U_i)}) = \mathsf{E}(g_1^{\delta - d(U, U_i)})$$

and then privately share $d = \delta - d(U, U_i)$ by running Algorithm 1. After executing Algorithm 1, assume that server S_k ($k = 1, 2, \ldots, n$) privately holds d_j such that $\sum_{k=1}^n d_j = d$, where $|d_j| < L$ for $j = 1, 2, \ldots, n-1$, $|d_n| < nL$, and $|d| \ll L$.

8.4 User Profile Matching Protocol

Private Comparison with Threshold For the purpose of security and efficiency, the original ElGamal encryption is used in this step. The n servers collaborate to select a multiplicative cyclic group $\mathbb{G}$ of prime order q', with a generator g_2, such that the discrete logarithm problem over $\mathbb{G}$ is difficult. Each server S_k chooses a private key sk_k randomly from $\mathbb{Z}_{q'}^* = \{1, 2, \ldots, q'-1\}$ and computes a public key $pk_k = g_2^{sk_k}$. The common public key is defined as $pk = \prod_{k=1}^{n} pk_i$.

Let $\lambda = \lceil \log_2 nL \rceil + 2$, and assume the two's complement of d_k is $\overline{d_k} = b_{\lambda-1}^{(k)} b_{\lambda-2}^{(k)} \ldots b_1^{(k)} b_0^{(k)}$, where $b_{\lambda-1}^{(k)} = 0$ if $d_k \geq 0$ and $b_{\lambda-1}^{(k)} = 1$ otherwise, according to [12].

To compare $d(U, U_i)$ and δ, the n servers collaborate to run Algorithm 3, where e represents the original ElGamal encryption with the public key pk.

Algorithm 3 Private comparison (PC) (n servers)

Input: $S_1 : \overline{d_1}, S_2 : \overline{d_2}, \ldots, S_n : \overline{d_n}$ (private), q', g_2, $\mathbb{G}$, pk (public)
Output: $e(1)$ if $d(U, U_i) \leq \delta$ and $e(g_2)$ otherwise.
1: Let $S = (e(g_2^{s_{\lambda-1}}), \ldots, e(g_2^{s_1}), e(g_2^{s_0}))$, where $s_j = 0$ for all j.
2: For $k = 1$ to n {
3: $\quad S_k$ encrypts $\overline{d_k}$ to get

$$e(\overline{d_k}) = (e(g_2^{b_{\lambda-1}^{(k)}}), \ldots, e(g_2^{b_1^{(k)}}), e(g_2^{b_0^{(k)}}))$$

4: The n servers collaborate to compute

$$S = S \boxplus e(\overline{d_k}) = (e(g_2^{s_{\lambda-1}}), \ldots, e(g_2^{s_0}))$$

$\quad$ as described in Sect. 5.3, where s_j is either 0 or 1.
5: }
6: **return** $e(g_2^{s_{\lambda-1}})$

Theorem 8.3 (Private Comparison Correctness) *In Algorithm 3, we have $s_{\lambda-1} = 0$ if $d(U, U_i) \leq \delta$ and 1 otherwise.*

Proof According to the binary addition described in Sect. 5.3, the sequence $(s_{\lambda-1} \cdots s_1 s_0)$ represents the two's complement of $d = \delta - d(U, U_i)$. Based on the properties of two's complement, we conclude that $s_{\lambda-1} = 0$ if $d \geq 0$ (i.e., $d(U, U_i) \leq \delta$) and 1 otherwise. Therefore, the theorem is proved. △

Private Generation of Response After obtaining $e(g_2^{s_{\lambda-1}})$, the n servers cooperate to decrypt it by executing Algorithm 4.

Theorem 8.4 (Collaborative Decryption Correctness) *In Algorithm 4, the output is α.*

Algorithm 4 Collaborative decryption (CD) (n servers)

Input: $S_1 : sk_1, S_2 : sk_2, \cdots, S_n : sk_n$ (private), $\mathsf{e}(g_2^\alpha)$ where $\alpha \in \{0, 1\}, q', g_2, \mathbb{G}, pk$ (public)
Output: α.
1: Let $(A_0, B_0) = \mathsf{E}(g^\alpha)$
2: For $k = 1$ to $n - 1$:
3: S_k randomly selects r_k from $\mathbb{Z}_p^*$, computes

$$A_k = A_{k-1}^{r_k}, B_k = \left(\frac{B_{k-1}}{A_{k-1}^{sk_k}}\right)^{r_k},$$

and sends (A_k, B_k) to the server S_{k+1}.
4: End for
5: The server S_n computes

$$Z = \frac{B_{n-1}}{A_{n-1}^{sk_n}}.$$

6: If $Z = 1$ then return 0.
7: Else return 1.

Proof When $\alpha = 0$, let $(A_0, B_0) = (g_2^{r_0}, pk^{r_0})$. We have the following:

$$A_1 = A_0^{r_1} = g_2^{r_0 r_1},$$

$$B_1 = \left(\frac{B_0}{A_0^{sk_1}}\right)^{r_1} = \left(\frac{(\prod_{l=1}^{n} pk_l)^{r_0}}{g_2^{r_0 sk_1}}\right)^{r_1} = \left(\prod_{l=2}^{n} pk_l\right)^{r_0 r_1}.$$

By the induction hypothesis, we assume that for $1 \leq k \leq n - 1$,

$$A_k = A_{k-1}^{r_k} = g_2^{r_0 r_1 \cdots r_k},$$

$$B_k = \left(\frac{B_{k-1}}{A_{k-1}^{sk_1}}\right)^{r_k} = \left(\prod_{l=k+1}^{n} pk_l\right)^{r_0 r_1 \cdots r_k}.$$

Therefore,

$$Z = \frac{B_{n-1}}{A_{n-1}^{sk_n}} = \frac{pk_n^{r_0 r_1 \cdots r_{n-1}}}{g_2^{r_0 r_1 \cdots r_{n-1} sk_n}} = 1,$$

and the output is 0. When $\alpha = 1$, we have $Z = g_2^{r_1 r_2 \cdots r_{n-1}} \neq 1$, and the output is 1. Thus, the theorem is proved. △

8.4 User Profile Matching Protocol

If the output of Algorithm 4 is 0, the n servers cooperate to encrypt CI_i with the public key PK_U of the query user U by executing Algorithm 5, where $\mathsf{E}(g_1^m, PK)$ and $\mathsf{E}(g_1^m, PK_U)$ represent encryptions of g_1^m under the public keys PK and PK_U, respectively.

Algorithm 5 Collaborative encryption (CE) (n servers)

Input: $S_1 : SK_1, S_2 : SK_2, \cdots, S_n : SK_n$ (private), $\mathsf{E}(g_1^m, PK), N, g, PK_U$ (public)
Output: $\mathsf{E}(g_1^m, PK_U)$.
1: Given $\mathsf{E}(g_1^m, PK)$, the n servers cooperate to execute Algorithm 1 to privately share m, such that each server S_k privately holds m_k with $\sum_{k=1}^n m_k = m$.
2: **For** $k = 1$ to n:
3: $\quad S_k$ randomly selects r_k from $\mathbb{Z}_p^*$, computes

$$A_k = g^{r_k} \bmod N^2, \quad B_k = g_1^{m_k} PK_U^{r_k} \bmod N^2,$$

$\quad$ and sends (A_k, B_k) to the service provider.
4: **End for**
5: The service provider computes

$$A = \prod_{k=1}^n A_k \bmod N^2, \quad B = \prod_{k=1}^n B_k \bmod N^2.$$

6: **return** (A, B)

Theorem 8.5 (Collaborative Encryption Correctness) *In Algorithm 5, (A, B) is an ElGamal encryption of $g_1^{m_1+m_2+\cdots+m_n}$ with the public key PK_U.*

Proof According to Algorithm 5, we have

$$A = \prod_{k=1}^n A_k = \prod_{k=1}^n g^{r_k} = g^{r_1+r_2+\cdots+r_n}$$

$$B = \prod_{k=1}^n B_k = \prod_{k=1}^n g_1^{m_k} PK_U^{r_k} = g_1^{m_1+\cdots+m_n} PK_U^{r_1+\cdots+r_n}$$

Therefore, the theorem is proved. △

After receiving $\mathsf{E}(g_1^{CI_i}, PK_U)$ from the service provider, the querying user U executes Algorithm 6 to retrieve CI_i.

Theorem 8.6 (Decryption Correctness) *In Algorithm 6, we have $x = y \bmod N$.*

Proof Assume that $A = g^r$ and $B = g_1^y PK_U^r$, we have

Algorithm 6 Decryption (User U)

Input: $\mathsf{E}(g_1^y, PK_U), N, g, PK_U$ (public), $U: SK_U$ (private)
Output: y.
1: Let $(A, B) = \mathsf{E}(g_1^y, PK_U)$
2: The querying user U computes

$$x = \frac{[B/A^{SK_U} \mod N^2 - 1]}{N}$$

3: **return** x

$$\begin{aligned}
x &= \frac{[B/A^{SK_U} \mod N^2 - 1]}{N} \\
&= \frac{[g_1^y PK_U^r / (g^r)^{SK_U} \mod N^2 - 1]}{N} \\
&= \frac{((1+N)^y \mod N^2 - 1)}{N} \\
&= \frac{(1 + y \mod NN - 1)}{N} = y \mod N
\end{aligned}$$

Thus, the theorem is proven. △

Since $CI_i < N$, the output of Algorithm 6 is CI_i when the input is $\mathsf{E}(g_1^{CI_i}, PK)$.

This process is repeated until t matching users are found in the database and returned to the querying user. If the total number of matching users from the database is fewer than t, all matching users are returned to the querying user.

8.5 Extended Profile Matching Protocol

User Profile Matching for Multiple Thresholds
In the user profile matching protocol outlined in Sect. 8.4, the querying user specifies only a single threshold δ. However, in some user matching queries, the user may wish to define different thresholds for various attribute groups. For instance, the querying user may seek individuals aged between 20 and 30 and with a height between 170 and 180 cm.

In this section, we extend the user profile matching protocol with a single threshold to allow the querying user to define multiple thresholds. Let us begin by considering two thresholds.

Assume that the querying user U selects two attribute groups: one group consisting of attributes $A_1, A_2, \ldots, A_m$ and the other containing attributes $A'_1, A'_2, \ldots, A'_{m'}$. The user wishes to find a matching user U_i such that

8.5 Extended Profile Matching Protocol

$$d_{A_1,A_2,\ldots,A_m}(U_i, U) = \sum_{j=1}^{m} w_j(a_{ij} - a_j)^2 \leq \delta,$$

$$d_{A'_1,A'_2,\ldots,A'_{m'}}(U_i, U) = \sum_{j=1}^{m'} w'_j(a'_{ij} - a'_j)^2 \leq \delta',$$

where a_{ij} and a'_{ij} represent the values of attributes A_j and A'_j, respectively, for the user U_i, and a_j, w_j, δ, a'_j, w'_j, and δ' are specified by U in the query.

First, the user U encrypts

$$g_1^{a_1}, \ldots, g_1^{a_m}, g_1^{a'_1}, \ldots, g_1^{a'_m}, g_1^{w_1}, \ldots, g_1^{w_m}, g_1^{w'_1}, \ldots, g_1^{w'_m}, g_1^{\delta}, g_1^{\delta'}$$

using the ElGamal encryption scheme and the public key PK as described in Sect. 5.3. Additionally, the user U randomly selects the private key SK_U and computes the public key $PK_U = g^{SK_U}$. The user then generates a query

$$Q = \{A_1, \mathsf{E}(g_1^{a_1}), \mathsf{E}(g_1^{w_1}), \ldots, A_m, \mathsf{E}(g_1^{a_m}), \mathsf{E}(g_1^{w_m}), \mathsf{E}(g_1^{\delta}),$$
$$A'_1, \mathsf{E}(g_1^{a'_1}), \mathsf{E}(g_1^{w'_1}), \ldots, A'_{m'}, \mathsf{E}(g_1^{a'_m}), \mathsf{E}(g_1^{w'_m}), \mathsf{E}(g_1^{\delta'}), PK_U\}$$

and submits Q to the service provider. Given Q, the n matching servers collaborate to execute Algorithm 1 repeatedly to privately share $a_1, w_1, \ldots, a_m, w_m, a'_1, w'_1, \ldots, a'_{m'}, w'_{m'}$.

For each user U_i in the database, the n matching servers work together to determine whether U_i meets the thresholds specified by the querying user U in three steps, as follows:

Step 1: For attributes $A_1, A_2, \ldots, A_m$, the n matching servers cooperate to compute $\mathsf{E}(g_1^{\delta - d_{A_1,\ldots,A_m}(U,U_i)})$ as described in Sect. 5.3. Then, the n servers collaborate to privately share $\delta - d_{A_1,\ldots,A_m}(U, U_i)$ without revealing it by running Algorithm 1, and to determine if $\delta - d_{A_1,\ldots,A_m}(U, U_i) \geq 0$ by executing Algorithm 3. Let the output of Algorithm 3 be $\mathsf{e}(g_2^{\alpha})$, where $\alpha = 0$ if $\delta - d_{A_1,\ldots,A_m}(U, U_i) \geq 0$ and $\alpha = 1$ otherwise.

Step 2: For attributes $A'_1, A'_2, \ldots, A'_{m'}$, the n matching servers collaborate to compute $\mathsf{E}(g_1^{\delta' - d_{A'_1,\ldots,A'_{m'}}(U,U_i)})$ as described in Sect. 5.3. Next, the n servers cooperate to privately share $\delta' - d_{A'_1,\ldots,A'_{m'}}(U, U_i)$ without revealing it by running Algorithm 1, and to determine if $\delta' - d_{A'_1,\ldots,A'_{m'}}(U, U_i) \geq 0$ by executing Algorithm 3. Let the output of Algorithm 3 be $\mathsf{e}(g_2^{\alpha'})$, where $\alpha' = 0$ if $\delta' - d_{A'_1,\ldots,A'_{m'}}(U, U_i) \geq 0$ and $\alpha' = 1$ otherwise.

Step 3: After obtaining $e(g_2^\alpha)$ and $e(g_2^{\alpha'})$, the n servers cooperate to compute

$$e(g_2^{\alpha\alpha'}) = e(g_2^{((2\alpha-1)\alpha'+\alpha')2^{-1}}) = (e(g_2^{(2\alpha-1)\alpha'})e(g_2^{\alpha'}))^{2^{-1}}.$$

Note that $e(g_2^{(2\alpha-1)\alpha'})$ can be computed with a conditional gate as described in Sect. 5.3. Then the n servers cooperate to compute

$$e(g_2^{(1-\alpha)(1-\alpha')}) = e(g_2)e(g_2^{\alpha\alpha'})/(e(g_2^\alpha)e(g_2^{\alpha'})),$$

and decrypt $e(g_2^{(1-\alpha)(1-\alpha')})$ by running Algorithm 4. If the output is 1 (i.e., $e(g_2^{(1-\alpha)(1-\alpha')}) = e(g_2)$, thus $\alpha = \alpha' = 0$), the n servers cooperate to encrypt CI_i by running Algorithm 5, and the social networking service provider returns $E(g_1^{CI_i}, PK_U)$ as the response to the query from user U. After receiving $E(g_1^{CI_i}, PK_U)$ from the service provider, the querying user U runs Algorithm 6 to obtain CI_i.

The above process is repeated until t matching users are found and returned to the querying user. If the total number of matching users is less than t, all matching users are returned to the querying user.

Now, we consider multiple thresholds for user matching in general. The query user selects several groups of attributes, specifies a dissimilarity threshold for each group of attributes, and defines logical relations (including NOT, AND, OR) for these thresholds, generating a query Q accordingly, which is then sent to the social service provider.

Given the query Q and the profile of a user U_i, the n matching servers cooperate to perform Step 1 for each group of attributes and each threshold to obtain $e(g_2^\alpha)$, where $\alpha = 0$ if the dissimilarity threshold is satisfied and 1 otherwise. Based on the logical relations specified by the query user U and the values $e(g_2^\alpha)$, $e(g_2^{\alpha'})$, ... obtained in Step 1 (and Step 2), the n matching servers cooperate to evaluate the logical conditions as in Step 3, using conditional gates described in Sect. 5.3.

When the output of Algorithm 5 satisfies the logical condition, the n matching servers cooperate to compute $E(g_1^{CI_i}, PK_U)$, and the social network provider returns it as the response to the querying user U, who then decrypts the response to obtain CI_i as in Step 3.

User Profile Matching for Categorical Attributes

For numerical attributes, dissimilarities between two users are defined as in Sect. 8.3. However, these dissimilarity definitions do not apply to categorical attributes, such as user nationality, user hobbies, etc. However, for specific queries, such as finding users with the same nationality, we can adapt our user profile matching protocol as follows.

Assume that $A_1, A_2, \ldots, A_m$ are categorical attributes. In the initialization phase, each user U_i maps the value of each categorical attribute A_j to the data a_{ij} of a fixed size using a hash function H, encrypts the values

$a_{i1}, a_{i1}^2, \ldots, a_{im}, a_{im}^2$ as described in Sect. 5.3, and then sends the ciphertexts $\mathsf{E}(g_1^{CI_i}), \mathsf{E}(g_1^{a_{i1}}), \mathsf{E}(g_1^{a_{i1}^2}), \ldots, \mathsf{E}(g_1^{a_{im}}), \mathsf{E}(g_1^{a_{im}^2})$ to the social networking service provider, which stores them in the user profile database DB.

When a user wishes to find people such that $a_{ij} = a_j$ for all $j = 1, 2, \ldots, m$, they send a query to the service provider:

$$Q = \{\mathsf{E}(g_1^{a_1}), w_1 = 2, \ldots, \mathsf{E}(g_1^{a_m}), w_m = 2, \delta = 1, PK_U\}.$$

As described in Sect. 8.4, the n matching servers cooperate to compute $\mathsf{E}(g_1^{(a_j-a_{ij})^2})$ for each $j = 1, 2, \ldots, m$ based on the query Q and the profile of user U_i in the database. Then the n servers compute

$$\prod_{j=1}^{m} \mathsf{E}(g_1^{(a_j-a_{ij})^2})^{w_j} = \mathsf{E}(g_1^{\sum_{j=1}^{m} 2(a_j-a_{ij})^2})$$

and

$$\mathsf{E}(g_1^{1-\sum_{j=1}^{m} 2(a_j-a_{ij})^2}).$$

Next, the n servers cooperate to share $d = 1 - \sum_{j=1}^{m} 2(a_j - a_{ij})^2$ by running Algorithm 1 and then determine if $d \geq 0$ by running Algorithm 3. Assume that the output of Algorithm 3 is $\mathsf{e}(g_2^\alpha)$. If $\alpha = 0$, then $d = 1 - \sum_{j=1}^{m} 2(a_j - a_{ij})^2 \geq 0$, and therefore $a_{ij} = a_j$ for $j = 1, 2, \ldots, m$.

The output of Algorithm 3, i.e., $\mathsf{e}(g_2^\alpha)$, may then be combined with other dissimilarity thresholds for numerical attributes as described in Sect. 8.4 to determine whether the profile of user U_i meets the search criteria.

8.6 Conclusion

In this chapter, we introduced a novel solution for privacy-preserving user profile matching using homomorphic encryption and multiple servers. Our approach enables users to find matching profiles with the assistance of several servers, while ensuring that neither the query nor the user profiles are disclosed.

Security analyses have demonstrated that the proposed protocol effectively safeguards both user profile privacy and query privacy. Additionally, the experimental results confirm that the protocol is both practical and feasible.

Future work will focus on enhancing the performance of conditional gate computation through parallel processing techniques.

References

1. R. Agrawal, A. Evfimievski, and R. Srikant, Information sharing across private databases, in SIGMOD 2003, pp. 86–97.
2. M. von Arb, M. Bader, M. Kuhn, and R. Wattenhofer, Veneta: Serverless friend-of-friend detection in mobile social networking, in IEEE WIMOB 2008, pp. 184–189.
3. B. H. Bloom, Space/time trade-offs in hash coding with allowable errors, Communications of the ACM 13 (7): 422–426, 1970.
4. D. Boneh, E. J. Goh, K. Nissim, Evaluating 2-DNF formulas on ciphertexts, in TCC 2006, pp 325–341.
5. D. Chaum, Blind signatures for untraceable payments, in Crypto 1982, pp. 199–203.
6. E. D. Cristofaro and G. Tsudik, Practical private set intersection protocols with linear complexity, in Financial Cryptography and Data Security 2010.
7. D. Dachman-Soled, T. Malkin, M. Raykova, and M. Yung, Efficient robust private set intersection, in ACNS 2009, pp. 125–142.
8. T. ElGamal, A public-key cryptosystem and a signature scheme based on discrete logarithms, IEEE Transactions on Information Theory 31 (4): 469–472, 1985
9. M. Freedman, K. Nissim, and B. Pinkas, Efficient private matching and set intersection, in EUROCRYPT 2004, pp. 1–19.
10. C. Gentry, Fully homomorphic encryption using ideal lattices, in STOC 2009, pp 169–178.
11. S. Goldwasser and S. Micali, Probabilistic encryption and how to play mental poker keeping secret all partial information, in Proc. 14th Symposium on Theory of Computing, 1982, pp. 365–377.
12. D. Harris, D. M. Harris and S. L. Harris, Digital Design and Computer Architecture, Morgan Kaufmann Publishers, 2007.
13. C. Hazay and Y. Lindell, Efficient protocols for set intersection and pattern matching with security against malicious and covert adversaries, in TCC 2008, pp. 155–175.
14. C. Hazay, G. L. Mikkelsen, T. Rabin, T. Toft, A. A. Nicolosi, Efficient RSA key generation and threshold paillier in the two-party setting, in CT-RSA 2012, pp. 313–331.
15. Y. Huang, L. Malka, D. Evans and J. Katz, Efficient privacy-preserving biometric identification, in NDSS 2011.
16. S. Jarecki and X. Liu, Efficient oblivious pseudorandom function with applications to adaptive OT and secure computation of set intersection, in TCC'09, 2009, pp. 577–594.
17. R. Li and C. Wu, An unconditionally secure protocol for multi-party set intersection, in ACNS 2007, pp. 226–236.
18. M. Li, N. Cao, S. Yu, and W. Lou, FindU: Privacy-preserving personal profile matching in mobile social networks, in IEEE INFOCOM 2010, pp. 1–9.
19. L. Kissner and D. Song, Privacy-preserving set operations, in CRYPTO 2005, pp. 241–257.
20. G. S. Narayanan, T. Aishwarya, A. Agrawal, A. Patra, A. Choudhary, and C. P. Rangan, Multi party distributed private matching, set disjointness and cardinality of set intersection with information theoretic security, in CANS 2009, pp. 21–40.
21. P. Paillier, Public-key cryptosystems based on composite degree residuosity classes, in Proc. 17th Int. Conf. Theory Appl. Cryptograph. Techn., 1999, pp. 223–238.
22. F. Y. Rao, On the security of a variant of ElGamal encryption scheme, IEEE Trans. Dependable and Secure Computing, 2017.
23. Y. Sang, H. Shen, and N. Xiong, Efficient protocols for privacy preserving matching against distributed datasets, in ICICS 2006, pp. 210–227.
24. S. F. Shahandashti, R. Safavi-Naini, P. Ogunbona, Private fingerprint matching, in ACISP 2012, pp 426–433.
25. A. Shamir, How to share a secret, Communications of the ACM 22 (11): 612–613, 1979.
26. L. Sun, L. Zhang, X. Ye, Randomized bit vector: Privacy-preserving encoding mechanism, in CIKM 2018, pp. 1263–1272.

References

27. J. Vaidya and C. Clifton, Secure set intersection cardinality with application to association rule mining, J. Comput. Secur. 13(4): 593–622, 2005.
28. Z. Yang, B. Zhang, J. Dai, A. Champion, D. Xuan, and D. Li, E-smalltalker: A distributed mobile system for social networking in physical proximity, in IEEE ICDCS, 2010.
29. A. C. Yao, Protocols for secure computations, in SFCS 1982.
30. Q. Ye, H. Wang, and J. Pieprzyk, Distributed private matching and set operations, in ISPEC 2008, pp. 347–360.
31. X. Yi, E. Bertino, F. Y. Rao, A. Bouguettaya, Practical privacy-preserving user profile matching in social networks, in ICDE 2016, pp. 373–384.
32. X. Yi, A. Bouguettaya, D. Georgakopoulos, A. Song. Privacy protection for wireless medical sensor data, IEEE Transactions on Dependable and Secure Computing, 13(3): 369–380, 2015.
33. Xun Yi, Elisa Bertino, Fang-Yu Rao, Kwok-Yan Lam, Surya Nepal, Athman Bouguettaya: Privacy-Preserving User Profile Matching in Social Networks. IEEE Trans. Knowl. Data Eng. 32(8): 1572–1585 (2020)

Chapter 9
Private k Nearest Neighbor Queries with Location Privacy

Abstract In mobile communication, spatial queries pose a significant threat to user location privacy, as the location of a query can reveal sensitive information about the mobile user. This chapter focuses on approximate k nearest neighbor (kNN) queries, where a mobile user queries a location-based service (LBS) provider to retrieve the approximate k nearest points of interest (POIs) based on their current location. We introduce two solutions: a basic solution and a generic solution, both aimed at preserving the user's location and query privacy in approximate kNN queries. The proposed solutions leverage the Paillier public-key cryptosystem and ensure both location and query privacy. To safeguard query privacy, the basic solution allows the mobile user to retrieve one type of POI, such as approximate k nearest car parks, without revealing the type of POI being queried to the LBS provider. The generic solution, on the other hand, extends this capability to support multiple discrete attributes in private location-based queries. Compared to existing methods for kNN queries with location privacy, our solutions demonstrate higher efficiency.

9.1 Introduction

The integration of positioning capabilities (e.g., GPS) into mobile devices has facilitated the emergence of location-based services (LBS), which are widely regarded as a transformative application in the wireless data market. LBS enables users to query service providers, such as Google or Bing Maps, in a ubiquitous manner to obtain detailed information about points of interest (POIs) in their vicinity, including restaurants, hospitals, and more.

LBS providers process spatial queries based on the location of the mobile user. However, location data collected from mobile users–either knowingly or unknowingly–can reveal much more than just latitude and longitude. Knowing a user's location can potentially expose their activities, such as attending a religious service, visiting a doctor, shopping for an engagement ring, or participating in a protest. Location data can also reveal who the user interacts with, how frequently, and their routines. When aggregated, this data can provide insights into regular habits and deviations from them, raising significant privacy concerns.

A survey conducted by Microsoft in 2010 across the United Kingdom, Germany, Japan, the United States, and Canada revealed that while 94

In this chapter, we examine approximate k nearest neighbor (kNN) queries, where a mobile user queries an LBS provider for the approximate k nearest POIs based on their current location. Typically, the user must share their location with the LBS provider, which then determines and returns the k nearest POIs by comparing distances between the user's location and nearby POIs. However, this process exposes the user's location to the provider.

Numerous techniques have been developed to provide varying degrees of location privacy, including:

- Information access control [13, 31],
- Mix zones [3],
- k-anonymity [2, 4, 12],
- "Dummy" locations [11, 23, 30],
- Geographic data transformation [9, 10, 27, 28],
- Private Information Retrieval (PIR) [6, 18, 19],
- Two LBS servers [5, 22].

Solutions based on access control, mix zones, and k-anonymity require a trusted third party to maintain user locations, making them vulnerable to misuse by untrusted entities. For instance, k-anonymity, originally designed for identity protection, is often inadequate for location privacy due to its reliance on the density and distribution of users, which are beyond the control of the technique.

Similarly, solutions relying on "dummy" locations require users to generate and transmit fake locations alongside their actual location, which incurs additional computational and communication overhead. For efficiency, users may choose fewer fake locations, which can allow the LBS provider to constrain the user's location to a small subspace, weakening privacy protections.

Geographic data transformation approaches are susceptible to access pattern attacks [26] because repeated queries often yield identical encoded results, potentially revealing query information based on ciphertext frequency. While PIR-based solutions offer strong cryptographic guarantees, they are computationally and communicationally intensive. To mitigate this, trusted hardware has been employed, but such approaches remain vulnerable to third-party misbehavior.

Providing practical solutions for kNN queries with location privacy remains a challenging task. In this chapter, we build upon our prior work [29] presented at ICDE 2014. Using the Paillier public-key cryptosystem, we propose enhanced solutions for private kNN queries. Our main contributions are as follows [32]:

- **Simplified PIR Process:** Unlike existing PIR-based solutions [6, 7, 19, 20], which require two stages of interaction, we propose a method where the user submits their encrypted location once and receives the encrypted k nearest POIs in response, requiring only a single PIR execution.
- **Query Privacy:** We introduce a solution that allows users to query for k nearest POIs of a specific type (e.g., car parks) without revealing the type of POI to the LBS provider, ensuring query privacy.

- **Support for Sequential Queries:** Our solution enables users to retrieve multiple POIs sequentially without executing the protocol multiple times, significantly improving efficiency.
- **Generic Support for Multiple Attributes:** For the first time, we extend PIR-based kNN queries to support multiple discrete type attributes (e.g., car park and daily parking fee categorized into "Low," "Middle," and "High").

We define a security model for private kNN queries and analyze the security of our solutions. The analysis demonstrates that our methods ensure location privacy, query privacy, and data privacy, limiting the LBS provider's knowledge to only the k nearest POIs per query.

We have implemented our solutions using a sample location-based database, and experimental results confirm their practicality. Compared to our prior work [29], the current chapter introduces several key enhancements, including variable k values, separate definitions for location and query privacy, and support for sequential queries and multiple attributes.

The remainder of this chapter is organized as follows. Section 9.2 reviews related work. Backgrounds are introduced in Sect. 9.3. We define our model in Sect. 9.4 and describe our solutions in Sects. ch9:sec5 and 9.6. Conclusions are drawn in the last section.

9.2 Related Works

Several prominent techniques have been developed to preserve location privacy in location-based services (LBS). These are summarized below:

Information Access Control
As proposed in [13, 31], user locations are sent directly to the LBS provider, which is tasked with restricting access to stored location data using rule-based policies. This technique supports three types of location-based queries: (1) *user location queries*, which seek the location of specific users identified by unique identifiers; (2) *enumeration queries*, which list users at specific locations, defined either geographically or symbolically; and (3) *asynchronous queries*, which request event-based information, such as when users enter or leave designated areas. However, this approach requires the LBS provider to maintain all user locations, rendering it vulnerable to potential misbehavior by the provider.

Mix Zones
The mix zone concept [3] employs a trusted middleware to mediate between mobile users and the LBS provider. Before forwarding user queries, the middleware anonymizes locations using pseudonyms. When a user enters a mix zone, a pseudonym is assigned, which is used for querying the LBS. The pseudonym is updated whenever the user re-enters a mix zone. This method, which has also been applied to road networks [17], relies on the middleware to anonymize user locations. However, it remains susceptible to misbehavior of the middleware.

k-Anonymity

The k-anonymity approach [24] ensures that a user's location cannot be distinguished from those of at least $k - 1$ others. Instead of sending an individual user's exact location, k-anonymity schemes aggregate k user locations and submit a minimal bounding region as the query parameter to the LBS. These schemes rely on either a trusted third party [2, 12] or peer-to-peer collaboration [4] for location aggregation. By achieving k-anonymity, the probability of identifying a specific user is limited to $1/k$. Nevertheless, the reliance on third-party or peer collaboration introduces vulnerabilities to their misbehavior.

Dummy Locations

In this method [11, 23], users send multiple fake locations alongside their actual location when querying the LBS. This obfuscates the real location, making it indistinguishable from the fake ones. Unlike k-anonymity, which uses other users' locations, this technique employs randomly generated or fixed locations, such as road intersections. Although this approach does not depend on third parties, it may allow the LBS provider to constrain users to a limited subspace, resulting in weaker privacy guarantees.

Private Information Retrieval (PIR)

PIR techniques [15] enable users to query a database without revealing which records are being retrieved. Protocols [6, 7, 19, 20] for point-of-interest (POI) queries typically consist of two stages: first, users privately determine the index of their location; second, they retrieve POIs using a PIR protocol. The distinction between the methods of Ghinita et al. [6, 7] and Paulet et al. [19, 20] lies in the first stage, where the former employs homomorphic encryption [16] and the latter uses oblivious transfer [14]. Trusted hardware [18] has also been used to support hardware-aided PIR [25]. Despite their benefits, these approaches rely on trusted third parties and remain vulnerable to their misbehavior.

Geographic Data Transformation

This technique [9, 27, 28] involves three parties: a data owner, an LBS provider, and a user. The data owner transforms the database using an encoding methodology before outsourcing it to the LBS provider. Users with transformation keys can issue encoded queries, preserving location privacy. For instance, Wong et al. [27] introduced secure point transformations that maintain relative distances, while Yao et al. [28] allowed approximate nearest neighbor (NN) searches directly on transformed data. However, these techniques are susceptible to access pattern attacks [26], as identical queries yield consistent encoded results.

Two LBS Servers

To mitigate access pattern attacks, solutions have been proposed using two non-colluding LBS servers. For instance, Elmehdwi et al. [5] presented a k-NN query solution based on semantically secure Paillier encryption [16], while Schlegel et al. [22] designed a method for continuous location-based services using separate query and service providers. These approaches assume that the two servers never collude.

Location-Based Alert Systems
Ghinita and Rughinis [8] proposed a location-based alert system where a mobile user continuously sends encrypted location data to the LBS server. The server decrypts and identifies the user's location only if they enter a predefined disaster area, thereby enabling timely alerts while preserving privacy.

9.3 Backgrounds

9.3.1 Paillier Public-Key Cryptosystem

Paillier public-key cryptosystem [16] is composed of three algorithms as follows.

- *Key Generation*: A user randomly chooses two large distinct primes p, q and an element g of $\mathbb{Z}_{N^2}^*$ whose order is a nonzero multiple of $N = pq$, publishes the public key $pk = (N, g)$, and keeps the private key $sk = (p, q)$ secret.
- *Encryption*: Given the public key pk of the user, one can encrypt a message m where m is a positive integer less than N by randomly choosing r from $\mathbb{Z}_{N^2}^*$ and computing

$$c = \mathsf{E}(m, pk) = g^m r^N \ (mod \ N^2) \tag{9.1}$$

where c is the ciphertext of m. Since r is randomly chosen, the ciphertext c of a message m is random. Therefore, Paillier cryptosystem is a probabilistic encryption.

- *Decryption*: The user can decrypt the ciphertext c with the private key sk by computing

$$m = \mathsf{D}(c, sk) = \frac{(c^\lambda (mod \ N^2) - 1)/N}{(g^\lambda (mod \ N^2) - 1)/N} (mod \ N) \tag{9.2}$$

where $\lambda = lcm(p-1, q-1)$.

Homomorphic Properties: Paillier cryptosystem has two homomorphic encryption properties as follows:

$$\mathsf{E}(m_1)\mathsf{E}(m_2) = \mathsf{E}(m_1 + m_2) \tag{9.3}$$

$$\mathsf{E}(m_1)^a = \mathsf{E}(am_1) \tag{9.4}$$

for any $m_1, m_2, m, a \in \mathbb{Z}_N$.

Suppose that $\mathsf{E}(m_i) = g^{m_i} r_i^N \ (mod \ N^2)$ for i=1, 2, it is easy to verify (3) and (4) because

$$\mathsf{E}(m_1)\mathsf{E}(m_2) = g^{m_1+m_2}(r_1 r_2)^N (mod \ N^2) = \mathsf{E}(m_1 + m_2),$$

$$\mathsf{E}(m_1)^a = g^{am_1}(r_1^a)^N (mod \ N^2) = \mathsf{E}(am_1).$$

9.3.2 RSA

RSA [21] is a public-key cryptosystem, composed of three algorithms as follows.

- *Key Generation*: A user randomly chooses two large distinct primes p, q and computes $N = pq$ and $\varphi(N) = (p-1)(q-1)$. Next, he chooses an integer e such that $1 < e < \varphi(N)$ and $gcd(e, \varphi(N)) = 1$, i.e., e and $\varphi(N)$ are coprime, and determines d such that $e \cdot d = 1 \ (mod \ \varphi(N))$ using the extended Euclidean algorithm. Then, he publishes the public key $pk = (e, N)$ and keeps the private key $sk = d$ secret. In addition, p, q, and $\varphi(N)$ must also be kept secret because they can be used to calculate d.
- *Encryption*: Given the public key (e, N) of the user, one can encrypt a message m where m is a positive integer less than N by computing

$$c = \mathsf{E}(m, pk) = m^e \ (mod \ N) \tag{9.5}$$

where c is the ciphertext of m.

- *Decryption*: The user can decrypt the ciphertext c with the private key d by computing

$$m = \mathsf{D}(c, sk) = c^d \ (mod \ N) \tag{9.6}$$

RSA is not a probabilistic encryption scheme. To transform RSA to a probabilistic encryption scheme, we need to add some random bits into the message m before encrypting m with RSA. Optimal Asymmetric Encryption Padding (OAEP) [1] is a padding scheme often used together with RSA encryption.

9.4 Model for Private k Nearest Neighbor Queries

This section discusses a model for location-based service scenarios in mobile environments, illustrated in Fig. 9.1. The primary entities involved are the mobile user, the location-based service (LBS) provider, the base station, and satellites, each serving distinct roles.

- The **mobile user** sends location-based queries to the LBS provider (also referred to as the LBS server) and receives location-based services in return.
- The **LBS provider** delivers location-based services to the mobile user.
- The **base station** facilitates communication between the mobile user and the LBS provider.
- **Satellites** provide location information to the mobile user.

It is assumed that the mobile user can acquire location information from satellites anonymously and that the base station and the LBS provider do not collude to compromise the user's location privacy. Alternatively, an anonymous channel,

9.4 Model for Private k Nearest Neighbor Queries

Fig. 9.1 Location-based service

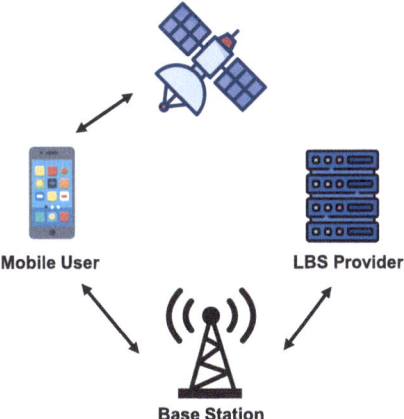

such as Tor,[1] can be used by the mobile user to send queries to and receive services from the LBS provider. This model focuses on protecting both the user's location and query privacy against the LBS provider. A private kNN query protocol (where $k \leq K$ and K is a constant) consists of the following three algorithms:

(1) **Query Generation** (QG): This algorithm takes as input a cloaking region CR with $n \times n$ cells, m distinct types of POIs, the location (i, j) of the mobile user, the type t of POIs, and the number of nearest neighbors k. It outputs a query Q (containing CR) and a secret s, denoted as $(Q, s) = \mathsf{QG}(CR, n, m, (i, j), t, k)$.

(2) **Response Generation** (RG): This algorithm takes as input the query Q and the location-based database D of POIs. It outputs a response R, denoted as $R = \mathsf{RG}(Q, D)$.

(3) **Response Retrieval** (RR): This algorithm takes as input the response R and the secret s of the mobile user. It outputs k nearest POIs of the type t, denoted as $kNN = \mathsf{RR}(R, s)$.

The private kNN query protocol, illustrated in Fig. 9.2, is deemed correct if $kNN = \mathsf{RR}(R, s)$ outputs the k nearest POIs of the type t corresponding to the cell at (i, j), where $(Q, s) = \mathsf{QG}(CR, n, m, (i, j), t, k)$ and $R = \mathsf{RG}(Q, D)$.

9.4.1 Location Privacy

The security of a private kNN query protocol involves ensuring location privacy. Intuitively, the mobile user $\mathcal{U}$ does not want to disclose their location (i, j) to the LBS server, which is considered an adversary.

---

[1] https://www.torproject.org/.

Fig. 9.2 Private kNN query

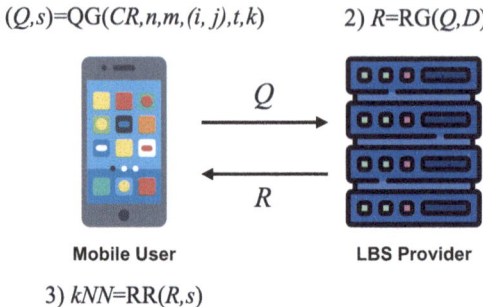

1) $(Q,s)=\text{QG}(CR,n,m,(i,j),t,k)$ 2) $R=\text{RG}(Q,D)$

Mobile User   LBS Provider

3) $kNN=\text{RR}(R,s)$

Location privacy is formally defined using the following game (Game 1):

Given a cloaking region CR with $n \times n$ cells, m types of POIs, and the number of nearest neighbors $k \leq K$, where K is a constant, consider a game between an adversary (the LBS provider) $\mathcal{A}$ and a challenger C. The game involves the following steps:

(1) For a given type of POIs, t, the adversary $\mathcal{A}$ selects two distinct tuples (i_0, j_0, t, k) and (i_1, j_1, t, k), where (i_b, j_b) represents the cell, from the cloaking region CR. These are sent to the challenger C.
(2) The challenger C selects a random bit $b \in \{0, 1\}$ and executes the Query Generation (QG) algorithm to obtain $(Q_b, s) = \text{QG}(CR, n, m, (i_b, j_b), t, k)$. The query Q_b is sent to the adversary $\mathcal{A}$.
(3) The adversary $\mathcal{A}$ analyzes Q_b (potentially using non-black-box methods) and finally outputs a bit $b' \in \{0, 1\}$.

The adversary wins if $b' = b$; otherwise, they lose. The adversary's advantage in this game is defined as:

$$\text{Adv}_{\mathcal{A}}(\kappa) = \left| \Pr(b' = b) - \frac{1}{2} \right|,$$

where κ is the security parameter.

Definition 9.1 (Location Privacy Definition) A kNN query protocol provides location privacy if, for any probabilistic polynomial-time (PPT) adversary $\mathcal{A}$, the advantage $\text{Adv}_{\mathcal{A}}(\kappa)$ is a negligible function, where the probability is taken over the randomness of both the challenger and the adversary.

Remark Location privacy ensures that the server cannot infer the precise location of the mobile user within the cloaking region CR.

9.4.2 Query Privacy

The security of a private kNN query protocol also encompasses query privacy. Here, the mobile user $\mathcal{U}$ seeks to prevent the LBS provider from identifying the type t of POIs the user is querying.

Now, we formally define query privacy using a game-based approach (Game 2) as follows.

Consider a cloaking region CR consisting of $n \times n$ cells, m types of POIs, and the number of nearest neighbors $k \leq K$. The game is played between an adversary (the LBS provider) $\mathcal{A}$ and a challenger C through the following steps:

(1) For any location (i, j), the adversary $\mathcal{A}$ selects two distinct tuples (i, j, t_0, k) and (i, j, t_1, k), where (i, j) represents the cell and t_b specifies the type of POIs. These tuples are then sent to the challenger C.
(2) The challenger C chooses a random bit $b \in \{0, 1\}$ and executes the Query Generation (QG) process to obtain $(Q_b, s) = \mathsf{QG}(CR, n, m, (i, j), t_b, k)$. The query Q_b is then returned to the adversary $\mathcal{A}$.
(3) The adversary $\mathcal{A}$ experiments with the code of Q_b in a non-black-box manner and ultimately outputs a bit $b' \in \{0, 1\}$.

The adversary wins the game if $b' = b$; otherwise, it loses. The advantage of the adversary $\mathcal{A}$ in this game is defined as:

$$\mathsf{Adv}_{\mathcal{A}}(\kappa) = |\mathsf{Pr}(b' = b) - 1/2|,$$

where κ denotes the security parameter.

Definition 9.2 (Query Privacy Definition) A kNN query protocol achieves query privacy if, for any probabilistic polynomial-time (PPT) adversary $\mathcal{A}$, the advantage $\mathsf{Adv}_{\mathcal{A}}(\kappa)$ is a negligible function, where the probability is taken over the random choices of the challenger and the adversary.

Remark Query privacy ensures that the server cannot infer the type of POIs involved in a kNN query from the mobile user.

The security of a private kNN query protocol also involves data privacy. Intuitively, the LBS provider S should release only the k nearest POIs of a single type to the mobile user $\mathcal{U}$ for each kNN query. In this context, the mobile user is considered the adversary.

Formally, data privacy is defined through the following game (Game 3):

Given a user location (i, j), where $1 \leq i, j \leq n$, one type t of POIs, and the number of nearest neighbors $k \leq K$, the game proceeds as follows:

(1) The adversary $\mathcal{A}$ selects two distinct cloaking regions CR_0 and CR_1 with $n \times n$ cells such that the k nearest POIs of type t in cell (i, j) are identical. The adversary generates a query Q to retrieve the k nearest POIs of type t in cell (i, j) and sends Q, CR_0, and CR_1 to the challenger C.

(2) The challenger C chooses a random bit $b \in \{0, 1\}$ and uses the response generation algorithm RG to compute $R_b = \text{RG}(Q, CR_b)$. The response R_b is then sent to the adversary $\mathcal{A}$.
(3) The adversary $\mathcal{A}$ experiments with the code of R_b in a non-black-box manner. If it retrieves the k nearest POIs of type t in cell (i, j) from R_b, it outputs its guess $b' \in \{0, 1\}$.

The adversary wins if $b' = b$; otherwise, it loses. The advantage of the adversary $\mathcal{A}$ is defined as:

$$\text{Adv}_{\mathcal{A}}(\kappa) = |\text{Pr}(b' = b) - 1/2|,$$

where κ is the security parameter.

Definition 9.3 (Data Privacy Definition) A kNN query protocol ensures data privacy if, for any probabilistic polynomial-time (PPT) adversary $\mathcal{A}$, the advantage $\text{Adv}_{\mathcal{A}}(\kappa)$ is a negligible function, where the probability is taken over the random choices of the challenger and the adversary.

Remark Data privacy ensures that the response distributions observed by the user are computationally indistinguishable for any two cloaking regions CR_1 and CR_2, provided the k nearest POIs of type t in cell (i, j) are identical in the two regions. This prevents a computationally bounded user from learning information beyond what is necessary for the query.

Unlike the model in [29], we assume that the LBS server encrypts all k nearest POIs using its public key pk. Consequently, during the Response Retrieval (RR) phase, the mobile user requires assistance from the LBS server to decrypt the k nearest POIs. This setup ensures the user obtains only one kNN POI per query. Additionally, if the user receives a sequence of encrypted k nearest POIs from the LBS server, it can repeatedly execute the RR algorithm to retrieve multiple POIs without needing to generate new queries, significantly enhancing the efficiency of private queries.

During the decryption process, the LBS server must remain unaware of the decrypted results. Otherwise, the server could deduce the location and query of the user. To achieve this, the mobile user employs a blind decryption algorithm with the LBS server as follows:

(1) Given an encrypted record of k nearest POIs, denoted as $C = \text{E}(m, pk)$, the user selects a random number r and computes a blinded ciphertext $C' = \text{F}(C, r)$ using a blinding operation F. The user then sends C' to the LBS server.
(2) The LBS server decrypts C' and returns the blinded plaintext $m' = \text{D}(C', sk)$, where D is the decryption algorithm, and sk is the server's private key.
(3) The user computes the unblinded plaintext $m = \text{G}(m', r)$ using an unblinding operation G.

The security of the blind decryption algorithm relies on its blindness property. Intuitively, this property ensures that the server provides decryption services without learning the actual input or output.

9.5 Basic Private k Nearest Neighbor Queries

Formally, blindness is defined through the following game (Game 4):

(1) The adversary $\mathcal{A}$ (the server) selects two distinct plaintexts m_0 and m_1, encrypts them, and sends the ciphertexts $C_0 = \mathsf{E}(m_0, pk)$ and $C_1 = \mathsf{E}(m_1, pk)$ to the challenger C.
(2) The challenger C (the user) selects a random bit $b \in \{0, 1\}$ and a random number r, then computes the blinded ciphertext $C'_b = \mathsf{F}(C_b, r)$. This ciphertext is returned to the adversary $\mathcal{A}$.
(3) The adversary $\mathcal{A}$ experiments with the code of C'_b in a non-black-box manner and outputs a guess $b' \in \{0, 1\}$.

The adversary wins the game if $b' = b$; otherwise, it loses. The advantage of the adversary $\mathcal{A}$ is defined as:

$$\mathsf{Adv}_{\mathcal{A}}(\kappa) = |\mathsf{Pr}(b' = b) - 1/2|,$$

where κ is the security parameter.

Definition 9.4 (Blindness Definition) The Response Retrieval (RR) algorithm satisfies blindness if, for any probabilistic polynomial-time (PPT) adversary $\mathcal{A}$, the advantage $\mathsf{Adv}_{\mathcal{A}}(\kappa)$ is a negligible function, where the probability is taken over the random choices of the challenger and the adversary.

9.5 Basic Private k Nearest Neighbor Queries

In this section, we present a basic construction for a private kNN query protocol based on our proposed model. This solution utilizes the Paillier encryption scheme [16] and RSA [21].

Initialization
Before initiating any private kNN protocol, an initialization phase takes place at the LBS server.

Initially, the LBS server partitions the location-based database D (a geographic map) into equal-sized cells, such as grids measuring 1 by 1 km. Each cell's center serves as the reference point, and for a given type of POI (Point of Interest), the server identifies K nearest POIs, denoted as $P_1, P_2, \cdots, P_K$. Figure 9.3 illustrates an example with $K = 8$, where each POI is represented as a tuple (x, y), corresponding to the latitude and longitude, respectively.

POI types are encoded numerically as $1, 2, \cdots, m$ and made publicly available. Examples of POI types include Churches, Schools, Post Offices/Postboxes, Telephone Boxes, Restaurants, Pubs, Car Parks, Speed Cameras, Tourist Attractions, among others.

For each cell (i, j) and each POI type t, the LBS server retains the K nearest POIs of type t, represented as an integer $d_{i,j,t}^K$ in a bitstream. The POIs are ordered by their distance from the cell center, with the first k POIs ($k \leq K$) denoted as

Fig. 9.3 k Nearest POIs for cells

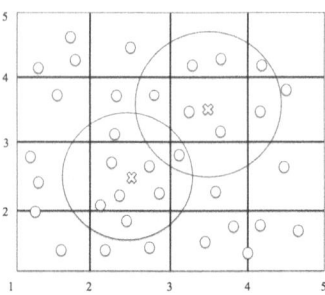

$d_{i,j,t}^k$, being the k nearest POIs. Each cell maintains m integers for different POI types. Additionally, $M(k)$ denotes the maximum length of any record, i.e., $M(k) = \max(d_{i,j,t}^k)$.

Remark The LBS server may construct distinct POI databases for various grid sizes. This approach maintains user and server privacy while offering flexibility. Smaller grids yield more accurate kNN query results but may reduce efficiency in response generation for a given cloaking region. A mobile user can select an appropriate grid size based on their query requirements.

Since the LBS provider identifies K nearest POIs based on the cell center (e.g., the intersections in Fig. 9.3), it returns identical k nearest POIs ($k \leq K$) to any two users within the same cell, irrespective of their exact location. For users near cell boundaries, querying multiple adjacent cells allows them to independently determine the k nearest POIs from the aggregated results. This design avoids private distance comparisons, which could inadvertently reveal the user's location.

The LBS server generates an RSA public and private key pair (pk, sk), where $pk = \{e, N\}$ and $sk = \{d\}$. For a user-specified k, the server encrypts $d_{i,j,t}^k$ to $d_{i,j,t}^{k\,\prime}$ for all i, j, t using the RSA encryption algorithm (described in Sect. 9.3) with Optimal Asymmetric Encryption Padding (OAEP) [1], as follows:

$$d_{i,j,t}^{k\,\prime} = \mathsf{E}(d_{i,j,t}^k, pk) = (d_{i,j,t}^k)^e \mod N \qquad (9.7)$$

Remark The LBS server's public key $pk = \{e, N\}$ is publicly available to all mobile users. It is required that $\log_2 N > 2\log_2 M(k)$ for the kNN query to ensure security. Different k values may necessitate distinct public keys. For instance, assuming a POI's location is represented by 32 bits, $d_{i,j,t}^k$ will typically be less than 1024 bits for $k = 5, 10, 20, 30$ and less than 2048 bits for unusually large k values such as $k = 40, 50$. The bit-length can be further reduced by using latitude and longitude relative to a reference point. For simplicity, we continue using $d_{i,j,t}^k$ to denote the k nearest POIs post-OAEP encryption.

Basic Private kNN Query Protocol

Assume that POI types are encoded as $1, 2, \cdots, m$ and made public. A mobile user $\mathcal{U}$ aims to find the k nearest POIs of type t around their location. The user $\mathcal{U}$ selects

9.5 Basic Private k Nearest Neighbor Queries

a cloaking region CR comprising $n \times n$ cells, where $\mathcal{U}$ resides in cell (i, j), and executes the kNN query protocol with the LBS provider $\mathcal{S}$. The protocol consists of Algorithms 1–3.

Algorithm 1 Query generation (user)

Input: $CR, n, m, (i, j), t, k, pk = \{e, N\}$
Output: Q, s
1: Randomly select two large primes p_1, q_1 such that $N_1 = p_1 q_1 > N$.
2: Randomly select two large primes p_2, q_2 such that $N_2 = p_2 q_2 > N$, with the condition $N_2^2 < N_1$.
3: Set $sk_1 = \{p_1, q_1\}$, $pk_1 = \{g_1, N_1\}$.
4: Set $sk_2 = \{p_2, q_2\}$, $pk_2 = \{g_2, N_2\}$.
5: For each $\ell \in \{1, 2, \cdots, n\}$, randomly choose an integer $r_\ell \in \mathbb{Z}^*_{N_1^2}$, and compute

$$c_\ell = \begin{cases} \mathsf{E}_1(1, pk_1) = g_1^1 r_\ell^{N_1} \pmod{N_1^2} & \text{if } \ell = i \\ \mathsf{E}_1(0, pk_1) = g_1^0 r_\ell^{N_1} \pmod{N_1^2} & \text{otherwise} \end{cases}$$

where E_1 denotes the Paillier encryption algorithm with the public key $pk_1 = \{g_1, N_1\}$, as described in Sect. 9.3.
6: For each $\ell \in \{1, 2, \cdots, m\}$, randomly select an integer $r'_\ell \in \mathbb{Z}^*_{N_2^2}$, and compute

$$c'_\ell = \begin{cases} \mathsf{E}_2(1, pk_2) = g_2^1 r'^{N_2}_\ell \pmod{N_2^2} & \text{if } \ell = t \\ \mathsf{E}_2(0, pk_2) = g_2^0 r'^{N_2}_\ell \pmod{N_2^2} & \text{otherwise} \end{cases}$$

where E_2 denotes the Paillier encryption algorithm with the public key $pk_2 = \{g_2, N_2\}$.
7: Let $Q = \{CR, n, m, k, c_1, c_2, \cdots, c_n, c'_1, c'_2, \cdots, c'_m, pk_1, pk_2\}$ and $s = \{sk_1, sk_2\}$.
8: **return** Q, s

Algorithm 2 Response generation RG (server)

Input: $D, Q = \{CR, n, m, k, c_1, c_2, \cdots, c_n, c'_1, c'_2, \cdots, c'_m, pk_1, pk_2\}$
Output: $R = \{C_1, C_2, \cdots, C_n\}$
1: Based on CR, n, and k, for each t in the cell (i, j), retrieve the first k points of $d^K_{i,j,t}$, denoted as $d^k_{i,j,t}$, and encrypt it using the RSA encryption algorithm described in Sect. 9.3. The result is denoted as $d^{k\,'}_{i,j,t}$.
2: Based on CR, n, and m, for each cell (α, β) in CR, compute

$$C_{\alpha,\beta} = \prod_{\ell=1}^{m} c'^{d^{k\,'}_{\alpha,\beta,\ell}}_\ell \pmod{N_2^2}$$

3: Based on CR and n, compute $R = \{C_1, C_2, \cdots, C_n\}$, where for $\beta \in \{1, 2, \cdots, n\}$,

$$C_\beta = \prod_{\alpha=1}^{n} c_\alpha^{C_{\alpha,\beta}} \pmod{N_1^2}.$$

4: **return** R

Algorithm 3 Response retrieval RR (user)

Input: $R = \{C_1, C_2, \cdots, C_n\}, s = \{sk_1, sk_2\}, sk = \{d\}$
Output: z
1: The user randomly selects an integer $r < N$ and computes and sends to the server the following:

$$w = r^e D_2(D_1(C_j, sk_1), sk_2) \mod N,$$

where D_1, D_2 represent the Paillier decryption algorithm, as described in Sect. 9.3.
2: The server computes and returns to the user:

$$v = D(w, sk) = w^d \mod N,$$

where D denotes the RSA decryption algorithm, as detailed in Sect. 9.3.
3: The user calculates:

$$z = r^{-1} v \mod N.$$

4: **return** z

Remark The CR may be defined by the coordinates (x, y) of an origin point and the order n of a square grid. The cell containing the origin point is labeled as $(1, 1)$, and the CR spans from the cell $(1, 1)$ to the cell (n, n).

Remark By making slight modifications to Algorithms 1–3, our protocol can be adjusted to preserve only location or query privacy.

If the user desires to maintain location privacy only (i.e., to keep (i, j) private in the query), the user executes Steps 1, 3, and 5 in Algorithm 1 and submits $Q = \{CR, n, m, k, c_1, c_2, \cdots, c_n, t, pk_1\}$ to the LBS server. Given t, the LBS server performs Steps 1 and 3 in Algorithm 2, where $C_{\alpha,\beta} = d^k_{\alpha,\beta,t}$ for $\beta = 1, 2, \cdots, n$, and returns $R = \{C_1, C_2, \cdots, C_n\}$ to the user. During the response retrieval phase, the user computes $w = r^e D_1(C_j, sk_1) \mod N$, where r is a random integer, and performs Steps 2 and 3 to retrieve the k nearest neighbors (kNN) of the type t.

If the user wants to preserve only query privacy (i.e., to keep the type t private in the query), given the location (i, j) of the user, they execute Steps 2, 4, and 6 in Algorithm 1 and submit $Q = \{CR, n, m, k, (i, j), c'_1, c'_2, \cdots, c'_m, pk_2\}$ to the LBS server. Since (i, j) is known, the LBS server performs Steps 1 and 2 in Algorithm 2, where $\alpha = i$ and $\beta = j$, and returns $R = \{C_{i,j}\}$ to the user. In the response retrieval phase, the user computes $w = r^e D_2(C_{i,j}, sk_2) \mod N$, where r is a random integer, and performs Steps 2 and 3 to retrieve the kNN of the type t.

Remark In Algorithm 3, when the mobile user receives the response, they can ignore C_ℓ for $\ell \neq j$ and only receive C_j because only C_j contains information about the k nearest POIs in the cell (i, j). In fact, the mobile user can retrieve the k nearest POIs of type t in any cell (i, γ), where $\gamma = 1, 2, \cdots, n$, by executing Algorithm 3 without needing to generate a new query or response. This functionality makes private queries highly efficient when the mobile user moves from one cell, such as (i, j), to another, such as $(i, j + 1)$ (where $j + 1 \leq n$) or $(i, j - 1)$ (where $j - 1 \geq 1$). Additionally, the LBS server can easily control the release of data by decryption, since each response retrieval invocation reveals only a single data point.

Theorem 9.1 (Correctness) *The basic kNN query protocol (Algorithms 1–3) is correct. Specifically, for any cloaking region CR with $n \times n$ grid and m types of POIs, for any cell (i, j) with $1 \leq i, j \leq n$, for any POI type t, and for any number of nearest neighbors $k \leq K$, we have:*

$$d^k_{i,j,t} = \mathsf{RR}(R, s),$$

where $d^k_{i,j,t}$ represents the k nearest POIs of type t for the center of the cell (i, j), and $(Q, sk) = \mathsf{QG}(CR, n, m, (i, j), t, k)$, $R = \mathsf{RG}(Q, D)$.

Proof Based on Algorithms 1 and 2, we have the following:

$$C_j = \prod_{\alpha=1}^{n} c_\alpha^{C_{\alpha,j}} = g_1^{C_{i,j}} \left(\prod_{\alpha=1}^{n} r_\alpha^{C_{\alpha,j}} \right)^{N_1} \mod N_1^2$$

which is a Paillier encryption of $C_{i,j}$. Therefore, $\mathsf{D}_1(C_j, sk_1) = C_{i,j}$.

Similarly, based on Algorithms 1 and 2, we have:

$$C_{i,j} = \prod_{\ell=1}^{m} c_\ell^{\prime d^{k'}_{i,j,\ell}} = g_2^{d^{k'}_{i,j,t}} \left(\prod_{\ell=1}^{m} r_\ell^{\prime d^{k'}_{i,j,\ell}} \right)^{N_2} \mod N_2^2$$

which is a Paillier encryption of $d^{k'}_{i,j,t}$. Therefore, $\mathsf{D}_2(\mathsf{D}_1(C_j, sk_1), sk_2) = d^{k'}_{i,j,t}$.

Finally, based on Algorithm 3, we have:

$$z = r^{-1} v = r^{-1} w^d$$
$$= r^{-1} \left(r^e d^{k}_{i,j,t}{}' \right)^d = r^{-1} r^{ed} d^{k}_{i,j,t}{}^d$$
$$= r^{-1} r d^{k}_{i,j,t} = d^{k}_{i,j,t} \mod N$$

Thus, we conclude that $d^k_{i,j,t} = \mathsf{RR}(R, s)$, and the theorem is proven. △

9.6 Genetic Private k Nearest Neighbor Queries

In this section, we extend our basic approach to present a generalized construction for a private kNN query protocol. Our generalized solution is designed for a multi-dimensional space, where each Point of Interest (POI) is characterized by location attributes (i, j) (where $1 \leq i, j \leq n$) and multiple discrete type attributes $(t_1, t_2, \cdots, t_T)$ (where t_λ is an integer and $1 \leq t_\lambda \leq m_\lambda$). For example, a car park (encoded as 3) located at (9,4) in the cell (8,5) with a "Mid" daily parking fee (encoded as 1) can be represented as $d^1_{8,5,3,1} = $ POI (9, 4), where the daily parking fee can be categorized as "Low" (<$10), "Mid" ($10-$30), or "High" (>$30)).

Initialization

As with our basic solution, the LBS server first divides the geographic database D into cells of equal size. For each cell centered at (i, j), the LBS server collects K nearest POIs of the specified types, $P_1, P_2, \ldots, P_K$, with each point represented by a tuple (x, y).

For each cell (i, j) and each POI type attribute $(t_1, t_2, \cdots, t_T)$, the LBS server retains K (e.g., $K = 20$) nearest POIs, represented as a stream of bits, denoted as an integer $d^K_{i,j,t_1,t_2,\cdots,t_T}$, where the points are ordered according to their distance from the center of the cell. The first k points, denoted as $d^k_{i,j,t_1,t_2,\cdots,t_T}$ (where $k \leq K$), represent the k nearest POIs. Each cell contains $\prod_{\lambda=1}^{T} m_\lambda$ integers corresponding to different type attributes of the POIs. We define $M(k)$ as the maximum value of $d^k_{i,j,t_1,t_2,\cdots,t_T}$, i.e., the longest record.

Next, the LBS server generates the RSA public and private key pair (pk, sk), where $pk = \{e, N\}$ and $sk = \{d\}$. Depending on the value of k specified by the user, the LBS server encrypts $d^k_{i,j,t_1,t_2,\cdots,t_T}$ to $d^{k'}_{i,j,t_1,t_2,\cdots,t_T}$ for all $i, j, t_1, t_2, \cdots, t_T$ according to the RSA encryption algorithm (described in Sect. 9.3) and Optimal Asymmetric Encryption Padding (OAEP) [1], as shown below:

$$d^{k'}_{i,j,t_1,t_2,\cdots,t_T} = \mathsf{E}(d^k_{i,j,t_1,t_2,\cdots,t_T}, pk)$$
$$= (d^k_{i,j,t_1,t_2,\cdots,t_T})^e \mod N \quad (9.8)$$

Generic Private kNN Query Protocol

We assume that the mobile user $\mathcal{U}$ wishes to find the k nearest POIs of type attributes $t_1, t_2, \cdots, t_T$ around their location. The user $\mathcal{U}$ chooses a cloaking region CR with $n \times n$ cells, where $\mathcal{U}$ is located in the cell (i, j), and then runs the kNN query protocol with the LBS provider $\mathcal{S}$, following the procedures outlined in Algorithms 4–6.

Theorem 9.2 (Correctness) *The correctness of our generic kNN query protocol (Algorithms 4–6) is as follows: For any cloaking region CR of size $n \times n$, the index i, j of any cell (where $1 \leq i, j \leq n$), any type attributes $(t_1, t_2, \cdots, t_T)$ of POIs, and any number of nearest neighbors $k \leq K$, the following holds:*

$$d^k_{i,j,t_1,t_2,\cdots,t_T} = \mathsf{RR}(R, s),$$

where $d^k_{i,j,t_1,t_2,\cdots,t_T}$ represents the k nearest POIs of the type attributes $(t_1, t_2, \cdots, t_T)$ with respect to the center of cell (i, j), and $(Q, s) = \mathsf{QG}(CR, n, m_1, m_2, \cdots, m_T, (i, j), (t_1, t_2, \cdots, t_T), k)$, and $R = \mathsf{RG}(Q, D)$.

9.6 Genetic Private k Nearest Neighbor Queries

Algorithm 4 Query generation (user)

Input: $CR, n, m_1, m_2, \cdots, m_T, k, (i, j), (t_1, t_2, \cdots, t_T), pk = \{e, N\}$
Output: Q, s

1: Randomly choose two large distinct primes p_1, q_1 such that $N_1 = p_1 q_1 > N$.
2: For each $\lambda \in \{1, 2, \cdots, T\}$, randomly choose two large distinct primes $p_{2\lambda}, q_{2\lambda}$ such that $N_{2\lambda} = p_{2\lambda} q_{2\lambda} > N$, where $N_{2\lambda}^2 < N_{2(\lambda-1)}$ and $N_{20} = N_1$.
3: Let $sk_1 = \{p_1, q_1\}$, $pk_1 = \{g_1, N_1\}$.
4: For each $\lambda \in \{1, 2, \cdots, T\}$, let $sk_{2\lambda} = \{p_{2\lambda}, q_{2\lambda}\}$, $pk_{2\lambda} = \{g_{2\lambda}, N_{2\lambda}\}$.
5: For each $\ell \in \{1, 2, \cdots, n\}$, pick a random integer $r_\ell \in \mathbb{Z}_{N_1^2}^*$, compute

$$c_\ell = \begin{cases} \mathsf{E}_1(1) = g_1{}^1 r_\ell^{N_1} \mod N_1^2 & \text{if } \ell = i \\ \mathsf{E}_1(0) = g_1{}^0 r_\ell^{N_1} \mod N_1^2 & \text{otherwise} \end{cases}$$

where E_1 denotes the Paillier encryption algorithm with public key $pk_1 = \{g_1, N_1\}$ as described in Sect. 9.3.

6: For each $\lambda \in \{1, 2, \cdots, T\}$ and each $\ell \in \{1, 2, \cdots, m_T\}$, pick a random integer $r'_{\lambda\ell} \in \mathbb{Z}_{N_{2\lambda}^2}^*$, compute

$$c'_{\lambda\ell} = \begin{cases} \mathsf{E}_{2\lambda}(1) = g_{2\lambda}{}^1 r'_{\lambda\ell}{}^{N_{2\lambda}} \mod N_{2\lambda}^2 & \text{if } \ell = t_\lambda \\ \mathsf{E}_{2\lambda}(0) = g_{2\lambda}{}^0 r'_{\lambda\ell}{}^{N_{2\lambda}} \mod N_{2\lambda}^2 & \text{otherwise} \end{cases}$$

where $\mathsf{E}_{2\lambda}$ denotes the Paillier encryption algorithm with public key $pk_{2\lambda} = \{g_{2\lambda}, N_{2\lambda}\}$ as described in Sect. 9.3

7: Let $Q = \{CR, n, m_1, m_2, \cdots, m_T, k, c_1, c_2, \cdots, c_n, c'_{11}, c'_{12}, \cdots, c'_{1m_1}, c'_{21}, c'_{22}, \cdots, c'_{2m_2}, \cdots, c'_{T1}, c'_{T2}, \cdots, c'_{Tm_T}, pk_1, pk_{21}, pk_{22}, \cdots, pk_{2T}\}$, $s = \{sk_1, sk_{21}, sk_{22}, \cdots, sk_{2T}\}$.
8: **return** Q, s

Proof Based on Algorithms 4–5, we can derive the following:

$$C_{\alpha,\beta,\ell_1,\cdots,\ell_{T-1}} = \prod_{\ell=1}^{m_T} c'_{T\ell}{}^{d^k_{\alpha,\beta,\ell_1,\cdots,\ell_{T-1},\ell}}$$

$$= c'_{Tt_T}{}^{d^k_{\alpha,\beta,\ell_1,\cdots,\ell_{T-1},t_T}} \pmod{N_{2T}{}^2}$$

$$= \mathsf{E}_{2T}(d^k_{\alpha,\beta,\ell_1,\cdots,\ell_{T-1},t_T})$$

for all possible $\ell_1, \ell_2, \cdots, \ell_{T-1}$.

$$C_{\alpha,\beta,\ell_1,\cdots,\ell_{T-2}} = \prod_{\ell=1}^{m_{T-1}} c'_{(T-1)\ell}{}^{C_{\alpha,\beta,\ell_1,\cdots,\ell_{T-2},\ell}}$$

$$= c'_{(T-1)t_{T-1}}{}^{C_{\alpha,\beta,\ell_1,\cdots,t_{T-1}}} \pmod{N_{2(T-1)}{}^2}$$

$$= \mathsf{E}_{2(T-1)}(\mathsf{E}_{2T}(d^k_{\alpha,\beta,\ell_1,\cdots,\ell_{T-2},t_{T-1},t_T}))$$

Algorithm 5 Response generation RG (server)

Input: $D, Q = \{CR, n, m_1, m_2, \cdots, m_T, k, c_1, c_2, \cdots, c_n, c'_{11}, c'_{12}, \cdots, c'_{1m_1}, c'_{21}, c'_{22}, \cdots, c'_{2m_2}, \cdots, c'_{T1}, c'_{T2}, \cdots, c'_{Tm_T}, pk_1, pk_{21}, pk_{22}, \cdots, pk_{2T}\}$

Output: $R = \{C_1, C_2, \cdots, C_n\}$

1: Based on $CR, n, m_1, m_2, \cdots, m_T, k$, for each cell (α, β) $(1 \leq \alpha, \beta \leq n)$ in CR:
2: Let $C_{\alpha,\beta,\ell_1,\cdots,\ell_T} = d^k_{\alpha,\beta,\ell_1,\cdots,\ell_{T-1},\ell_T}{}'$ for all possible $\ell_1, \ell_2, \cdots, \ell_T$.
3: For $\lambda = 1$ to $T - 1$:
4: For all possible $\ell_1, \ell_2, \cdots, \ell_{T-\lambda}$, compute

$$C_{\alpha,\beta,\ell_1,\cdots,\ell_{T-\lambda}} = \prod_{\ell=1}^{m_{T-\lambda+1}} c'_{(T-\lambda+1)\ell}{}^{C_{\alpha,\beta,\ell_1,\cdots,\ell_{T-\lambda},\ell}} \mod N^2_{2(T-\lambda+1)}$$

5: Compute

$$C_{\alpha,\beta} = \prod_{\ell=1}^{m_1} c'_{1\ell}{}^{C_{\alpha,\beta,\ell}} \mod N^2_{21}$$

6: Based on CR and n, compute $R = \{C_1, C_2, \cdots, C_n\}$, where for $\beta \in \{1, 2, \cdots, n\}$,

$$C_\beta = \prod_{\alpha=1}^{n} c_\alpha{}^{C_{\alpha,\beta}} \mod N^2_1$$

7: **return** R

Algorithm 6 Response retrieval RR (user)

Input: $R = \{C_1, C_2, \cdots, C_n\}, s = \{sk_1, sk_{21}, sk_{22}, \cdots, sk_{2T}\}, sk = \{d\}$
Output: z

1: The user randomly chooses an integer $r < N$ and computes, then sends to the server:

$$w = r^e \mathsf{D}_{2T}(\cdots(\mathsf{D}_{21}(\mathsf{D}_1(C_j, sk_1), sk_{21})\cdots), sk_{2T}) \mod N$$

where $\mathsf{D}_1, \mathsf{D}_{21}, \cdots, \mathsf{D}_{2T}$ are the Paillier decryption algorithms described in Sect. 9.3.

2: The server computes and replies to the user:

$$v = \mathsf{D}(w, sk) = w^d \mod N$$

where D denotes the RSA decryption algorithm described in Sect. 9.3.

3: The user computes:

$$z = r^{-1} v \mod N$$

4: **return** z

9.6 Genetic Private k Nearest Neighbor Queries

for all possible $\ell_1, \ell_2, \cdots, \ell_{T-2}$.

$$\cdots\cdots\cdots\cdots\cdots\cdots\cdots\cdots\cdots\cdots\cdots\cdots$$

$$C_{\alpha,\beta,\ell_1} = \prod_{\ell=1}^{m_2} c'^{C_{\alpha,\beta,\ell_1,\ell}}_{2\ell}$$

$$= c'^{C_{\alpha,\beta,\ell_1,t_2}}_{2t_2} \pmod{N_{22}^2}$$

$$= \mathsf{E}_{22}(\mathsf{E}_{23}(\cdots(\mathsf{E}_{2T}(d^{k'}_{\alpha,\beta,\ell_1,t_2,\cdots,t_T}))))$$

for all possible ℓ_1.

$$C_{\alpha,\beta} = \prod_{\ell=1}^{m_1} c'^{C_{\alpha,\beta,\ell}}_{1\ell}$$

$$= c'^{C_{\alpha,\beta,t_1}}_{1t_1} \pmod{N_{21}^2}$$

$$= \mathsf{E}_{21}(\mathsf{E}_{22}(\cdots(\mathsf{E}_{2T}(d^{k'}_{\alpha,\beta,t_1,t_2,\cdots,t_T}))))$$

In addition, we have

$$C_j = \prod_{\alpha=1}^{n} c_\alpha^{C_{\alpha,j}} = g_1^{C_{i,j}} \left(\prod_{\alpha=1}^{n} r_\alpha^{C_{\alpha,j}}\right)^{N_1} \pmod{N_1^2}$$

which represents a Paillier encryption of $C_{i,j}$. Therefore, $\mathsf{D}_1(C_j, sk_1) = C_{i,j} = \mathsf{E}_{21}(\mathsf{E}_{22}(\cdots(\mathsf{E}_{2T}(d^{k'}_{i,j,t_1,t_2,\cdots,t_T}))))$.

Based on Algorithm 6, we derive

$$w = r^e d^{k'}_{\alpha,\beta,t_1,t_2,\cdots,t_T}$$

Thus, we obtain

$$z = r^{-1}v = r^{-1}w^d$$

$$= r^{-1}\left(r^e d^{k'}_{i,j,t_1,t_2,\cdots,t_T}\right)^d = r^{-1}r^{ed}d^{k}_{i,j,t_1,t_2,\cdots,t_T}{}^d$$

$$= r^{-1}rd^{k}_{i,j,t_1,t_2,\cdots,t_T} = d^{k}_{i,j,t_1,t_2,\cdots,t_T} \pmod{N}$$

Hence, we conclude that $d^{k}_{i,j,t_1,t_2,\cdots,t_T} = \mathsf{RR}(R, s)$, and the theorem is proven. △

Remark Similar to our basic solution, the generic solution can protect location and query privacy with only slight modifications to Algorithms 4–6.

Specifically, the generic solution can be adapted to preserve query privacy for partial type attributes. For instance, if a user wishes to find the k nearest POIs around their location (i, j) with type attributes $(t_1, t_2, \cdots, t_S, t_{S+1}, \cdots, t_T)$, where the location (i, j) and partial type attributes $(t_1, t_2, \cdots, t_S)$ need to remain private from the LBS server, the user can run Algorithm 1. In this case, T is replaced by S, and the following query is sent to the LBS provider:

$Q = \{CR, n, m_1, m_2, \cdots, m_S, k, c_1, c_2, \cdots, c_n, c'_{11}, c'_{12}, \cdots, c'_{1m_1}, c'_{21}, c'_{22},$
$\cdots, c'_{2m_2}, \cdots, c'_{S1}, c'_{S2}, \cdots, c'_{Sm_S}, t_{S+1}, \cdots, t_T, pk_1, pk_{21}, pk_{22}, \cdots, pk_{2S}\}$

Based on Q, the LBS server computes:

$$C_{\alpha,\beta,\ell_1,\cdots,\ell_{S-1}} = \prod_{\ell=1}^{m_S} c'^{d^k_{\alpha,\beta,\ell_1,\cdots,\ell_{S-1},\ell,t_{S+1},\cdots,t_T}}_{S\ell} = c'^{d^k_{\alpha,\beta,\ell_1,\cdots,\ell_{S-1},t_S,\cdots,t_T}}_{S t_S} \pmod{N_{2S}^2}$$

$$\cdots$$

$$C_{\alpha,\beta} = \prod_{\ell=1}^{m_1} c'^{C_{\alpha,\beta,\ell}}_{1\ell}$$

$$C_\beta = \prod_{\alpha=1}^{n} c_\alpha^{C_{\alpha,\beta}} \pmod{N_1^2}$$

The LBS server then returns $(C_1, C_2, \cdots, C_n)$ to the user, who computes:

$$w = r^e D_{2S}(\cdots(D_{21}(D_1(C_j, sk_1), sk_{21})\cdots sk_{2S})) \pmod{N}$$

Finally, the user executes Steps 2 and 3 of Algorithm 6 and obtains $d^k_{i,j,t_1,t_2,\cdots,t_T}$.

9.7 Conclusion

In this chapter, we have introduced both a basic and a generic approximate kNN query protocol. The security analysis demonstrates that our protocols ensure location privacy, query privacy, and data privacy. From a performance perspective, our basic protocol outperforms existing PIR-based LBS query protocols, particularly in terms of parallel computation and communication overhead. Experimental evaluations further confirm that our basic protocol is practical and feasible. In future work, we plan to implement our protocol on mobile devices.

References

1. M. Bellare and P. Rogaway. Optimal asymmetric encryption - how to encrypt with RSA. In Proc. Eurocrypt 1994.

2. B. Bamba, L. Liu, P. Pesti, and T. Wang. Supporting anonymous location queries in mobile environments with PrivacyGrid. In Proc. WWW 2008.
3. A. R. Beresford and F. Stajano. Location privacy in pervasive computing. IEEE Pervasive Computing 2(1), 2003.
4. C. Y. Chow, M. F. Mokbel, and X. Liu. A peer-to-peer spatial cloaking algorithm for anonymous location-based services. In Proc. ACM GIS 2006.
5. Y. Elmehdwi, B. K. Samanthula, W. Jiang. Secure k-nearest neighbor query over encrypted data in outsourced environments. In Proc. ICDE 2014.
6. G. Ghinita, P. Kalnis, A. Khoshgozaran, C. Shahabi, and K.-L. Tan. Private queries in location-based services: Anonymizers are not necessary. In Proc. ACM SIGMOD 2008.
7. G. Ghinita, P. Kalnis, M. Kantarcioglu, and E. Bertino. Approximate and exact hybrid algorithms for private nearest-neighbor queries with database protection. GeoInformatica 15(14): 699–726, 2010.
8. G. Ghinita, R. Rughinis. An efficient privacy-reserving system for monitoring mobile users: making searchable encryption practical. in Proc. ACM CODASPY 2014.
9. Haibo Hu, Jianliang Xu, Chushi Ren, and Byron Choi, Processing Private Queries over Untrusted Data Cloud through Privacy Homomorphism, In Proc. ICDE 2011.
10. A. Khoshgozaran and C. Shahabi. Blind evaluation of nearest neighbor queries using space transformation to preserve location privacy. In Proc. SSTD 2007.
11. H. Kido, Y. Yanagisawa, and T. Satoh. An anonymous communication technique using dummies for location-based services. In Proc. ICPS 2005, pages 88–97.
12. M. F. Mokbel, C.-Y. Chow, and W. G. Aref. The new casper: query processing for location services without compromising privacy. In Proc. VLDB 2006.
13. G. Myles, A. Friday, and N. Davies. Preserving privacy in environments with location-based applications. IEEE Pervasive Computing 2(1):56–64, 2003.
14. M. Naor and B. Pinkas. Oblivious transfer with adaptive queries. In Proc. CRYPTO 1999, pages 791–791.
15. R. Ostrovsky and W. Skeith. A survey of single-database private information retrieval: techniques and applications. In Proc. PKC 2007, pages 393–411.
16. P. Paillier. Public key cryptosystems based on composite degree residue classes. In Proc. EUROCRYPT 1999, pages 223–238.
17. B. Palanisamy and L. Liu. Mobimix: Protecting location privacy with mix-zones over road networks. In Proc. ICDE 2011, pages 494–505.
18. S. Papadopoulos, S. Bakiras, D. Papadias. Nearest neighbor search with strong location privacy. In Proc. VLDB 2010.
19. R. Paulet, M. Golam Kaosar, X. Yi, and E. Bertino. Privacy-preserving and content-protecting location based queries. In Proc. ICDE 2012, pages 44–53.
20. R. Paulet, M. Golam Kaosar, X. Yi, and E. Bertino. Privacy-preserving and content-protecting location based queries. IEEE Transactions on Knowledge and Data Engineering, accepted in 2013.
21. R. Rivest, A. Shamir and L. Adleman. A method for obtaining digital signatures and public-key cryptosystems. Communications of the ACM 21 (2): 120–126, 1978.
22. R. Schlegel, C. Chow, Q. Huang, D. Wong. User-defined privacy grid system for continuous location-based services. IEEE Transactions on Mobile Computing, Jan. 2015.
23. P. Shankar, V. Ganapathy and L. Iftode. Privately querying location-based services with SybilQuery. In Proc. Ubicomp 2009, pages 31–40.
24. L. Sweeney. k-anonymity: a model for protecting privacy. Int. J. Uncertain. Fuzziness Knowl.-Based Syst. 10: 557–570, 2002.
25. S. Wang, X. Ding, R. H. Deng, and F. Bao. Private information retrieval using trusted hardware. In Proc. ESORICS 2006.
26. P. Williams and R. Sion. Usable PIR. In NDSS, 2008.
27. W. K. Wong, D. W. Cheung, B. Kao, and N. Mamoulis. Secure kNN computation on encrypted databases. In Proc. SIGMOD 2009.
28. B. Yao, F. Li, and X. Xiao. Secure nearest neighbor revisited. In Proc. ICDE 2013.

29. X. Yi, R. Paulet, E. Bertino, V. Varadharajan. Practical k nearest neighbor queries with location privacy. In Proc. ICDE 2014. pages 640–651.
30. M. L. Yiu, C. Jensen, X. Huang, and H. Lu. SpaceTwist: Managing the trade-offs among location privacy, query performance, and query accuracy in mobile systems. In Proc. ICDE 2008.
31. M. Youssef, V. Atluri, and N. R. Adam. Preserving mobile customer privacy: An access control system for moving objects and custom problems. In Proc. MDM 2005.
32. X Yi, R Paulet, E Bertino, V Varadharajan. Practical approximate k nearest neighbor queries with location and query privacy IEEE Transactions on Knowledge and Data Engineering 28 (6), 1546–1559, 2016.

Chapter 10
Differentially Private Distributed Frequency Estimation

Abstract In order to remain competitive, Internet companies often collect and analyze user data to enhance user experiences. Frequency estimation is a widely used statistical technique, but it can potentially conflict with privacy regulations. To address this, privacy-preserving analytical methods based on differential privacy have been proposed. These methods typically require either a large user base or a trusted server. While these requirements may be feasible for larger companies, they may present challenges for smaller organizations. To overcome this limitation, we introduce a distributed, privacy-preserving, sampling-based frequency estimation method in this chapter. This method maintains high accuracy, even with a small number of users, and does not require a trusted server. The approach combines multi-party computation and sampling techniques to achieve this. Additionally, we establish a relationship between the privacy guarantee, output accuracy, and the number of participants. Unlike most existing methods, our approach offers a *centralized* differential privacy guarantee without the need for a trusted server. Our results demonstrate that, even with a small number of participants, our method can produce estimates with high accuracy. This provides smaller companies with greater opportunities for growth through privacy-preserving statistical analysis. Furthermore, we propose an architectural model to support weighted aggregation, which improves the accuracy of estimates by accommodating users with varying privacy requirements. Our method outperforms unweighted aggregation, delivering more precise estimates. Extensive experimental results confirm the effectiveness of the proposed methods.

10.1 Introduction

The rapid growth of the digital economy has led to a significant increase in the adoption of Internet applications and services, which are now used by billions of people worldwide. To remain competitive and sustain growth in this highly dynamic digital landscape, service and application providers must continuously adapt their offerings to improve user experience. One way to achieve this is by analyzing usage data using various data analytical and statistical techniques. Frequency estimation is a fundamental statistical tool that is commonly employed to analyze categorical data within a fixed domain space [20]. It allows companies to estimate the distribution of user data, providing valuable insights into data trends and user preferences. Frequency estimation has been widely utilized by many companies. For example, Google analysts have used frequency estimation to study users' web browsing behavior [3], and Apple researchers have applied it to analyze user preferences regarding various system features, such as emojis, health data types, and media playback [38].

Frequency estimation also serves as a critical building block for many advanced data analysis and machine learning techniques, including ranking [41], feature selection [40], and natural language processing [42]. While frequency estimation is valuable for decision-making processes, improper use of this technique can compromise user privacy, especially when handling sensitive information. Ensuring the privacy of user data is a growing concern, and various privacy protection regulations have been established to address this issue. In order to meet privacy expectations while still providing accurate estimates, a well-designed privacy-preserving mechanism must be employed.

Differential privacy (DP) has become a standard privacy measure widely used to quantify the level of privacy protection provided by various mechanisms. Two main types of differential privacy have been proposed: centralized DP and local DP. Local DP has garnered more attention recently and is widely deployed by large companies. In a local setting, each user perturbs their private data before transmitting it to the server. This ensures that no user's private data ever leaves their device, which aligns with the privacy expectations of users when contributing their data. However, local DP mechanisms typically require a large number of participants in order to produce useful results while maintaining a satisfactory level of privacy guarantee [3–5, 30, 37].

In contrast, centralized DP mechanisms can achieve high accuracy even with a smaller number of participants. However, these mechanisms typically require a trusted server that stores users' private data in its raw form, performing privacy-preserving operations to prevent information leakage to clients querying the dataset. This requirement implies that small companies may struggle to perform accurate privacy-preserving frequency estimation on their users' data due to either an insufficient number of users or users' reluctance to trust the server.

In this work, we aim to design a *centrally* differentially private mechanism that addresses the frequency estimation problem, and operates effectively without the need for a large user base or a trusted server [1]. This approach empowers smaller companies to conduct user analysis, thereby leveling the playing field and providing them with greater growth opportunities. Specifically, we propose a sampling-based frequency estimation solution that ensures high accuracy while preserving user privacy, even in scenarios with a limited number of users and without the need for a trusted server. To achieve this, we combine multiparty computation techniques with differential privacy mechanisms, thus providing a centralized differential privacy guarantee without relying on a trusted server.

10.1 Introduction

In selecting the differential privacy mechanism for this study, we focus on sampling methods due to their efficiency advantages. Sampling-based mechanisms are generally simpler than other differential privacy methods, such as the Gaussian mechanism, which is more commonly used in such applications. As a result, sampling-based mechanisms typically have lower computational and communication complexities. Moreover, as demonstrated in our analysis, when the privacy level is kept constant, the efficiency improvements offered by sampling-based mechanisms often result in better estimation accuracy. These considerations motivate the exploration of sampling-based frequency estimation mechanisms in this work.

Considering the approach discussed above, this work establishes a theoretical relationship between the privacy level, accuracy requirement, and the number of users in a classical sampling-based frequency estimation method. This relationship demonstrates that, for a limited number of participants and a fixed privacy level, the sampling-based mechanism outperforms traditional perturbation-based methods, such as the Gaussian mechanism. Specifically, to achieve the same level of privacy and accuracy, the sampling-based solution requires fewer participants than its counterparts based on the Gaussian mechanism.

Furthermore, due to the less stringent participant requirements compared to local DP mechanisms, as well as the reduced computational demands of the sampling method, the use of multiparty computation (MPC) schemes becomes more practical. With this consideration, we extend the sampling-based mechanism to a decentralized setting by employing MPC schemes. To reduce the communication requirements imposed by the MPC scheme, we introduce an additional sampling step, ensuring that users report only a fraction of the data required by the initial mechanism. Combined with the insights gained from our earlier analysis, this demonstrates that sampling-based differentially private mechanisms can be used for data analysis in the absence of a trusted server, especially when the number of users is limited.

This approach provides a stronger privacy guarantee compared to the original centralized setting, as no confidential data leaves the data owners' devices. This offers a practical privacy-preserving aggregation technique, enabling companies with fewer users to conduct privacy-preserving data analysis. Lastly, we propose an architectural model to support weighted aggregation, aimed at achieving more accurate estimates in scenarios where users have varying sensitivities to the privacy of their data. The model is designed to reduce statistical error by prioritizing reports that exhibit higher statistical accuracy. Compared to the conventional unweighted aggregation used in previous studies [14], our weighted aggregation method provides more statistically accurate estimates. Extensive experiments have been conducted to demonstrate the effectiveness of the proposed methods.

The rest of this chapter is organized as follows. Section 10.2 discusses related work. Section 10.3 introduces the preliminaries. The proposed solution is detailed in Sect. 10.4. The chapter concludes in Sect. 10.5.

10.2 Related Work

Frequency Estimation with Local Differential Privacy Frequency estimation mechanisms that ensure local differential privacy against the server have been extensively studied in the literature [22]. One of the simplest methods for achieving this is the randomized response technique, which was later generalized by Kairouz et al. [25] to apply to data with higher-dimensional attributes. Subsequently, Wang et al. [24] proposed an optimized unary encoding (OUE) method, which encodes the true value using a one-hot encoding scheme, followed by applying randomized response to each bit of the resulting vector. To handle higher-dimensional attributes, Erlingsson et al. [3] and Wang et al. [24] introduced a hash-based method that maps user data to a much smaller domain before performing the randomized response, thereby reducing statistical variance. Bassily and Smith [4] proposed a transformation-based method that reduces the user's data from d bits to just one bit. This transformation not only results in an estimation with smaller statistical variance but also significantly reduces communication costs. Additionally, Wang et al. [31] suggested a subset selection method in which users randomly choose some of the items to report instead of reporting all the real items. This approach performs well in the intermediate privacy region when compared to other methods. However, as noted, all of these local differential privacy techniques require millions of participants to ensure statistical accuracy due to the large variance introduced by the perturbation method.

Combination of MPC and Differential Privacy Multi-party computation (MPC) provides a means to securely compute functions on users' private data without disclosing the data itself. However, MPC does not protect the privacy of any information that may be inferred from the output. To address this, it is natural to incorporate differential privacy techniques into the MPC framework to safeguard the privacy of the output.

In general, the combination of MPC and differential privacy methods discussed in the literature commonly utilizes Laplace or Gaussian mechanisms to provide differential privacy guarantees. For instance, Pettai and Laud [11] studied and analyzed the overhead associated with adding Laplace noise to secure multiparty computations. Bindschaedler et al. [10] developed secure aggregation protocols that introduce Laplace and Gaussian noise to ensure differential privacy for participants in a star network. The works in [8, 9] focus on collaborative model training, where the Gaussian mechanism is incorporated into the training process to produce aggregated parameters that are differentially private. Furthermore, secure aggregation techniques are introduced to reduce the required amount of added noise. In addition, Hu et al. [7] developed a secure computation protocol based on the Flajolet-Martin sketches to solve the Private Distributed Cardinality Estimation problem. In their approach, they utilize the uncertainty introduced by the intrinsic estimation variance of the FM sketch to produce a differentially private output.

Sampling Privacy Sampling methods offer a level of privacy protection and have been shown to perform more effectively when the number of participants is small. Typically, they are employed as an amplification technique to strengthen the privacy guarantees in distributed settings [28, 29]. Additionally, Joy and Gerla et al. [27] introduced a sampling privacy mechanism for distributed environments. However, instead of focusing solely on a pure sampling mechanism, they proposed that some data owners occasionally provide reports that contradict their expected responses. Husain et al. [26] assumed that each user has multiple records and applied the sampling process to each local dataset. To the best of our knowledge, this paper is the first to explore the performance of a pure sampling mechanism for frequency estimation in a distributed environment.

Heterogeneous Privacy Setting Privacy-preserving schemes that offer multiple levels of privacy are typically referred to as personalized differential privacy settings. These schemes are designed to provide personalized privacy protections, catering to users with varying privacy requirements.

Most of the research in this area [15–19] has focused on centralized settings, targeting different applications such as matrix factorization and crowdsourcing task assignment. In contrast, personalized differential privacy in distributed systems has not been as extensively studied. Akter et al. [14] proposed a numerical perturbation method for estimating the data average under a personalized differential privacy setting, inspired by the approach of Duchi et al. [12]. However, in their method, estimates from different privacy settings are directly aggregated together. Ye et al. [13] proposed a similar weighted aggregation method for frequency estimation, though the proposed weights depend solely on group size, which is not an optimal solution.

10.3 Preliminaries

10.3.1 Problem Definition

In this paper, we consider the fundamental primitive for computing item frequencies. Formally, let $U = \{u_1, \ldots, u_n\}$ represent a set of users with size n, where each user $u_i \in U$ has a value v_i within a domain $\mathcal{I} = \{I_1, \ldots, I_N\}$ of size N and reports it to an aggregator. The goal of the aggregator is to determine the number of users holding different items in $\mathcal{I}$.

We assume that each user's private data can be efficiently and uniquely encoded into a data space $\mathcal{D}$ of size N. Let q denote the smallest prime such that $q > n$, and let $\mathbb{F}_q$ represent the finite field containing q elements. Thus, we have $\mathcal{D} \subseteq \mathbb{F}_q^N$.

In our approach, we encode each user's value v_i as a vector of length N, denoted by $\mathbf{e}_j \in \mathbb{F}_q^N$, where the j-th entry is set to 1, and all other entries are set to 0.

Additionally, a default encoding with no private value is represented by $\mathbf{0} \in \mathbb{F}_q^N$, the zero vector of length N. Therefore, we have $\mathcal{D} = \{\mathbf{0}, \mathbf{e}_1, \ldots, \mathbf{e}_N\}$.

Let $[N]$ denote the set $\{1, 2, \ldots, N\}$. We define $S_{N,M} = \{A \subseteq [N] : |A| = M\}$, which will be useful in the discussion of two-stage sampling. Here, M and N are positive integers.

To protect the user's data, we propose a randomized algorithm that takes an encoded private data value from $\mathcal{D}$ and outputs a report in a report space $\mathcal{R}$ for transmission to the aggregator. This solution allows the aggregator to collect the users' private data, providing only an estimation of the number of users holding different items in $\mathcal{I}$, without revealing other information about the users' values. The goal is to minimize the error in these estimations.

This work is based on the following assumptions:

- A secure communication channel exists between any pair of participants, including the server(s) and users.
- An honest-but-curious adversary model, where the adversary observes the data and communication of corrupted parties but does not maliciously alter any data.
- The set of corrupted parties contains at most one of the servers (no collusion among servers) or some users, but not both.

10.3.2 Differential Privacy

Differential privacy is a privacy concept introduced by Dwork et al. [32] in 2006. It protects users' private information by introducing randomness, as defined below.

Definition 10.1 ((ϵ, δ)-Differential Privacy [23]) Let $\mathcal{M}$ be a probabilistic algorithm, $\mathcal{M} : \mathcal{D} \to \mathcal{R}$. $\mathcal{M}$ is (ϵ, δ)-differentially private if, for all possible non-empty sets of outputs $S \subseteq \mathcal{M}(\mathcal{D})$ and for all neighboring data sets D and $D' \in \mathcal{D}$, the following condition holds:

$$Pr[\mathcal{M}(D) \in S] \leq e^\epsilon Pr[\mathcal{M}(D') \in S] + \delta. \tag{10.1}$$

If $\delta = 0$, we say that $\mathcal{M}$ is ϵ-differentially private.

Intuitively, this definition ensures that adding, removing, or changing a record in the data set cannot significantly affect the final statistics.

Definition 10.2 (ϵ-Local Differential Privacy [37]) Let $\mathcal{M}$ be a randomized algorithm, $\mathcal{M} : \mathcal{D} \to \mathcal{R}$. $\mathcal{M}$ satisfies ϵ-local differential privacy if and only if, for any pair of distinct input values $v, v' \in \mathcal{D}$ and for any $\emptyset \subsetneq S \subseteq \mathcal{M}(\mathcal{D})$, the following condition holds:

$$Pr[\mathcal{M}(v) \in S] \leq e^\epsilon Pr[\mathcal{M}(v') \in S].$$

10.3 Preliminaries

Compared to centralized differential privacy, the perturbation mechanism $\mathcal{M}$ for local differential privacy is applied independently to each data record.

In the centralized setting, where a trusted server stores the users' raw data, sampling is often used to achieve (ϵ, δ)-differential privacy for mean value estimation [2, 6]. In this approach, the server randomly samples some of the records to estimate the overall statistics. To preserve privacy, the sampling result should not be disclosed along with the final statistics. In this work, we focus on using sampling methods to design differentially private mechanisms, rather than relying on more traditional techniques such as Laplace and Gaussian perturbations, due to the following practical efficiency considerations:

- We use population and data sampling from uniform or Bernoulli distributions, which is simpler than sampling from more complex distributions, such as Laplace or Gaussian.
- Sampling can be directly applied to an N-dimensional object, while Laplace and Gaussian perturbations must be applied independently to each of the N dimensions.
- Population or data sampling produces a vector response over $\mathbb{F}_q$, whereas the Laplace or Gaussian mechanism produces a real-number vector response. In this work, we extend the sampling-based mechanism to a decentralized setting through secret sharing. This can be done without altering the underlying field, while maintaining information-theoretical security. In contrast, an information-theoretically secure decentralized protocol based on the Gaussian or Laplace mechanism requires additional encoding to finite field elements using techniques like fixed-point encoding [39]. This leads to a larger finite field, which increases the storage, communication, and computation requirements for each user.

We also demonstrate that our mechanism satisfies (ϵ, δ)-differential privacy and provides more accurate estimates with smaller population sizes compared to the estimates produced by the traditional Gaussian mechanism, which also achieves (ϵ, δ)-differential privacy.

If any part of the mechanism depends solely on the output of other mechanisms, without any direct reliance on the private data, such a sub-protocol can be considered post-processing and does not consume any privacy budget, as formalized in Proposition 10.1.

Proposition 10.1 (Post-processing [23]) Let $\mathcal{M} : \mathcal{D} \to \mathcal{R}$ be a randomized algorithm that is (ϵ, δ)-differentially private. Let f be an arbitrary randomized mapping, $f : \mathcal{R} \to \mathcal{R}'$. Then $f \circ \mathcal{M} : \mathcal{D} \to \mathcal{R}'$ is (ϵ, δ)-differentially private. □

10.3.3 Additive Secret Sharing

Secret sharing schemes (SSS) are privacy-preserving techniques designed to enable a dealer with a secret value s to distribute it among a group of participants in the form of shares. This concept was first independently introduced by Shamir and Blakley in 1979. The shares are probabilistically generated from the secret, ensuring that a sufficient number of shares can be used to reconstruct the original secret. One of the most widely used secret sharing schemes is the additive secret sharing scheme. The formal definition of this scheme is as follows.

Definition 10.3 (Additive Secret Sharing Scheme) Let m be a positive integer and $\mathbb{F}$ a finite field. Given a secret value $s \in \mathbb{F}$, an additive secret sharing scheme with m parties over $\mathbb{F}$ generates the m shares in the following manner. First, we independently sample $s_1, \ldots, s_{m-1}$ from $\mathbb{F}$ uniformly at random. Then, having $s_1, \ldots, s_{m-1}$ and s, define $s_m = s - (s_1 + \cdots + s_{m-1})$. The m shares are then defined as $(s_1, \ldots, s_m)$.

It is straightforward to see that, with knowledge of all m shares, the secret s can be recovered by calculating $s_1 + \cdots + s_m$. However, if an adversary learns at most $m - 1$ of the shares, the distribution of s remains uniform, independent of the knowledge of the $m - 1$ revealed shares. In other words, the adversary gains no information about s.

A key property of the additive secret sharing scheme, which is essential to this paper, is its additive nature. This property is summarized in the following proposition.

Proposition 10.2 Suppose we are using an additive secret sharing scheme with m parties over a finite field $\mathbb{F}$. For two secret values $a, b \in \mathbb{F}$, let a and b be additively shared with shares $(a_1, \ldots, a_m)$ and $(b_1, \ldots, b_m)$, respectively. Then, $(a_1+b_1, a_2+b_2, \ldots, a_m+b_m)$ is a valid additive secret sharing of $a + b$. □

The additive property demonstrates that, to compute a valid additive secret sharing of the sum of M private values, we can individually apply additive secret sharing to each private value across the m parties. Each party can then sum up the M shares they receive to obtain a share for the sum of the M private values. It can be shown, through simulation-based security, that such an additive secret-sharing-based aggregation method provides information-theoretic security against an honest-but-curious adversary who controls up to $m - 1$ of the computing parties. A more detailed discussion of simulation-based security and the adversarial setting can be found, for example, in [21].

This security guarantee illustrates that, by using the additive secret-sharing-based aggregation method described above, we can simulate the existence of a trusted server that assists the users in the calculation with perfect security, without actually requiring such a trusted server. This demonstrates that we can achieve a centralized differential privacy guarantee without revealing any private data to other entities.

10.4 Proposed Solution

In this section, we present our proposed solution for privacy-preserving frequency estimation, aimed at mitigating the fairness gap between companies of different

10.4 Proposed Solution

sizes. Our solution enables organizations with smaller user bases to perform accurate statistical analysis on their users' data without the need for any user data to leave their respective devices. The workflow of the proposed methods is outlined below.

1. Users are grouped based on their privacy preferences. This step is assumed to be completed prior to the protocol, and we omit further discussion of this process due to its simplicity.
2. Each privacy group executes a privacy-preserving frequency estimation mechanism according to its chosen privacy level. This mechanism is used to estimate the distribution of the data held by users within the group.
3. After obtaining the data distribution estimates from all privacy groups, the server performs a weighted aggregation to derive an overall estimate of the data distribution. The aggregation is carried out using the different weights assigned to each privacy group.

The following sections provide a more detailed discussion of the mechanisms involved in Steps 2 and 3.

10.4.1 Differential Privacy with Distributed Sampling

We consider a privacy-preserving solution to the frequency estimation problem based on a sampling method. Initially, we examine a common approach in which sampling is used to design a privacy-preserving frequency estimation mechanism, which is then extended to a decentralized setting. To reduce the communication burden imposed on users due to the use of secret sharing, we propose a variant of this approach, which we refer to as the two-stage sampling method.

10.4.1.1 General Sampling

In general sampling, we refer to the method of calculating the frequency estimate from the private values of a randomly selected subset of users. Specifically, in a centralized setting, the server randomly selects a subset of records and performs statistical analysis on these selected records. Algorithm 1 provides the specification for the general sampling-based frequency estimation mechanism, denoted as DPCS.

Privacy Analysis

We demonstrate that DPCS provides (ϵ, δ)-differential privacy, as discussed in Theorem 10.1.

Theorem 10.1 *Let $\epsilon, \delta > 0$ be given. Suppose there exists a positive real number β such that for each $i = 1, \ldots, N$, there are at least βn users who own item I_i, and*

$$\beta \geq \frac{1}{2\pi n(e^{-\epsilon} - e^{-2\epsilon})} \max\left(\left(\frac{2\pi}{\delta}\right)^{\frac{2}{N+1}}, \left(\frac{1}{\delta}\right)^{\frac{2}{N}}\right). \tag{10.2}$$

Algorithm 1 Differential privacy with centralized sampling (DPCS)

Input: The trusted server holds the encoded private values of n users, $\mathbf{t}_i \in \{\mathbf{e}_1, \ldots, \mathbf{e}_N\} \subseteq \mathcal{D}$, sampling probability p, and privacy budget ϵ;
Output: Estimate of the normalized frequency of each value, $\widehat{\psi}$;
1: The trusted server samples from the n users to select the participants, with each user having a probability p of being chosen. Let $\mathcal{A} \subseteq U$ be the set of participating users;
2: The trusted server computes $\widehat{\tau} = \sum_{u_i \in \mathcal{A}} \mathbf{t}_i$;
3: The server computes and outputs the normalized frequency estimation $\widehat{\psi} = \frac{1}{pn} \widehat{\tau}$ as a vector of real numbers;

Then, setting $p = 1 - e^{-\epsilon}$, the mechanism DPCS *provided in Algorithm 1 is (ϵ, δ)-differentially private.*

Discussion According to Theorem 10.1, we observe that the privacy level is determined not only by the privacy parameters but also by the population size and data dimension. Theorem 10.1 provides a more relaxed bound for privacy. Regardless of changes in the data set, as long as it satisfies Eq. (10.2), it guarantees fixed privacy for a given sampling probability p. To further understand the impact of the sampling mechanism on the privacy guarantee, we provide a tighter bound in Theorem 10.2. For the same sampling probability, Theorem 10.2 offers a stronger privacy guarantee. However, it is important to note that this stronger guarantee comes with a more stringent requirement on the data distribution.

Theorem 10.2 *Let $\epsilon, \delta > 0$ be given. Suppose there exist two positive real numbers, β and z, such that for each $i = 1, \ldots, N$, there are at least βn users who own item I_i. We require β and z to satisfy the following conditions. First, we require that*

$$\beta \geq \frac{e^{z+\epsilon}}{2\pi n(1 - e^{-\frac{\epsilon}{2}})^2} \max\left(\frac{\left(\frac{2\pi}{\delta}\right)^{\frac{2}{N+1}}}{(1 + e^{-\frac{\epsilon}{2}})^2}, \left(\frac{4\pi}{\delta}\right)^{\frac{2}{N}} \right).$$

Additionally, we require that

$$z \leq \ln\left(1 + n(1 - \beta(N-1))(1 - e^{-\frac{\epsilon}{2}})\right).$$

Then, setting $p = 1 - e^{-z-\epsilon}$, the mechanism DPCS *in Algorithm 1 preserves (ϵ, δ)-differential privacy.*

Utility Analysis
We now provide a statistical analysis of $\widehat{\psi}$ to demonstrate the accuracy of the sampling.

Lemma 10.1 *Suppose that for each $i = 1, \ldots, N$, there are Π_i users whose true value is $\mathbf{e}_i$. In other words, the target value is $\psi = \left(\frac{\Pi_1}{n}, \ldots, \frac{\Pi_N}{n}\right)$. Then, $\widehat{\psi}$ is an*

10.4 Proposed Solution

unbiased estimator of ψ, and the variance of $\widehat{\psi}$ is given by

$$\text{Var}\left(\widehat{\psi}\right) = \frac{1-p}{pn^2} V,$$

where V is a diagonal matrix of size $N \times N$ with the entry in row i and column i being Π_i. Furthermore, the expected squared L^2 distance of $\widehat{\psi}$ from the true value ψ is

$$\mathbb{E}\left(\left\|\widehat{\psi} - \psi\right\|_2^2\right) = \frac{1-p}{pn}.$$

10.4.1.2 General Sampling (Decentralized)

We begin by designing the protocol DPDS, which implements DPCS in a decentralized setting. The complete specification of the protocol DPDS can be found in Algorithm 2.

Algorithm 2 Differential privacy with distributed sampling (DPDS)

Input: Encoded private values of n users $\mathbf{t}_i \in \{\mathbf{e}_1, \ldots, \mathbf{e}_N\} \subseteq \mathcal{D}$, sampling probability p, privacy budget ϵ;
Output: Estimate of the normalized frequency of each value $\widehat{\psi}$;
1: *[User side]*
2: User u_i samples a random Bernoulli random variable with success probability p to decide whether they actively participate in the count;
3: **if** u_i actively participates, **then**
4:    u_i encodes their response as $\widehat{\mathbf{t}}_i = \mathbf{t}_i \in \mathcal{D}$;
5: **else**
6:    u_i encodes their response as $\widehat{\mathbf{t}}_i = \mathbf{0} \in \mathcal{D}$;
7: **end if**
8: u_i samples $n-1$ random vectors of length N, $\widehat{\mathbf{s}}_{i,1}, \ldots, \widehat{\mathbf{s}}_{i,n-1} \in \mathbb{F}_q^N$ and sets $\widehat{\mathbf{s}}_{i,n} = \widehat{\mathbf{t}}_i - \sum_{j=1}^{n-1} \widehat{\mathbf{s}}_{i,j}$;
9: u_i sends $\widehat{\mathbf{s}}_{i,j}$ to u_j for $j = 1, \ldots, n$;
10: u_i computes $\widehat{\boldsymbol{\tau}}_i = \sum_{j=1}^n \widehat{\mathbf{s}}_{j,i}$ and sends it to the server;
11: *[Server side]*
12: The server computes $\widehat{\boldsymbol{\tau}} = \sum_{i=1}^n \widehat{\boldsymbol{\tau}}_i \in \mathbb{F}_q^N$ and treats the sum as a vector over real numbers of length N;
13: The server computes and outputs the normalized frequency estimation $\widehat{\psi} = \frac{1}{pn}\widehat{\boldsymbol{\tau}}$ as a vector over real numbers;

As shown in Algorithm 2, each user's value is represented using one-hot encoding. Each user independently decides whether to participate in the reporting process with probability p (Line 2). Here, p represents the sampling probability, which directly influences the privacy level. If the user decides to actively participate, they retain their true value. If the user does not actively participate, their value is perturbed as a zero vector of length N (Lines 3–7). Subsequently, the secret-sharing-based aggregation scheme is applied, where the n users act as the n computing parties, and they send shares of the aggregated value to the server, which serves as the aggregator (Lines 8–10). Finally, the server estimates the frequency for each item by calculating $\widehat{\psi} = \frac{1}{pn} \sum_{i=1}^{n} \widehat{\tau}_i$ (Line 12).

Privacy and Utility Analysis

To facilitate the transition from the centralized to the decentralized setting, we introduce a slight modification to DPCS, which we refer to as DPCS*. The primary difference between DPCS and DPCS* lies in their outputs. Instead of directly outputting $\widehat{\psi}$ as in DPCS, the trusted server secretly shares $\widehat{\tau}$ as n vectors of length N over $\mathbb{F}_q$. With these shares, the receiver can recover the value of $\widehat{\tau}$, treat it as a vector over real numbers, and normalize it by calculating $\widehat{\psi} = \frac{1}{pn}\widehat{\tau}$.

Note that, with this modification, the only additional information that the receiver may gain is the number of users, n, which is publicly known. Therefore, DPCS* offers the same privacy and accuracy guarantees as DPCS, and the proofs in Theorems 10.1 and 10.2, as well as Lemma 10.1, are equally applicable to DPCS*. It is evident that, from the perspective of the receiver, DPDS is indistinguishable from DPCS*. Hence, the privacy and accuracy guarantees of DPCS* also apply to DPDS.

Moreover, since we employ an additive secret sharing scheme with n parties, we can ensure that any collusion involving up to $n - 1$ parties does not reveal any information about the private value of the remaining party. Therefore, DPDS guarantees privacy against an honest-but-curious adversary, whether they control up to $n - 1$ users or the server, but not both.

Complexity Analysis

It is straightforward to observe that in the DPDS protocol, each user communicates nN elements from $\mathbb{F}_q$. Regarding computational cost, each user performs one Bernoulli sampling, $(n - 1)N$ uniform samplings from $\mathbb{F}_q$, and $2(n - 1)N$ field addition operations. On the other hand, the server incurs no communication cost and performs $(n - 1)N$ field addition operations, along with N real-number multiplication operations.

Comparison with the Distributed Gaussian Mechanism

Recall that our mechanism ensures (ϵ, δ)-differential privacy. A comparable mechanism that provides the same privacy guarantee is the Gaussian mechanism. In this section, we extend Gaussian-based frequency estimation to a distributed setting using secret sharing. The perturbation can be applied by allowing each user to perform a Gaussian sampling step before participating in the secret-sharing-based aggregation.

For a dataset containing information from n users, where Π_i users hold item I_i for $i = 1, \ldots, N$, the goal is to compute the output $\Psi = \frac{1}{n}(\Pi_1, \ldots, \Pi_N)$. It is clear that the sensitivity of the query is $G_f = \frac{\sqrt{2}}{n}$. To achieve (ϵ, δ)-differential privacy, the data can be perturbed by Gaussian noise sampled from $\mathcal{N}(0, G_f^2 \sigma_G^2)$,

10.4 Proposed Solution

where $\sigma_G = \frac{\sqrt{2\ln\left(\frac{1.25}{\delta}\right)}}{\epsilon}$. By the additivity property of identically and independently distributed random variables from a fixed Gaussian distribution, such noise can be generated by having each user add Gaussian noise sampled from $\mathcal{N}(0, (\frac{G_f \sigma_G}{n})^2)$, followed by additive secret sharing to distribute the shares among the n users, with the aggregated result revealed to the server. We denote this mechanism as DPDG.

It is easy to observe that DPDG is (ϵ, δ)-differentially private and produces an unbiased estimator of Ψ with variance $V_G = \frac{4\ln\left(\frac{1.25}{\delta}\right)}{n^2 \epsilon^2}$. The following section discusses the advantages and disadvantages of DPDG in comparison with DPDS.

In contrast to DPDS, which requires that $\Pi_i \geq \beta n$ for some $\beta \in (0, 1)$ to ensure privacy, DPDG does not impose such constraints on the actual data distribution. Additionally, DPDG provides a smaller variance, $V_G < V$, when n exceeds a threshold, i.e., when $n > \frac{4(e^\epsilon - 1)\ln\left(\frac{1.25}{\delta}\right)}{\epsilon^2}$. This demonstrates that in situations with smaller populations, DPDS generally outperforms DPDG. Furthermore, DPDG generally requires higher computational and communication complexity.

Compared to DPDS, the variant DPDG replaces one Bernoulli sampling, which outputs a single bit, with one multi-dimensional Gaussian sampling, which outputs a vector of length N, with each entry being a real number. Since this Gaussian sampling produces a vector of real numbers, we cannot directly store any element as a field element. Secret-sharing schemes over real numbers are typically defined by encoding real numbers as field elements through fixed-point arithmetic [39]. For k-bit decimal precision, the field size must be increased by at least k bits, resulting in significantly higher computational and communication complexity compared to DPDS for achieving the same level of accuracy.

10.4.1.3 Two-Stage Sampling

Although the mechanism DPDS discussed in the previous section provides highly accurate frequency estimations, it requires distributing n vectors of length N across n different users. This requirement may become impractical as the number of users increases. To mitigate the communication cost, a natural solution is to reduce both the number of items reported and the number of shares generated for each item. Based on this observation, we propose a two-stage sampling method. Intuitively, after performing the same sampling method as in DPDS to determine each user's participation level, each user conducts a second round of sampling to decide which items to report. Specifically, we consider two types of sampling for the second stage.

Uniform Sampling In this method, to determine which items to report, we fix the number of items and sample each item uniformly at random. By reporting only a subset of the values instead of the entire set $\widehat{\mathbf{t}}_i \in \mathcal{D}$, the overall communication and computational costs can be significantly reduced. However, this comes at the cost of reduced statistical accuracy. Formally, the uniform sampling process involves randomly selecting αN items from the N items for each user to report, where α is a predetermined fraction. In other words, for any user, for any $A \in S_{N,\alpha N}$, the probability that a user reports the αN items corresponding to the elements of A is given by $\frac{1}{\binom{N}{\alpha N}}$. We define this reporting distribution as χ_U.

Adaptive Sampling To limit the decrease in statistical accuracy resulting from reporting fewer values, the sampling probability can be adjusted. Instead of uniformly sampling the values, we increase the probability that an actively participating user reports the value corresponding to the actual item they hold. Specifically, we adopt the approach proposed in [33] for item sampling. For a fixed constant $\alpha \in (0, 1)$, the sampling process generates a random subset of $[N]$ of size αN, with sets containing the user's true value having a higher probability of being sampled. We formally define the adaptive sampling process as follows.

Definition 10.4 (Adaptive Sampling) Let N be a positive integer, $\alpha \in (0, 1)$, and $\gamma > 1$. For $i = 1, \ldots, n$, let $Y_i^{(N,\alpha,\gamma)}$ be a random variable with values from $S_{N,\alpha N}$, representing the set of items that user u_i reports. We assume that after the first stage of the sampling process, the response of u_i is encoded as $\widehat{\mathbf{t}_i} \in \mathcal{D}$. Then, for $A \in S_{N,\alpha N}$, we define the probability that $Y_i^{(N,\alpha,\gamma)} = A$ as ζ_A:

$$\zeta_A = \begin{cases} \frac{1}{\binom{N}{\alpha N}}, & \text{if } \widehat{\mathbf{t}_i} = \mathbf{0}, \\ \frac{\gamma}{\gamma\binom{N-1}{\alpha N-1}+\binom{N-1}{\alpha N}}, & \text{if } \mathbf{v} = \mathbf{e}_j \text{ and } j \in A, \\ \frac{1}{\gamma\binom{N-1}{\alpha N-1}+\binom{N-1}{\alpha N}}, & \text{if } \mathbf{v} = \mathbf{e}_j \text{ and } j \notin A. \end{cases}$$

We denote this reporting distribution as χ_A.

Remark The second stage of sampling can reduce the communication and computational burdens on each user. However, it also introduces some drawbacks. First, each user needs to submit a preliminary report to identify the items they will be reporting. This ensures that the shares are correctly labeled and that the aggregation process does not begin until all shares have been submitted. Moreover, this preliminary reporting process can lead to some information leakage. Specifically, the server can identify a smaller group of users from whom the reported counts originate. In some cases, such additional information could be used to infer more details about the users' private values. To minimize this risk of information leakage, we introduce an additional server responsible for receiving the preliminary reports and performing further processing to reduce the information exposed to the aggregating server. We refer to this scheme as TSS, and its full specification is provided in Algorithm 3.

As illustrated in Algorithm 3, rather than reporting the entire encoded value, users report the set of items they intend to report to server $\mathcal{S}_1$ following the encoding and sampling processes (Lines 2–9). Consequently, the only information that $\mathcal{S}_1$ obtains regarding the users' private values is the sets A_i, which contain the αN items that

10.4 Proposed Solution

Algorithm 3 Two-stage sampling (TSS)

Input: Private values of n users $\mathbf{t}_i \in \{\mathbf{e}_1, \cdots, \mathbf{e}_N\} \subseteq \mathbb{F}_q^N$, sampling probability p, reporting proportion α, reporting distribution $\chi \in \{\chi_U, \chi_A\}$ (with possible reporting parameter γ if necessary) and privacy budget ϵ;
Output: Estimate of normalized frequency of each value $\widehat{\psi}$;

1: *[User side]*
2: User u_i samples a random Bernoulli random variable with success probability p to determine whether they actively participate in the count;
3: **if** u_i actively participates, **then**
4:    u_i encodes their response as $\widehat{\mathbf{t}_i} = \mathbf{t}_i \in \mathcal{D}$;
5: **else**
6:    u_i encodes their response as $\widehat{\mathbf{t}_i} = \mathbf{0} \in \mathcal{D}$;
7: **end if**
8: u_i samples $A_i \in S_{N,\alpha N}$ using distribution χ with possible implicit parameters γ and $\widehat{\mathbf{t}_i}$;
9: u_i defines a binary vector $\mathbf{a}_i$ where its j-th entry is 1 if $j \in A_i$ and 0 otherwise, which is sent to $\mathcal{S}_1$;
10: *[$\mathcal{S}_1$ Side]*
11: $\mathcal{S}_1$ compiles $\mathbf{a}_1, \cdots, \mathbf{a}_n$ to obtain the numbers $m_1, \cdots, m_N$ where m_j is the number of users reporting the count for item I_j;
12: **for** $j = 1, \cdots, N$ **do**
13:    $\mathcal{S}_1$ randomly selects m_j users $u_{j,1}, \cdots, u_{j,m_j}$;
14:    $\mathcal{S}_1$ publishes $(j, u_{j,1}, \cdots, u_{j,m_j})$ to all users and the aggregating server $\mathcal{S}_2$;
15:    **for** i such that $j \in A_i$ **do**
16:      u_i with encoded response for item I_j, $\widehat{t_{i,j}}$ samples $m_j - 1$ random elements from $\mathbb{F}_q$, denoted by $s_{j,i,1}, \cdots, s_{j,i,m_j-1} \in \mathbb{F}_q$ and sets $s_{j,i,m_j} = \widehat{t_{i,j}} - \sum_{t=1}^{m_j-1} s_{j,i,t} \in \mathbb{F}_q$;
17:      u_i sends $(j, s_{j,i,t})$ to $u_{j,t}$ for $t = 1, \cdots, m_j$;
18:    **end for**
19:    *[For selected users $u_{j,1}, \cdots, u_{j,m_j}$]*
20:    Upon receiving m_j values $s_{j,i,t}$ for i such that $j \in A_i$, each $u_{j,t}$ computes $s_{j,t} = \sum_{i:j \in A_i} s_{j,i,t}$ and sends $(j, s_{j,t})$ to $\mathcal{S}_2$;
21:    *[$\mathcal{S}_2$ Side]*
22:    $\mathcal{S}_2$ calculates the estimate of the count for item j as $s_j = \sum_{t=1}^{m_j} s_{j,t}$ and treats it as a real number;
23:    $\mathcal{S}_2$ performs post-processing to obtain the unbiased estimator for the count of item I_j as $\widehat{\psi_j} = \frac{1}{q_\chi n} s_j$ where $q_\chi = p p_\chi$ and

$$p_\chi = \begin{cases} \alpha, & \text{if } \chi = \chi_U \\ \frac{\alpha\gamma}{\alpha\gamma + 1 - \alpha}, & \text{if } \chi = \chi_A \end{cases}.$$

24: **end for**

---

user u_i intends to report, thereby revealing the number of users reporting different items I_j (Line 11).

For $j = 1, \cdots, N$, once the number m_j of users reporting item I_j is determined, server $\mathcal{S}_1$ randomly selects m_j users to assist in reporting item I_j to server $\mathcal{S}_2$ and notifies all participants about the identity of the elected users (Lines 13 and 14). The original users who intend to report item I_j then privately share their encoded response for item I_j with the m_j elected users using additive secret sharing (Lines

15 – 18). The elected users aggregate the shares they receive for item I_j and send the sum to server S_2 (Line 20). After receiving the m_j reports from the elected users, server S_2 estimates the distribution of item I_j for all $j = 1, \cdots, N$ (Lines 22 – 23).

With the assistance of server S_1 and the elected users, although server S_2 may learn m_j—the total number of users that could influence the count s_j—the identities of these m_j users remain concealed. In the following, we analyze the privacy protection guarantees of TSS against the two servers.

Privacy Analysis Recall that server S_1 only receives an indicator from the user to determine which items they intend to report. Thus, the information received by S_1 consists solely of the items each user plans to report, without any further details about the actual reports. Intuitively, such information does not reveal the user's private value if the reporting distribution χ is uniform. However, if the indicator is selected adaptively, this could potentially disclose some information about the user's private value. This is because the distribution design may cause sets containing the user's actual private value to be more likely to be sampled. As shown in Theorem 10.3, this potential privacy leak can be effectively bounded.

Theorem 10.3 *Let $\mathcal{M}$ be the mechanism that follows TSS from the perspective of S_1. In other words, for each user, the mechanism takes $\mathbf{v}_i$ as input and outputs A_i. Let α be the reporting proportion for the second stage, while χ_U and χ_A represent the uniform and adaptive reporting distributions, respectively. Then, the mechanism $\mathcal{M}$ provides an ϵ^* local differential privacy against S_1, where*

$$\epsilon^* = \begin{cases} 0, & \text{if } \chi = \chi_U; \\ \log \gamma, & \text{if } \chi = \chi_A; \end{cases} \tag{10.3}$$

Theorem 10.3 states that, in the case of uniform sampling, the user's information is perfectly concealed from S_1. However, when adaptive sampling is used, the user's true item has a higher likelihood of being selected, which could potentially be used to infer statistical information about the user's true item. Nonetheless, the sampling process is shown to satisfy the local differential privacy definition, with the privacy disclosure bounded by $\log \gamma$. Next, we consider the privacy provided by TSS against S_2.

Theorem 10.4 *The mechanism TSS provides at least the same level of privacy as DPDS against S_2, given the same sampling probability, regardless of the chosen reporting distribution χ.*

Utility Analysis
Let $\widehat{\psi}_j$ be the random variable that estimates the actual count of item I_j. It is important to note that since it is impossible for holders of other items to influence the count s_j, we can treat each random variable $\widehat{\psi}_j$ independently. Lemma 10.2 provides the distribution of the final estimation.

Lemma 10.2 *Fix $i \in \{1, \cdots, N\}$ and assume that a user u_j holds item I_i and decides to actively participate in the TSS mechanism. Suppose further that he uses*

10.4 Proposed Solution

the distribution χ to determine A_j, the αN items he will report. Then the probability that $i \in A_j$ is p_χ, where

$$p_\chi = \begin{cases} \alpha, & \text{if } \chi = \chi_U; \\ \frac{\alpha\gamma}{\alpha\gamma+1-\alpha}, & \text{if } \chi = \chi_A; \end{cases}$$

Therefore, since the probability that a user u_j actively participates in TSS is p, we have that for any user u_j holding item I_i, the probability that he can influence the count of s_j is $q_\chi \triangleq p_\chi \cdot p$.

Such a distribution can then be used to provide the statistical accuracy analysis for TSS.

Theorem 10.5 *For TSS, suppose that for $i = 1, \ldots, N$, there are Π_i users holding the value $\mathbf{e}_i$. In other words, the target value we aim to achieve is $\boldsymbol{\psi} = \left(\frac{\Pi_1}{n}, \ldots, \frac{\Pi_N}{n}\right)$. Then, $\widehat{\boldsymbol{\psi}}$ is an unbiased estimator of $\boldsymbol{\psi}$ with variance given by*

$$\mathrm{Var}\left(\widehat{\boldsymbol{\psi}}\right) = \frac{1-q_\chi}{n^2 q_\chi} V,$$

where V is an $N \times N$ diagonal matrix, with the entry in row i and column i being Π_i. Furthermore, the expected squared L^2 distance between $\widehat{\boldsymbol{\psi}}$ and its true value $\boldsymbol{\psi}$ is

$$\mathbb{E}\left(\|\widehat{\boldsymbol{\psi}} - \boldsymbol{\psi}\|_2^2\right) = \frac{1-q_\chi}{n q_\chi}.$$

Proof This theorem can be proven using the same structure as the proof of Lemma 10.1, with the only difference being that for any $j = 1, \ldots, N$, the binomial random variable $\widehat{\psi}_j$ has a success probability of q_χ instead of p. By following the reasoning from the previous proof, we obtain the desired result. □

Complexity Analysis

Finally, we discuss the computational and communication costs for different groups of participants in TSS. A summary of these requirements, along with a comparison to those for DPDS, can be found in Table 10.1. A detailed discussion of the complexity calculations for TSS is provided in the Appendix. As observed, although the complexity for the server increases–especially for the newly introduced server S_1– the average computational and communication complexity incurred by an individual user is reduced by approximately a factor of $\frac{1}{p_\chi \alpha}$, thus achieving the objective of introducing the second stage of sampling.

Security Against Collusion of Users

It is important to note that since participating users secretly share their private values with a randomly selected set of parties, there exists a non-zero probability

Table 10.1 Expected complexity requirements for DPDS and TSS

	DPDS		TSS		
	Server	User	S_1	S_2	User
Computation	FA:$((n-1)N)$ RM: N	FA:$(2(n-1)N)$ BS:1 FS:$(n-1)N$	nA:$((n-1)N)$ nS:$\alpha n N$	FA:$(\alpha n - 1)N$ RM:N	FA:$(p_\chi n - 1)(p_\chi + \alpha)N$ BS:1, aS:1, FS: $(p_\chi n - 1)(\alpha N)$
Communication	0	qD:nN in 2 rounds	nD:$\alpha n N(n+1)$ ND:$(n+1)N$ in 1 round	0	bD:N ND:$\alpha N(1 + p_\chi n)$ qD: $\alpha N(1 + p_\chi n)$ in 3 rounds

FA: Field Addition; RM: Real Number Multiplication; BS: Bernoulli Sampling; aS: Sampling from $S_{N,\alpha N}$ following χ; FS: Field Sampling; nA: Addition of elements in $[n]$; nD: Sending elements of $[n]$; ND: Sending elements of $[N]$; bD: Sending binary elements; qD: Sending elements of $\mathbb{F}_q$.

10.4 Proposed Solution

that all the chosen parties may collude, especially if the set size is insufficient. In such cases, colluding parties could potentially recover all the private values of the reporting users. To mitigate this risk, we propose a minor adjustment to Algorithm 3 to safeguard against user collusion. Let ϕ denote a positive integer representing the maximum number of colluding parties. We can guarantee privacy by ensuring that for each item, at least $\phi + 1$ users are elected to participate in the reporting process to S_2. This can be achieved by modifying the steps from Lines 13 to 20 of the algorithm as follows:

Algorithm 4 Two-stage sampling (TSS$'$)

Input: Private values of n users $\mathbf{t}_i \in \{\mathbf{e}_1, \cdots, \mathbf{e}_N\} \subseteq \mathbb{F}_q^N$, sampling probability p, reporting proportion α, reporting distribution $\chi \in \{\chi_U, \chi_A\}$ (with an optional reporting parameter γ if needed), reporter bound ϕ, and privacy budget ϵ;
Output: Estimate of normalized frequency for each value $\widehat{\psi}$;

1: {Follow the first 11 lines of TSS in Algorithm 3}
2: **for** $j = 1, \cdots, N$ **do**
3:   S_1 sets $m_j' = \max(\phi + 1, m_j)$;
4:   S_1 randomly selects m_j' users $u_{j,1}, \cdots, u_{j,m_j'}$;
5:   S_1 publishes $(j, m_j, u_{j,1}, \cdots, u_{j,m_j'})$ to all users and the aggregating server S_2;
6:   **for** i such that $j \in A_i$ **do**
7:     u_i with encoded response for item I_j, $\widehat{t_{i,j}}$ samples $m_j' - 1$ random elements from $\mathbb{F}_q$, denoted by $s_{j,i,1}, \cdots, s_{j,i,m_j'-1} \in \mathbb{F}_q$ and sets $s_{j,i,m_j'} = \widehat{t_{i,j}} - \sum_{t=1}^{m_j'-1} s_{j,i,t} \in \mathbb{F}_q$;
8:     u_i sends $(j, s_{j,i,t})$ to $u_{j,t}$ for $t = 1, \cdots, m_j'$;
9:   **end for**
10: *[For elected users $u_{j,1}, \cdots, u_{j,m_j'}$]*
11: Upon receiving m_j values $s_{j,i,t}$ for i such that $j \in A_i$, $u_{j,t}$ computes $s_{j,t} = \sum_{i: j \in A_i} s_{j,i,t}$ and sends $(j, s_{j,t})$ to S_2;
12: *[S_2 Side]*
13: S_2 calculates the estimate of the count for item j as $s_j = \sum_{t=1}^{m_j'} s_{j,t}$ and treats it as a real number;
14: S_2 performs post-processing to obtain the unbiased estimator for the count of item I_j as $\widehat{\psi}_j = \frac{1}{q_\chi n} s_j$ where $q_\chi = p p_\chi$ and

$$p_\chi = \begin{cases} \alpha, & \text{if } \chi = \chi_U \\ \frac{\alpha \gamma}{\alpha \gamma + 1 - \alpha}, & \text{if } \chi = \chi_A \end{cases}.$$

15: **end for**

It is easy to observe that these minor adjustments do not impact the privacy protection against any of the servers nor the utility of TSS. However, they do provide a privacy guarantee for any user's private value against collusion involving up to ϕ other users. This ensures that TSS$'$ maintains privacy in the presence of an honest-but-curious adversary controlling either S_1, S_2, or any subset of ϕ users from the total n users. It is important to note that this privacy guarantee no longer holds if the adversary controls both servers, or any one server in combination with the ϕ colluding users.

Furthermore, the adjustments introduced in TSS' affect the complexity for the participants. Specifically, for any j such that $m_j < \phi + 1$, there is an additional $\phi + 1 - m_j$ sampling required over $[n]$, as well as the publication of the value of $m_j \in [n]$ to $\mathcal{S}_1$. Similarly, for any j such that $m_j < \phi + 1$, an extra $\phi + 1 - m_j$ field addition operations are required at $\mathcal{S}_2$. Lastly, for any j such that $m_j < \phi + 1$, users reporting the j-th value will incur additional sampling, field additions, and transmission of $\phi + 1 - m_j$ field elements. Users selected to assist in the reporting process for item j will need to perform additional $\phi + 1 - m_j$ field additions.

10.4.2 Weighted Aggregation for Hybrid Privacy Levels

Providing a uniform level of privacy for all users within a mechanism may simplify its analysis. However, this approach could lead to suboptimal statistical accuracy, as it may offer a higher level of privacy than required for some users. To address this, we propose a more flexible approach by introducing multiple privacy levels such as *Very Strong*, *Strong*, *Normal*, and *Weak*, allowing users to select the level that best matches their privacy preferences. This approach is not entirely new, as prior studies have explored the impact of multiple privacy levels on the statistical accuracy of mechanisms [13, 14]. However, in these studies, aggregation is performed without considering the varying distortion levels chosen by individual users. Although such aggregation methods can achieve satisfactory statistical accuracy, they may not be optimal, particularly when assuming that the private data follows a fixed distribution. Intuitively, the statistical accuracy of aggregation can be enhanced by assigning larger weights to reports with lower distortion. To formalize this approach, we propose a general theoretical framework for data aggregation with different levels of perturbation.

We assume that users' data are independent and identically distributed random variables with an expected value equal to the desired average $\bar{v}$. This assumption holds in scenarios where users' data follow a fixed distribution, such as human weight data (which typically follows a Gaussian distribution) [34], web page click frequencies, or node degrees in social networks, which often follow a power-law distribution [35, 36]. We further assume that the output of each user is an unbiased estimator of their true value, which can be achieved through any unbiased perturbation mechanism.

The general idea is to assign different weights to each group of users in order to better calibrate the statistical mean. We assume that there are μ different privacy levels, $\{\epsilon_1, \cdots, \epsilon_\mu\}$, with the j-th privacy level being selected by n_j users. For each $j = 1, \cdots, \mu$ and $k = 1, \cdots, n_j$, we denote the k-th user choosing the j-th privacy level as $u_{j,k}$. We assign a weight $w_j \in \mathbb{R}_{>0}$ to the j-th privacy level. If the output of $u_{j,k}$ is $\tilde{\mathbf{v}}_{j,k}$, the average $\bar{\mathbf{v}}$ is estimated as

10.4 Proposed Solution

$$\widehat{\bar{\mathbf{v}}} = \frac{\sum_{j=1}^{\mu} w_j \sum_{k=1}^{n_j} \tilde{\mathbf{v}}_{j,k}}{\sum_{j=1}^{\mu} n_j w_j}.$$

We determine the values of w_j that provide high accuracy, which largely depends on the expected squared L^2 distance of each estimation $\tilde{v}_{j,k}$ from its expected value $\bar{v}$. For simplicity, we assume that the expected squared L^2 error of $\tilde{v}_{j,k}$ is the same for all users with the same privacy level ϵ_j, and we denote this error by V_j. This assumption is applicable to most existing differential privacy (DP) mechanisms.

To improve the accuracy of the estimation, we assign more weight to users who select a larger privacy budget. We evaluate the statistical error by calculating its expected squared error, $\Delta = \mathbb{E}\left[\left\|\left(\frac{\sum_{j=1}^{\mu} w_j \sum_{k=1}^{n_j} \tilde{v}_{j,k}}{\sum_{j=1}^{\mu} n_j w_j} - \bar{v}\right)\right\|_2^2\right]$. By simple algebraic manipulation, and noting that $V_j = \mathbb{E}\left(\|\tilde{v}_{j,k} - \bar{v}\|_2^2\right)$ and that the values from any two different users are uniformly and identically distributed (implying zero correlation between them), we can calculate Δ as

$$\Delta = \frac{\sum_{j=1}^{\mu} w_j^2 n_j V_j}{\left(\sum_{j=1}^{\mu} n_j w_j\right)^2}. \tag{10.4}$$

To correct the statistical error and achieve a more accurate statistical mean, we propose two potential solutions:

Solution 1 We model the weight assignment process as an optimization problem:

$$\min \left\{ \frac{\sum_{j=1}^{\mu} n_i w_i^2 V_j}{\left(\sum_{i=1}^{\mu} n_i w_i\right)^2} : \sum_{i=1}^{\mu} w_i = 1, 0 < w_i < 1, \forall i \in [\mu] \right\}.$$

This problem can be solved numerically using existing optimization techniques, such as gradient descent.

Solution 2 We treat the expected squared error formula as a multi-variable function of the weights and consider its critical points to identify possible local minima. One such critical point occurs at $(w_1, \cdots, w_\mu)$, where for $i = 1, \cdots, \mu$,

$$w_i = \frac{1/V_i}{\sum_{j=1}^{\mu} \frac{1}{V_j}}. \tag{10.5}$$

Lemma 10.3 *Define a function* $f : \mathbb{R}^\mu \to \mathbb{R}$ *such that* $f(w_1, \cdots, w_\mu) = \frac{1}{(\sum_{j=1}^{\mu} n_j w_j)^2} \sum_{j=1}^{\mu} w_j^2 n_j V_j$. *The critical points of f are of the form* $c \cdot \left(\frac{1}{V_1}, \cdots, \frac{1}{V_\mu}\right)$ *for some constant* $c \in \mathbb{R}$.

Proof For $t = 1, \cdots, \mu$, define $f_t \triangleq \frac{\partial f}{\partial w_t}$ as the partial derivative of f with respect to w_t. Then,

$$f_t = \frac{2n_t}{\left(\sum_{j=1}^{\mu} w_j n_j\right)^2} \left[w_t V_t - \frac{\sum_{j=1}^{\mu} w_j^2 n_j V_j}{\sum_{j=1}^{\mu} n_j w_j} \right].$$

Setting $f_t = 0$ and subtracting $f_{t'}$ from f_t for any $t' \neq t$, we get

$$(w_t V_t - w_{t'} V_{t'}) \left(\sum_{j=1}^{\mu} w_j n_j\right) = 0. \tag{10.6}$$

Since $\sum_{j=1}^{\mu} w_j n_j > 0$, Eq. (10.6) implies that $w_t V_t = w_{t'} V_{t'}$ for any $t \neq t'$. This shows that $w_t V_t$ is constant for all t. Let this constant be denoted as c. Therefore, we conclude that $f_t = 0$ for all t if and only if there exists a constant c such that $w_t = \frac{c}{V_t}$. Hence, the critical points of f are of the form $c \cdot \left(\frac{1}{V_1}, \cdots, \frac{1}{V_\mu}\right)$. □

Note that although the given values $(w_1, \cdots, w_\mu)$ in Eq. (10.5) represent a critical point of the function f, it only indicates that $(w_1, \cdots, w_\mu)$ could be a local minimum of the function. However, this does not guarantee that using these weights results in a smaller expected error compared to an unweighted estimator, where $w_i = 1$ for all $i = 1, \cdots, \mu$. The following lemma confirms that when we set $w_i = \frac{1/V_i}{\sum_{j=1}^{\mu} 1/V_j}$, the expected error is indeed smaller than the expected error when all weights are set to 1, i.e., $w_i = 1$ for all $i = 1, \cdots, \mu$.

Lemma 10.4 *For $i = 1, \cdots, \mu$, let w_i be defined as in Eq. (10.5). Then, $f(w_1, \cdots, w_\mu) \leq f(1, \cdots, 1)$. Furthermore, equality holds if and only if $V_i = V_j$ for any $i, j \in \{1, \cdots, \mu\}$.*

Proof Let $E_1 = f(w_1, \cdots, w_\mu)$ and $E_2 = f(1, \cdots, 1)$. It is straightforward to verify that $E_1 = \frac{1}{\sum_{j=1}^{\mu} \frac{n_j}{V_j}}$ and $E_2 = \frac{\sum_{j=1}^{\mu} n_j V_j}{\left(\sum_{j=1}^{\mu} n_j\right)^2}$. Then, $E_1 \leq E_2$ if and only if

$$\sum_{1 \leq i < j \leq \mu} 2 n_i n_j \leq \sum_{1 \leq i < j \leq \mu} n_i n_j \left(\frac{V_i}{V_j} + \frac{V_j}{V_i}\right). \tag{10.7}$$

Note that for any $i = 1, \cdots, \mu$ and $j = i + 1, \cdots, \mu$, since $(V_i - V_j)^2 \geq 0$ and it equals zero if and only if $V_i = V_j$, we have $\frac{V_i}{V_j} + \frac{V_j}{V_i} \geq 2$ for any positive real numbers V_i and V_j. Equality holds if and only if $V_i = V_j$. This completes the proof. □

Lemma 10.4 demonstrates that, unless the variance is identical across all privacy levels, using a weighted average with weights defined in Eq. (10.5) yields a statistically more accurate estimate compared to the unweighted average. Since we assume the estimator is unbiased, a smaller mean squared error implies a reduction in the variance of the estimator. This further reinforces the conclusion that the weighted average provides improved statistical accuracy.

Remark In addition to guaranteeing better statistical accuracy compared to unweighted aggregation, weighted aggregation is performed as a post-processing step that does not consume any additional privacy budget. It is also important to note that this post-processing technique requires only μ real number multiplications to apply the corresponding weights to the aggregated data for different privacy settings. This operation is asymptotically negligible, particularly when μ is assumed to be small.

10.5 Conclusion

In this chapter, we have examined the frequency estimation problem and introduced a solution designed to offer fair opportunities for companies of varying sizes to conduct user data analysis. Specifically, our approach is based on sampling methods.

We initially provided a theoretical bound for the sampling privacy and utility of the proposed method. Subsequently, we extended the mechanism to a decentralized setting using secret sharing, allowing accurate analysis to be performed without accessing the users' original data. To enhance the feasibility of this mechanism in Multi-Party Computation (MPC) settings, we introduced a two-stage sampling method, which further alleviates the communication and computational burdens on users.

Additionally, we addressed scenarios where users have different levels of privacy concerns. To accommodate this, users were partitioned into several disjoint groups, each operating independently to provide frequency estimates for their respective groups. We also proposed a weighted aggregation framework, which enables a more accurate overall frequency estimate to be derived from the individual estimates of different privacy groups.

References

1. M Yang, I Tjuawinata, KY Lam, T Zhu, J Zhao, Differentially Private Distributed Frequency Estimation IEEE Transactions on Dependable and Secure Computing 20 (5), 3910–3926, 2022.
2. Lichao Sun and Lingjuan Lyu. Federated model distillation with noise-free differential privacy. *arXiv preprint arXiv:2009.05537*, 2020.
3. Úlfar Erlingsson, Vasyl Pihur, and Aleksandra Korolova. Rappor: Randomized aggregatable privacy-preserving ordinal response. In *Proceedings of the 2014 ACM SIGSAC Conference on Computer and Communications Security*, pages 1054–1067, 2014.
4. Raef Bassily and Adam Smith. Local, private, efficient protocols for succinct histograms. In *Proceedings of the Forty-Seventh Annual ACM Symposium on Theory of Computing*, pages 127–135, 2015.
5. Wennan Zhu, Peter Kairouz, Brendan McMahan, Haicheng Sun, and Wei Li. Federated heavy hitters discovery with differential privacy. In *International Conference on Artificial Intelligence and Statistics*, pages 3837–3847, 2020. PMLR.
6. Ninghui Li, Wahbeh Qardaji, and Dong Su. On sampling, anonymization, and differential privacy or, k-anonymization meets differential privacy. In *Proceedings of the 7th ACM Symposium on Information, Computer and Communications Security*, pages 32–33, 2012.

7. Changhui Hu, Jin Li, Zheli Liu, Xiaojie Guo, Yu Wei, Xuan Guang, Grigorios Loukides, and Changyu Dong. How to make private distributed cardinality estimation practical, and get differential privacy for free. In *30th USENIX Security Symposium*, 2021. Newcastle University.
8. Lichao Sun, Jianwei Qian, Xun Chen, and Philip S. Yu. Ldp-fl: Practical private aggregation in federated learning with local differential privacy. *arXiv preprint arXiv:2007.15789*, 2020.
9. Yuanxiong Guo and Yanmin Gong. Practical collaborative learning for crowdsensing in the internet of things with differential privacy. In *2018 IEEE Conference on Communications and Network Security (CNS)*, pages 1–9, 2018. IEEE.
10. Vincent Bindschaedler, Shantanu Rane, Alejandro E. Brito, Vanishree Rao, and Ersin Uzun. Achieving differential privacy in secure multiparty data aggregation protocols on star networks. In *Proceedings of the Seventh ACM on Conference on Data and Application Security and Privacy*, pages 115–125, 2017.
11. Martin Pettai and Peeter Laud. Combining differential privacy and secure multiparty computation. In *Proceedings of the 31st Annual Computer Security Applications Conference*, pages 421–430, 2015.
12. John C. Duchi, Michael I. Jordan, and Martin J. Wainwright. Minimax optimal procedures for locally private estimation. *Journal of the American Statistical Association*, 113(521):182–201, 2018. Taylor & Francis.
13. Yutong Ye, Min Zhang, Dengguo Feng, Hao Li, and Jialin Chi. Multiple privacy regimes mechanism for local differential privacy. In *International Conference on Database Systems for Advanced Applications*, pages 247–263, 2019. Springer.
14. Mousumi Akter and Tanzima Hashem. Computing aggregates over numeric data with personalized local differential privacy. In *Australasian Conference on Information Security and Privacy*, pages 249–260, 2017. Springer.
15. Mengmeng Yang, Tianqing Zhu, Yang Xiang, and Wanlei Zhou. Personalized privacy preserving collaborative filtering. In *International Conference on Green, Pervasive, and Cloud Computing*, pages 371–385, 2017. Springer.
16. Hui Cai, Yanmin Zhu, Jie Li, and Jiadi Yu. A profit-maximizing mechanism for query-based data trading with personalized differential privacy. *The Computer Journal*, 64(2):264–280, 2021. Oxford University Press.
17. Zhibo Wang, Jiahui Hu, Ruizhao Lv, Jian Wei, Qian Wang, Dejun Yang, and Hairong Qi. Personalized privacy-preserving task allocation for mobile crowdsensing. *IEEE Transactions on Mobile Computing*, 18(6):1330–1341, 2018. IEEE.
18. Ben Niu, Yahong Chen, Boyang Wang, Jin Cao, and Fenghua Li. Utility-aware exponential mechanism for personalized differential privacy. In *2020 IEEE Wireless Communications and Networking Conference (WCNC)*, pages 1–6, 2020. IEEE.
19. Shun Zhang, Laixiang Liu, Zhili Chen, and Hong Zhong. Probabilistic matrix factorization with personalized differential privacy. *Knowledge-Based Systems*, 183:104864, 2019. Elsevier.
20. Graham Cormode, Tejas Kulkarni, and Divesh Srivastava. Answering range queries under local differential privacy. *Proceedings of the VLDB Endowment*, 12(10):1126–1138, 2019. VLDB Endowment.
21. Yehuda Lindell. Secure Multiparty Computation (MPC). Cryptology ePrint Archive, Report 2020/300, 2020. Available at https://eprint.iacr.org/2020/300.
22. Mengmeng Yang, Lingjuan Lyu, Jun Zhao, Tianqing Zhu, and Kwok-Yan Lam. Local differential privacy and its applications: A comprehensive survey. *arXiv preprint arXiv:2008.03686*, 2020.
23. Cynthia Dwork, Aaron Roth, et al. The algorithmic foundations of differential privacy. *Foundations and Trends in Theoretical Computer Science*, 9(3-4):211–407, 2014.
24. Tianhao Wang, Jeremiah Blocki, Ninghui Li, and Somesh Jha. Locally differentially private protocols for frequency estimation. In *26th USENIX Security Symposium (USENIX Security 17)*, pages 729–745, 2017.
25. Peter Kairouz, Keith Bonawitz, and Daniel Ramage. Discrete distribution estimation under local privacy. In *Proceedings of the 33rd International Conference on International Conference on Machine Learning - Volume 48*, pages 2436–2444, 2016. JMLR.org.

26. Hisham Husain, Borja Balle, Zac Cranko, and Richard Nock. Local Differential Privacy for Sampling. In *International Conference on Artificial Intelligence and Statistics*, pages 3404–3413, 2020. PMLR.
27. Josh Joy and Mario Gerla. Differential privacy by sampling. *arXiv preprint arXiv:1708.01884*, 2017.
28. Borja Balle, Gilles Barthe, and Marco Gaboardi. Privacy profiles and amplification by subsampling. *Journal of Privacy and Confidentiality*, 10(1), 2020.
29. Stacey Truex, Ling Liu, Ka-Ho Chow, Mehmet Emre Gursoy, and Wenqi Wei. LDP-Fed: Federated learning with local differential privacy. In *Proceedings of the Third ACM International Workshop on Edge Systems, Analytics and Networking*, pages 61–66, 2020.
30. T-H. Hubert Chan, Elaine Shi, and Dawn Song. Optimal lower bound for differentially private multi-party aggregation. In *Algorithms – ESA 2012*, pages 277–288. Springer Berlin Heidelberg, 2012.
31. Shaowei Wang, Liusheng Huang, Pengzhan Wang, Yiwen Nie, Hongli Xu, Wei Yang, Xiang-Yang Li, and Chunming Qiao. Mutual information optimally local private discrete distribution estimation. *arXiv preprint arXiv:1607.08025*, 2016.
32. Cynthia Dwork. Differential privacy. In *Proc of the 33rd International Colloquium on Automata, Languages and Programming*, 2006. Springer, Berlin.
33. M. Yang, I. Tjuawinata, K.-Y. Lam, J. Zhao, and L. Sun. Secure hot path crowdsourcing with local differential privacy under fog computing architecture. *IEEE Transactions on Services Computing*, 2020.
34. Brian A'hearn, Franco Peracchi, and Giovanni Vecchi. Height and the normal distribution: Evidence from Italian military data. *Demography*, 46(1):1–25, 2009. Springer.
35. Neil Zhenqiang Gong, Wenchang Xu, Ling Huang, Prateek Mittal, Emil Stefanov, Vyas Sekar, and Dawn Song. Evolution of social-attribute networks: Measurements, modeling, and implications using Google+. In *Proceedings of the 2012 Internet Measurement Conference*, pages 131–144, 2012.
36. Aaron Clauset, Cosma Rohilla Shalizi, and Mark EJ Newman. Power-law distributions in empirical data. *SIAM Review*, 51(4):661–703, 2009. SIAM.
37. John C. Duchi, Michael I. Jordan, and Martin J. Wainwright. Local privacy and statistical minimax rates. In *2013 IEEE 54th Annual Symposium on Foundations of Computer Science*, pages 429–438, 2013.
38. Differential Privacy Team, Apple. Learning with Privacy at Scale. Available at: https://docs-assets.developer.apple.com/ml-research/papers/learning-with-privacy-at-scale.pdf.
39. Octavian Catrina and Amitabh Saxena. Secure computation with fixed-point numbers. In *Financial Cryptography and Data Security*, pages 35–50. Springer Berlin Heidelberg, 2010. ISBN 978-3-642-14577-3.
40. Amirali Aghazadeh, Ryan Spring, Daniel LeJeune, Gautam Dasarathy, Anshumali Shrivastava, and Richard G. Baraniuk. Mission: Ultra large-scale feature selection using count-sketches. In *International Conference on Machine Learning*, 2018.
41. Fabon Dzogang, Thomas Lansdall-Welfare, Saatviga Sudhahar, and Nello Cristianini. Scalable preference learning from data streams. In *Proceedings of the 24th International Conference on World Wide Web*, pages 885–890, ACM, 2015.
42. Amit Goyal, Hal Daumé III, and Graham Cormode. Sketch algorithms for estimating point queries in NLP. In *Proceedings of the 2012 Joint Conference on Empirical Methods in Natural Language Processing and Computational Natural Language Learning*, pages 1093–1103, Association for Computational Linguistics, 2012.

Chapter 11
Privacy and Digital Trust in VANETs

Abstract Privacy and digital trust are foundational to the successful development and deployment of Vehicular Ad Hoc Networks (VANETs). These networks rely on secure communication between vehicles and infrastructure to ensure road safety and efficient traffic management. However, the dynamic, decentralized, and open architecture of VANETs introduces significant challenges in maintaining privacy and security, which are integral to fostering digital trust. This chapter uses VANETs as a case study to examine the critical interplay between privacy and digital trust, emphasizing how authentication mechanisms, such as Public Key Infrastructure (PKI)-based and blockchain-based approaches, exemplify the principles of digital trust. The chapter evaluates these schemes by analyzing their strengths, limitations, and applicability within the VANET environment. Additionally, it reviews current research advancements in trust management, highlighting the role of cryptographic techniques, decentralized trust models, and emerging technologies such as blockchain and 5G in enhancing privacy and security. Finally, the chapter outlines future research directions to address scalability, efficiency, and interdisciplinary challenges, advancing privacy and digital trust within vehicular networks.

11.1 Introduction

Privacy and digital trust are foundational pillars in the modern interconnected world, shaping how individuals and organizations interact online by ensuring the responsible handling of sensitive information [1]. Digital trust, at its core, encompasses the confidence users place in digital platforms, services, and technologies to protect their personal data. This confidence is built on five fundamental principles:

- **Security:** Safeguarding against unauthorized access, breaches, and cyber threats.
- **Privacy:** Guaranteeing that personal information is collected, stored, and utilized in compliance with regulations and user expectations.
- **Integrity:** Ensuring data accuracy and protection from unauthorized alterations.
- **Transparency:** Providing clear, honest communication about data usage, storage, and security measures, empowering informed user decisions.

- **Reliability:** Delivering dependable service functionality, free from unexpected failures or vulnerabilities.

Despite these pillars, digital trust remains fragile. A McKinsey survey [2] reports that while 70% of consumers believe companies protect their data, 57% of executives acknowledged experiencing at least one significant data breach in the past three years. Such breaches erode trust, with nearly half of U.S. consumers discontinuing services due to privacy concerns [3].

Data privacy is a critical component of digital trust, focusing on safeguarding personal information through robust policies, practices, and technologies. Effective privacy protection not only prevents identity theft and financial harm but also strengthens trust between users and organizations, supports regulatory compliance, mitigates risks, and enhances reputations. In today's digital landscape, prioritizing privacy is essential for sustaining trust and reducing liabilities.

This chapter delves into the dynamic interplay between privacy and digital trust within the context of Vehicular Ad Hoc Networks (VANETs), offering a case study to illustrate these concepts in practice. VANETs, a key component of intelligent transportation systems (ITS), enable communication between vehicles and infrastructure to improve road safety, optimize traffic flow, and enhance transportation management. However, the open, decentralized, and dynamic nature of VANETs presents significant privacy and trust challenges, demanding innovative solutions to secure communication and maintain data integrity.

VANETs rely heavily on Vehicle-to-Vehicle (V2V) and Vehicle-to-Infrastructure (V2I) communication, utilizing advanced radio access technologies such as IEEE 802.11bd. Cooperative Awareness Messages (CAMs) play a vital role in facilitating real-time data exchange, including vehicle speed, location, and trajectory, to prevent accidents and manage traffic [14]. Public Key Infrastructure (PKI) underpins the security of these communications, managing public and private key pairs and issuing certificates through trusted Certificate Authorities (CAs) for Road Side Units (RSUs) and On-Board Units (OBUs).

To protect privacy, pseudonym-based authentication schemes anonymize vehicle identities during communication. Frequent pseudonym updates are essential for mitigating risks such as tracking attacks, as recommended by ETSI standards [15, 34]. However, these systems face challenges like linkability, where adversaries could correlate pseudonyms to track vehicle movements. Robust mechanisms are required to ensure unlinkability and protect user privacy effectively.

Trust management is equally critical in VANETs, enabling vehicles to securely and reliably exchange sensitive data such as safety messages. Effective trust management must balance anonymity to protect privacy and accountability to ensure safety and deter malicious behavior. Temporary identifiers for safety messages are one solution that preserves privacy while enabling accountability in cases of misbehavior.

Additionally, blockchain technology has emerged as a promising solution to the intertwined challenges of privacy and trust in VANETs. Blockchain offers a decentralized, tamper-resistant, and transparent framework for managing data integrity, ensuring authenticated and immutable transmissions. Through digital signatures, blockchain enables vehicles to securely validate transactions without relying on a central authority. Furthermore, blockchain supports secure and private communication using digital wallets, enhancing both privacy and digital trust in vehicle-to-vehicle interactions.

This chapter explores the privacy and trust challenges in VANETs, emphasizing how advanced technologies such as PKI and blockchain exemplify the principles of digital trust. By examining these technologies and their applications, the chapter highlights potential solutions to enhance privacy and trust in VANETs, while identifying future research opportunities in this evolving field.

11.2 Privacy and Trust Requirements

In Vehicular Ad Hoc Networks (VANETs), ensuring both privacy and trust is indispensable for secure communication and maintaining system integrity. VANETs enable real-time data exchange between vehicles and infrastructure, which is crucial for applications like traffic management and accident prevention. However, the openness and dynamic nature of these networks introduce significant challenges in safeguarding sensitive information and ensuring the reliability of exchanged data.

Privacy and trust are interdependent in VANETs, forming the foundation for secure interactions. Trust relies on robust security measures to protect communication, while privacy ensures users' data and identities remain shielded from misuse or unauthorized access. The primary security requirements in VANETs–confidentiality, data integrity, availability, non-repudiation, and unforgeability–play essential roles in establishing trust. Similarly, privacy requirements such as anonymity, pseudonymity, and unlinkability are critical for protecting user identities and preventing tracking. This section examines these key requirements, offering a comprehensive understanding of the challenges involved in ensuring privacy and digital trust in VANETs.

11.2.1 Security Requirements in VANETs

Security is a cornerstone of digital trust in VANETs, ensuring that interactions between vehicles and infrastructure are both reliable and protected. The following requirements underpin secure communication in these networks:

- **Confidentiality:** Safeguarding the network ensures that only authorized entities can access sensitive resources. Cryptographic protocols, such as encryption and trust management, prevent unauthorized access to data transmitted across the network, ensuring privacy and trustworthiness [24].

- **Data Integrity:** Ensuring that transmitted data remains accurate and unaltered is critical. Unauthorized modification or retransmission of data can result in significant harm. Mechanisms to detect and prevent replay attacks bolster trust in the accuracy and authenticity of network communications [25, 31, 40].
- **Availability:** Availability guarantees that legitimate users have on-demand access to network services, even under attack. Mechanisms must ensure that essential services remain operational, safeguarding trust in the network's reliability [40].
- **Non-repudiation:** Non-repudiation ensures that neither the sender nor the recipient can deny their involvement in a transaction. Digital signatures provide verifiable proof of origin, enhancing accountability and trust among network participants [5].
- **Unforgeability:** Only authorized entities with valid credentials can generate secure messages. Ensuring that certificates or credentials cannot be forged strengthens trust in the authenticity of network participants [16].
- **Identity and Authentication Management:** Robust authentication systems verify the identities of participants, forming the first line of defense against adversaries. Mutual authentication and message integrity checks are essential for maintaining trust within the network [11, 22].
- **Revocability:** Certificate revocation mechanisms prevent compromised or unauthorized vehicles from participating in the network. Maintaining forward unlinkability ensures that revoked credentials do not expose additional sensitive information [6, 38].

11.2.2 Privacy Requirements in VANETs

Privacy is equally essential to fostering digital trust in VANETs. Users must have confidence that their personal information is protected from misuse or unauthorized access. The following privacy requirements are critical:

- **Anonymity and Pseudonymity:** Protecting users' identities requires anonymity, which ensures that a vehicle cannot be directly identified within a group [29]. Pseudonym-based schemes further enhance privacy by periodically changing identifiers to prevent linkability [28, 44].
- **Conditional Privacy:** While anonymity is essential, mechanisms must allow authorized entities, such as Certification Authorities (CAs), to trace the origin of malicious activities. Conditional privacy strikes a balance between protecting users and ensuring accountability [25].
- **Homomorphic Encryption:** This advanced cryptographic technique enables computations on encrypted data without revealing the underlying information, ensuring data remains private even during processing.
- **Rerandomization:** Rerandomization ensures that encrypted data cannot be correlated over time, preventing pattern recognition and safeguarding privacy.

- **Zero-Knowledge Proofs (ZKPs):** ZKPs enable participants to prove knowledge of specific information without revealing the actual data, supporting privacy-preserving authentication and access control.
- **Secure Multi-Party Computation (SMPC):** SMPC facilitates collaborative computations while keeping individual inputs confidential. This is particularly useful in scenarios requiring joint analysis of sensitive data, such as traffic or vehicle usage statistics.
- **Federated Privacy:** Federated privacy models decentralize data processing, allowing entities to maintain control over their data while enabling secure collaboration. This reduces risks associated with centralized storage.
- **Secure Enclaves (e.g., SGX):** Hardware-based secure enclaves provide isolated environments for executing sensitive operations, ensuring the integrity and confidentiality of private data.
- **Unlinkability:** Unlinkability ensures that interactions cannot be associated with a single vehicle, even if multiple messages are sent. This prevents tracking and reinforces privacy [4].
- **Anonymity Techniques:** Advanced techniques, including onion routing, data masking, and k-anonymity, enhance privacy by concealing user identities and interactions within the network.

This section underscores how privacy and security requirements work in tandem to establish digital trust in VANETs. By addressing these requirements, VANETs can provide a secure and reliable communication environment, enabling the adoption of privacy-preserving and trust-enhancing technologies.

11.3 Privacy and Trust Challenges: A Review of Current Research

Vehicular Ad Hoc Networks (VANETs) exemplify the complexities of maintaining privacy and digital trust in dynamic, decentralized systems. Over the years, several surveys have explored the diverse facets of privacy-preserving authentication in VANETs, including routing protocols, security concerns, privacy requirements, and associated risks. These surveys provide valuable insights into the evolving landscape of VANET security but also highlight critical gaps that must be addressed to foster digital trust.

For instance, [13] presents a comprehensive analysis of pseudonym change strategies, categorizing and comparing various approaches used in VANETs. However, the lack of focus on decentralized pseudonym management represents a significant limitation in addressing the privacy needs of modern VANETs. Similarly, [19] reviews simulation tools for vehicular network security, trust, and privacy. While it identifies transmission patterns and trends in VANET security, privacy, and trust-based attacks, it does not compare its findings with other surveys or provide a comprehensive analysis of the current research landscape.

In another study, [8] evaluates the performance of existing authentication and privacy-preserving schemes, categorizing privacy-preserving systems in VANETs. However, it falls short in addressing specific security threats and privacy concerns. Likewise, [9] investigates VANET security attacks, including those targeting cloud computing, communication protocols, and architecture, but omits a detailed exploration of trust management models, which are fundamental to building reliable communication systems. Furthermore, [32] offers an overview of cryptographic schemes, such as symmetric and asymmetric cryptography, identity-based cryptography, and signature methods used in VANETs. However, it fails to examine trust management, which is crucial for ensuring the authenticity and reliability of VANET communications.

The survey by Manivannan et al. [23] reviews VANET security, privacy, and message dissemination, focusing on secure authentication and privacy-preserving systems. Nevertheless, it does not address emerging cryptographic methods like symmetric-key cryptography, ring signatures, and blockchain-based approaches that are gaining momentum in VANET security. Additionally, the study primarily references research conducted between 2009 and 2016, leaving out more recent advancements. Similarly, [4] discusses the benefits and challenges of software-defined networks (SDN) in VANETs, focusing on SDN layers and routing. However, it limits its scope to V2V and V2I communication patterns, overlooking broader implications for privacy and trust.

More recent work by Mundhe et al. [26] categorizes current authentication and privacy-preserving schemes, offering an updated view of the field. However, it does not address critical issues like single points of failure and fault tolerance. Similarly, [10] summarizes the applications of 5G, 5G-SDN, and blockchain in VANET authentication and privacy. While promising, this study lacks an in-depth analysis of trust model management, a key component of reliable VANET systems. The survey by [12] offers insights into machine learning-based misbehavior detection systems in vehicular networks, but it gives limited attention to trust model management or decentralization.

Blockchain Technology for Enhancing Privacy and Trust in VANETs
Blockchain has emerged as a transformative solution for addressing privacy and trust challenges in VANETs. Its decentralized, transparent, and immutable framework provides robust mechanisms for enhancing data integrity and mitigating risks associated with centralized systems. By eliminating single points of failure, blockchain ensures that VANETs can operate securely and efficiently in highly dynamic environments.

Blockchain facilitates distributed consensus mechanisms, enabling all nodes to participate in maintaining the network ledger. This decentralization reduces the dependency on central authorities, making VANETs more resilient to attacks targeting centralized trust models. For example, blockchain can securely manage vehicle identity authentication and message integrity without relying on a single trusted entity.

When integrated with 5G technology, blockchain's potential for improving VANET security is significantly amplified. 5G's low latency, high bandwidth, and network slicing capabilities enable faster and more reliable inter-vehicle communication. Combined with blockchain's tamper-resistant architecture, this synergy ensures that safety-critical data, such as vehicle-to-vehicle (V2V) and vehicle-to-infrastructure (V2I) communications, are processed securely and efficiently.

This section highlights how blockchain and other advanced technologies are reshaping the landscape of VANET security and trust management, underscoring the need for continued innovation in these areas to strengthen digital trust in vehicular networks.

11.4 Survey of Authentication Schemes for Privacy and Trust

Authentication mechanisms are integral to fostering privacy and digital trust within Vehicular Ad-hoc Networks (VANETs). These mechanisms not only secure communication but also establish a foundation of trust between vehicles and roadside infrastructure, which is crucial for maintaining a reliable and secure network environment. Trust management techniques lie at the heart of these authentication schemes, providing a structured approach to assessing trustworthiness and facilitating secure interactions. This section examines the classification of trust management techniques commonly employed in VANETs, highlighting their role in addressing privacy and digital trust challenges. Trust can be broadly categorized into three primary types: subject trust, trust-based services, and the origin of trust. Additionally, trust can be assessed through two key dimensions: entity trust and content trust. These classifications form the basis for understanding the methodologies used in existing VANET authentication schemes.

- **Entity Trust:** Entity trust evaluates the trustworthiness of network participants, such as vehicles and infrastructure. It leverages advanced technologies, including cryptography, game theory, fuzzy logic, social networking, and machine learning, to provide a robust assessment of an entity's reliability within the network.
- **Content Trust:** Content trust focuses on validating the reliability of the information exchanged in the network. Techniques such as data analytics, plausibility assessments, watermarking, and evidence-based methodologies are employed to ensure the integrity and accuracy of the transmitted data.
- **Trust-Based Services:** Trust-based services utilize established trust metrics to enhance network operations. These services encompass secure routing, effective relay selection for message dissemination, and optimized information sharing. Trust scores assigned to network nodes play a critical role in ensuring the efficient and secure performance of these processes.

The origin of trust within VANETs can be classified into three distinct categories:

- **Direct Trust:** Direct trust arises from interactions between nodes, such as Vehicle-to-Vehicle (V2V) and Vehicle-to-Infrastructure (V2I) communications. These interactions allow vehicles to evaluate trustworthiness based on real-time exchanges.
- **Indirect Trust:** Indirect trust is built on recommendations or feedback provided by neighboring nodes. A vehicle can rely on trust evaluations from other vehicles to determine the reliability of a specific node or entity.
- **Aggregated Trust:** Aggregated trust integrates both direct and indirect assessments to deliver a comprehensive evaluation of a node's or entity's trustworthiness.

Trust management models form the cornerstone of VANET security frameworks. Trust is a multifaceted characteristic that applies to both the participating entities and the information they exchange. Numerous methods have been proposed in the literature to evaluate and manage trust in VANETs, focusing on the paths used to establish trust and the underlying methodologies. Depending on the use case, trust may be established through direct, indirect, or hybrid approaches [17].

This section concentrates on two prominent authentication schemes: Public Key Infrastructure (PKI)-based schemes and Blockchain-based schemes. These approaches represent significant advancements in addressing the unique privacy and security challenges inherent in VANETs. PKI offers a hierarchical and well-established framework for managing vehicle authentication and privacy, but it encounters challenges such as reliance on centralized authorities and inefficiencies in certificate revocation processes. Blockchain, on the other hand, provides a decentralized and tamper-resistant architecture that aligns well with the distributed and dynamic nature of VANETs. However, blockchain-based solutions face scalability and resource constraints despite their strengths in transparency and trust management. Together, these schemes highlight the progress made in improving the security and privacy of vehicular networks. Further research into integrating or optimizing these technologies offers promising opportunities to advance digital trust in VANETs. For additional insights into other authentication schemes, such as identity-based cryptography, pseudonym systems, and group or ring signature-based methods, readers may refer to [43].

11.4.1 Public Key Infrastructure-Based Schemes

Public Key Infrastructure (PKI) serves as a cornerstone for establishing privacy and digital trust in Vehicular Ad-hoc Networks (VANETs). The European Telecommunications Standards Institute (ETSI) Intelligent Transport Systems (ITS) framework emphasizes the hierarchical nature of PKI, where the Certificate Authority (CA) occupies the top level, supported by intermediate CAs, with vehicles at the base of the hierarchy. This structure enables VANETs to meet stringent security and privacy requirements through advanced encryption techniques. PKI-based systems

11.4 Survey of Authentication Schemes for Privacy and Trust

leverage public key cryptography, wherein the Trusted Authority (TA) oversees key pair distribution and authenticates vehicles within the network to ensure secure communication. Authentication is achieved using a vehicle's public key alongside the CA's digital signature, creating a robust framework for both vehicle authentication and privacy preservation. Furthermore, CA certificates enhance traceability and facilitate pseudonym-based authentication, which bolsters privacy protections [30].

Lu et al. proposed a system that employs short-lifetime pseudonyms and certificates distributed via Roadside Units (RSUs) to address the challenges of centralized administration. Although this approach provides anonymity, it also introduces several limitations. For instance, the Efficient Conditional Privacy Preservation (ECPP) system suffers from high latency and necessitates tamper-resistant hardware for RSUs to mitigate physical attacks. Additionally, ECPP lacks a comprehensive pseudonym revocation mechanism, leaving malicious vehicles unchecked. Compromised RSUs exacerbate the problem by using locally retained, unchanged pseudonyms to track vehicle movements, undermining unlinkability.

Wasef and Shen [37] enhanced authentication protocols using HMAC with a global key. While this method improves security, frequent key updates impose significant resource demands. To mitigate the inefficiencies associated with large Certificate Revocation Lists (CRLs), Simplicio Jr. et al. [18] introduced the Activation Code-based Certificate (ACPC) system, which assigns short-bit activation codes to vehicles, reducing the size of CRLs. However, managing CRLs, especially for pseudonym revocation in large-scale networks, remains a challenge.

To address RSU overhead, [18] proposed a bilinear pairing-based authentication scheme employing proxy vehicles. These proxies authenticate multiple signatures simultaneously and relay the results to RSUs. Although resilient to proxy compromise, this approach lacks well-defined criteria for selecting proxy vehicles and fails to provide mechanisms for excluding malicious entities from the network.

Zhang et al. [30] introduced a hierarchical privacy-preserving authentication framework, where vehicles self-register with a Key Generation Center (KGC). The KGC issues temporary credentials, such as Session Tag Public (STP) and Session Tag Key (STK), enabling vehicles to sign messages securely. Despite its promise, the comparative effectiveness of this method against alternative schemes has not been comprehensively evaluated.

Vijayakumar et al. [37] proposed an anonymous authentication scheme focused on computational efficiency. Following offline registration, the TA provides vehicles with a password, public-private key pair, re-encryption key, and license. Traceability is maintained through a tracking list of vehicle identities. Authentication is performed using a combination of the vehicle's private key, short-lived public key, and anonymous certificate. However, the process of managing and revoking pseudonym certificates is resource-intensive and time-consuming.

Azees et al. [30] developed a bilinear pairing-based anonymous authentication mechanism, enabling mutual verification between RSUs and vehicles without involving the TA. While scalable, this method incurs high transmission and computational costs associated with certificate management and revocation. Vehicles must preload multiple anonymous certificates, and the TA must store these records, resulting in substantial overhead.

In [36], an anonymous message authentication scheme called LIAP is proposed, wherein vehicles obtain master keys from local RSUs. These keys enable the creation of anonymous identities and allow vehicles to sign messages. This approach ensures message integrity, authentication, and privacy while protecting against replay and collusion attacks. Nevertheless, managing CRLs, certificate distribution, and membership revocation presents significant communication and computational challenges.

Alrawais et al. [18] proposed a fog computing-based revocation system utilizing a Merkle hash tree in place of traditional CRLs. In this framework, fog nodes act as trusted gateways, issuing certificates to local vehicles and updating revocation information through the Merkle tree. While this approach eliminates CRL verification overhead, maintaining synchronized and up-to-date Merkle trees across all nodes poses logistical difficulties. Furthermore, implementing this system requires significant infrastructure investments to support its operation.

11.4.2 Blockchain-Based Schemes

Blockchain technology has emerged as a promising solution to address the intertwined challenges of privacy and digital trust in Vehicular Ad-hoc Networks (VANETs). Traditional authentication methods often face issues such as a lack of transparency in Trusted Authority (TA) operations, reliance on Certificate Revocation Lists (CRLs) that impose significant computational and storage burdens, and inefficiencies in identity and message verification under high-speed and high-traffic conditions. Blockchain provides a decentralized, tamper-resistant, trustworthy, transparent, and anonymous framework, making it highly compatible with the dynamic and distributed nature of VANETs [27].

Blockchain-based Approaches Several blockchain-based techniques have been proposed to enhance privacy-preserving authentication in VANETs. Key approaches include:

- **Blockchain-based Privacy-Preserving Authentication (BPPA):** Lu et al. [20] introduced the BPPA scheme, which integrates standard blockchain with a Chronological Merkle Tree (CMT) and Merkle-Patricia Tree (MPT). This design improves authentication efficiency while preserving privacy. However, the inclusion of Certificate Authorities (CAs) and Local Execution Agents (LEAs) introduces additional computational and communication overhead.
- **Certificate-based Blockchain Scheme:** A blockchain scheme utilizing bilinear pairing public key signatures was proposed by [7]. This approach ensures transparency and security by enabling pseudo-identity revocation via blockchain. However, the complexity of batch signature aggregation and verification remains a significant limitation.

11.4 Survey of Authentication Schemes for Privacy and Trust

- **Blockchain-based Traffic Event Validation (BTEV):** Yang et al. [39] proposed a BTEV mechanism that employs Proof-of-Event (PoE) consensus instead of the more resource-intensive Proof-of-Work (PoW). While PoE facilitates efficient validation of traffic events and trust verification, the need for all network members to verify both PoE and PoW can introduce processing delays.
- **Secure Data Sharing and Storage:** Zhang and Chen [42] presented a blockchain-based framework for secure data sharing and storage. Intelligent contracts distribute data coins to participating vehicles, while elliptic curve signatures ensure message security. However, combining bilinear pairing with blockchain results in substantial computational overhead.
- **Blockchain-based Secure Authentication and Key Management:** Tan and Chung [33] introduced a blockchain and edge-computing-integrated authentication and key management approach. This system provides conditional privacy and unforgeability while addressing network members' storage and computational constraints. Nevertheless, generating and verifying signatures remain resource-intensive.
- **Distributed Blockchain-based Key Management Mechanism (DBKMM):** Ma et al. [21] proposed a DBKMM using a bivariate polynomial-based mutual authentication and key agreement protocol. This approach updates public-private key pairs through smart contracts and defends against attacks such as collusion, denial-of-service, and public key tampering. However, its complexity poses challenges for lightweight VANET devices.
- **Blockchain-based Trust and Secure Communication Authentication (BTSCA):** Wang et al. [35] developed a BTSCA scheme to evaluate vehicle trustworthiness and secure data transfers between Roadside Units (RSUs). While resistant to replay and Man-in-the-Middle (MITM) attacks, the scheme lacks comparative analyses to demonstrate advantages over existing methods.
- **Optimized Blockchain Techniques:** Ali et al. [41] proposed an optimized blockchain system to alleviate throughput bottlenecks while combating fake messages. However, the reliance on conventional signature techniques leaves the system vulnerable to rogue key attacks.

Key Benefits and Challenges Blockchain-based schemes present substantial advantages for VANETs by enhancing privacy, security, and digital trust through transparency, tamper resistance, and immutability. These features empower any network member to verify operations conducted by the TA [20, 21, 33, 35, 39, 41, 42]. However, these approaches face notable challenges:

- **Resource Constraints:** The computational and storage requirements for blockchain operations can overwhelm lightweight devices commonly used in VANETs.
- **Scalability Issues:** Blockchain size becomes increasingly challenging to manage as vehicles dynamically enter and exit the network.

- **High Complexity:** Advanced cryptographic techniques, such as bilinear pairing, batch verification, and hierarchical methods, introduce additional computational overhead.

Despite these limitations, blockchain provides a resilient and decentralized framework for improving privacy and security in VANETs. It addresses many vulnerabilities inherent in traditional centralized systems while offering an infrastructure that aligns with the requirements of modern vehicular networks. By leveraging blockchain, VANETs can foster a higher degree of digital trust and reliability, paving the way for secure and privacy-preserving communication in intelligent transportation systems.

11.5 Conclusion

This chapter has delved into the critical interplay between privacy and digital trust within the framework of Vehicular Ad Hoc Networks (VANETs). As demonstrated, achieving a balance between privacy and trust is fundamental to facilitating secure, efficient, and reliable communication between vehicles and infrastructure. Using VANETs as a case study, we identified and analyzed the key privacy and security challenges posed by their open, dynamic, and decentralized nature.

We explored various authentication schemes that are central to addressing these challenges, with a particular emphasis on Public Key Infrastructure (PKI)-based and blockchain-based approaches. PKI, as a hierarchical and well-established framework, continues to serve as a cornerstone for vehicle authentication and privacy management. However, its reliance on centralized authorities and inefficiencies in certificate revocation remain significant hurdles. Blockchain, in contrast, offers a decentralized and tamper-resistant model that aligns closely with the principles of digital trust by eliminating single points of failure and enhancing transparency. Despite its advantages, blockchain-based solutions face their own limitations, such as scalability and resource demands, which must be addressed to enable widespread adoption.

Trust management models play a pivotal role in strengthening digital trust within VANETs. This chapter highlighted the potential of decentralized trust mechanisms to enhance the reliability and security of vehicular networks. Emerging technologies such as 5G, with its low latency and high bandwidth capabilities, complement blockchain's decentralized ledger, paving the way for innovative solutions to longstanding challenges in privacy and security.

Future research must prioritize addressing the scalability, efficiency, and resource constraints of these technologies to ensure their seamless integration into VANETs. Moreover, interdisciplinary research will be critical for tackling the socio-technical dimensions of trust in VANETs, such as user behavior, regulatory frameworks, and ethical considerations. By aligning technical advancements with broader societal and organizational contexts, we can foster an environment of enhanced digital trust that supports the sustainable development of intelligent transportation systems.

As VANETs continue to evolve as a cornerstone of modern transportation networks, advancing privacy-preserving and trust-enabling technologies will be essential. These innovations will not only strengthen the security and reliability of vehicular communications but also contribute to building the public confidence necessary for the success of future intelligent transportation systems.

References

1. Jodi Daniels. The Convergence Of Data Privacy And Digital Trust. 2023. Available at: https://www.forbes.com/councils/forbesbusinesscouncil/2023/03/17/the-convergence-of-data-privacy-and-digital-trust/. Accessed: 27 Nov 2024.
2. Jim Boehm, Liz Grennan, Alex Singla, and Kate Smaje. Consumer faith in cybersecurity, data privacy, and responsible AI hinges on what companies do today–and establishing this digital trust just might lead to business growth. 2024. Available at: https://www.mckinsey.com/capabilities/quantumblack/our-insights/why-digital-trust-truly-matters. Accessed: 27 Nov 2024.
3. Andrew Perrin. Half of Americans have decided not to use a product or service because of privacy concerns. 2020. Available at: https://www.pewresearch.org/short-reads/2020/04/14/half-of-americans-have-decided-not-to-use-a-product-or-service-because-of-privacy-concerns/. Accessed: 27 Nov 2024.
4. Othman S. Al-Heety, Zahriladha Zakaria, Mahamod Ismail, Mohammed Mudhafar Shakir, Sameer Alani, and Hussein Alsariera. A comprehensive survey: Benefits, services, recent works, challenges, security, and use cases for SDN-VANET. *IEEE Access*, 8:91028–91047, 2020.
5. Mishri Saleh Al-Marshoud, Ali H. Al-Bayatti, and Mehmet Sabir Kiraz. Improved chaff-based cMix for solving location privacy issues in VANETs. *Electronics*, 10(11):1302, 2021.
6. Mahmood A. Al-Shareeda and Selvakumar Manickam. Security methods in Internet of Vehicles. *arXiv preprint arXiv:2207.05269*, 2022.
7. Ikram Ali, Mwitende Gervais, Emmanuel Ahene, and Fagen Li. A blockchain-based certificateless public key signature scheme for vehicle-to-infrastructure communication in VANETs. *Journal of Systems Architecture*, 99:101636, 2019.
8. Ikram Ali, Alzubair Hassan, and Fagen Li. Authentication and privacy schemes for vehicular ad hoc networks (VANETs): A survey. *Vehicular Communications*, 16:45–61, 2019.
9. Muhammad Arif, Guojun Wang, Md Zakirul Alam Bhuiyan, Tian Wang, and Jianer Chen. A survey on security attacks in VANETs: Communication, applications and challenges. *Vehicular Communications*, 19:100179, 2019.
10. Farooque Azam, Sunil Kumar Yadav, Neeraj Priyadarshi, Sanjeevikumar Padmanaban, and Ramesh C. Bansal. A comprehensive review of authentication schemes in vehicular ad-hoc networks. *IEEE Access*, 9:31309–31321, 2021.
11. Maria Azees, Pandi Vijayakumar, and Lazarus Jegatha Deborah. Comprehensive survey on security services in vehicular ad-hoc networks. *IET Intelligent Transport Systems*, 10(6):379–388, 2016.
12. Abdelwahab Boualouache and Thomas Engel. A survey on machine learning-based misbehavior detection systems for 5G and beyond vehicular networks. *IEEE Communications Surveys & Tutorials*, 25(2):1128–1172, 2023.

13. Abdelwahab Boualouache, Sidi-Mohammed Senouci, and Samira Moussaoui. A survey on pseudonym changing strategies for vehicular ad-hoc networks. *IEEE Communications Surveys & Tutorials*, 20(1):770–790, 2017.
14. Zouina Doukha and Samira Moussaoui. An SDMA-based mechanism for accurate and efficient neighborhood-discovery link-layer service. *IEEE Transactions on Vehicular Technology*, 65(2):603–613, 2015.
15. ETSI ITS. Intelligent Transport Systems (ITS); Security; Pre-standardization study on pseudonym change management, 2018. Available at https://www.etsi.org/deliver/etsi_tr/103400_103499/103415/01.01.01_60/tr_103415v010101p.pdf. Cited 27 Nov 2024.
16. Shi-Jinn Horng, Shiang-Feng Tzeng, Po-Hsian Huang, Xian Wang, Tianrui Li, and Muhammad Khurram Khan. An efficient certificateless aggregate signature with conditional privacy-preserving for vehicular sensor networks. *Information Sciences*, 317:48–66, 2015.
17. Rasheed Hussain, Jooyoung Lee, and Sherali Zeadally. Trust in VANET: A survey of current solutions and future research opportunities. *IEEE Transactions on Intelligent Transportation Systems*, 22(5):2553–2571, 2020.
18. Yiliang Liu, Liangmin Wang, and Hsiao-Hwa Chen. Message authentication using proxy vehicles in vehicular ad hoc networks. *IEEE Transactions on Vehicular Technology*, 64(8):3697–3710, 2014.
19. Zhaojun Lu, Gang Qu, and Zhenglin Liu. A survey on recent advances in vehicular network security, trust, and privacy. *IEEE Transactions on Intelligent Transportation Systems*, 20(2):760–776, 2018.
20. Zhaojun Lu, Qian Wang, Gang Qu, Haichun Zhang, and Zhenglin Liu. A blockchain-based privacy-preserving authentication scheme for VANETs. *IEEE Transactions on Very Large Scale Integration (VLSI) Systems*, 27(12):2792–2801, 2019.
21. Zhuo Ma, Junwei Zhang, Yongzhen Guo, Yang Liu, Ximeng Liu, and Wei He. An efficient decentralized key management mechanism for VANET with blockchain. *IEEE Transactions on Vehicular Technology*, 69(6):5836–5849, 2020.
22. Khalid Mahmood, Shehzad Ashraf Chaudhry, Husnain Naqvi, Taeshik Shon, and Hafiz Farooq Ahmad. A lightweight message authentication scheme for smart grid communications in power sector. *Computers & Electrical Engineering*, 52:114–124, 2016.
23. Dakshnamoorthy Manivannan, Shafika Showkat Moni, and Sherali Zeadally. Secure authentication and privacy-preserving techniques in Vehicular Ad-hoc NETworks (VANETs). *Vehicular Communications*, 25:100247, 2020.
24. Mohamed Nidhal Mejri, Jalel Ben-Othman, and Mohamed Hamdi. Survey on VANET security challenges and possible cryptographic solutions. *Vehicular Communications*, 1(2):53–66, 2014.
25. Alfred J. Menezes, Paul C. Van Oorschot, and Scott A. Vanstone. *Handbook of Applied Cryptography*. CRC Press, 2018.
26. Pravin Mundhe, Shekhar Verma, and SJCSR Venkatesan. A comprehensive survey on authentication and privacy-preserving schemes in VANETs. *Computer Science Review*, 41:100411, 2021.
27. Satoshi Nakamoto. Bitcoin: A peer-to-peer electronic cash system. *Satoshi Nakamoto*, 2008.
28. Jonathan Petit, Florian Schaub, Michael Feiri, and Frank Kargl. Pseudonym schemes in vehicular networks: A survey. *IEEE Communications Surveys & Tutorials*, 17(1):228–255, 2014.
29. Andreas Pfitzmann and Marit Köhntopp. Anonymity, unobservability, and pseudonymity–a proposal for terminology. In *Designing Privacy Enhancing Technologies: International Workshop on Design Issues in Anonymity and Unobservability, Berkeley, CA, USA, July 25–26, 2000 Proceedings*, pages 1–9. Springer, 2001.
30. Maxim Raya and Jean-Pierre Hubaux. Securing vehicular ad hoc networks. *Journal of Computer Security*, 15(1):39–68, 2007.
31. Kamile Nur Sevis and Ensar Seker. Survey on data integrity in cloud. In *2016 IEEE 3rd International Conference on Cyber Security and Cloud Computing (CSCloud)*, pages 167–171. IEEE, 2016.

32. Muhammad Sameer Sheikh, Jun Liang, and Wensong Wang. A survey of security services, attacks, and applications for vehicular ad hoc networks (VANETs). *Sensors*, 19(16):3589, 2019.
33. Haowen Tan and Ilyong Chung. Secure authentication and key management with blockchain in VANETs. *IEEE Access*, 8:2482–2498, 2019.
34. IEEE Vehicular Technology Society. IEEE Standard for Wireless Access in Vehicular Environments (WAVE) – Networking Services. *IEEE Std 1609.3-2016 (Revision of IEEE Std 1609.3-2010)*, pages 1–160, 2016. doi:10.1109/IEEESTD.2016.7458115.
35. Chen Wang, Jian Shen, Jin-Feng Lai, and Jianwei Liu. B-TSCA: Blockchain assisted trustworthiness scalable computation for V2I authentication in VANETs. *IEEE Transactions on Emerging Topics in Computing*, 9(3):1386–1396, 2020.
36. Shibin Wang and Nianmin Yao. LIAP: A local identity-based anonymous message authentication protocol in VANETs. *Computer Communications*, 112:154–164, 2017.
37. Albert Wasef and Xuemin Shen. EMAP: Expedite message authentication protocol for vehicular ad hoc networks. *IEEE Transactions on Mobile Computing*, 12(1):78–89, 2011.
38. Yanjiang Yang, Xuhua Ding, Haibing Lu, Jian Weng, and Jianying Zhou. Self-blindable credential: Towards anonymous entity authentication upon resource constrained devices. In *Information Security: 16th International Conference, ISC 2013, Dallas, Texas, November 13-15, 2013, Proceedings*, pages 238–247. Springer, 2015.
39. Yao-Tsung Yang, Li-Der Chou, Chia-Wei Tseng, Fan-Hsun Tseng, and Chien-Chang Liu. Blockchain-based traffic event validation and trust verification for VANETs. *IEEE Access*, 7:30868–30877, 2019.
40. Ezer Osei Yeboah-Boateng. Cyber-Security Challenges with SMEs in Developing Economies: Issues of Confidentiality, Integrity & Availability (CIA). *Institut for Elektroniske Systemer, Aalborg Universitet*, 1st edition, 2013.
41. Lei Zhang, Mingxing Luo, Jiangtao Li, Man Ho Au, Kim-Kwang Raymond Choo, Tong Chen, and Shengwei Tian. Blockchain-based secure data sharing system for Internet of Vehicles: A position paper. *Vehicular Communications*, 16:85–93, 2019.
42. Xiaohong Zhang and Xiaofeng Chen. Data security sharing and storage based on a consortium blockchain in a vehicular ad-hoc network. *IEEE Access*, 7:58241–58254, 2019.
43. Mishri AlMarshoud, Mehmet Sabir Kiraz, and Ali H. Al-Bayatti. Security, privacy, and decentralized trust management in VANETs: A review of current research and future directions. *ACM Computing Surveys*, 56(10):1–39, 2024.
44. Xu Yang, Fei Zhu, Xuechao Yang, Junwei Luo, Xun Yi, Jianting Ning, and Xinyi Huang. Secure reputation-based authentication with malicious detection in VANETs. *IEEE Transactions on Dependable and Secure Computing*, 2024.

Chapter 12
Conclusions and Acknowledgements

Abstract In this chapter, we provide a summary of this book and acknowledge financial supports of research grants.

12.1 Summary

In this book, we have introduced state-of-the-art Privacy Enhancing Techniques (PET), including

- Homomorphic encryption techniques (Chap. 2)
- Secure multiparty computation techniques (Chap. 3)
- Differential privacy techniques (Chap. 4)

In particular, we have discussed tools and libraries with some examples about how to implement PET in this book.

Then we have described the applications of these techniques in data privacy protection, such as

- Privacy-Preserving Data Mining (Chap. 5)
- Privacy-Preserving Single Database PIR (Chap. 6)
- Privacy-Preserving Machine Learning (Chap. 7)
- Privacy-Preserving Social Network (Chap. 8)
- Privacy-Preserving Location-Based Queries (Chap. 9)
- Privacy-Preserving Distributed Frequency Estimation (Chap. 10)
- Privacy and Digital Trust in VANETs (Chap. 11).

12.2 Acknowledgements

At last, we would like to acknowledge financial supports of the following research grants when writing this book.

- ARC Discovery Project: DP180103251, Privacy-Preserving Online User Matching, Australia
- ARC Discovery Project: DP190102835, Private Searching on Streaming Data, Australia
- ARC Discovery Project: DP220102803, Privacy-Preserving Location Based Queries, Australia
- ARC Discovery Project: DP240102140, Data Privacy Protection in Wireless Sensor Networks, Australia
- ARC Linkage Project: LP160101766, Privacy-Preserving Cloud Data Mining-as-a-Service, Australia
- ARC Linkage Project: LP180101062, Development of Cryptographic Library and Support System, Australia
- ARC Linkage Project: LP220200649, Privacy-Preserving Collaborative Analytics on Sensitive Data, Australia
- Digital Trust Centre, funded by Infocomm Media Development Authority (S$49,999,892) and supported by National Research Foundation, Singapore, 1 October 2022 - 30 September 2027, Singapore
- Strategic Centre for Research in Privacy-Preserving Technologies & Systems, funded by Infocomm Media Development Authority (S$15,300,000) and supported by National Research Foundation, Singapore, 1 November 2018 - 31 March 2025, Singapore
- NSF Award#2112471: AI Institute for Future Edge Networks and Distributed Intelligence (AI-EDGE), USA
- NSF Award#2229876: AI Institute for Agent-based Cyber Threat Intelligence and Operation, USA

In addition, we would like to thank Xinqian Wang for his help with multiparty computation libraries and implementation codes.

This research is supported by the National Research Foundation, Singapore and Infocomm Media Development Authority under its Trust Tech Funding Initiative and Strategic Capability Research Centres Funding Initiative. Any opinions, findings and conclusions or recommendations expressed in this material are those of the author(s) and do not reflect the views of National Research Foundation, Singapore and Infocomm Media Development Authority.

The manufacturer's authorised representative in the EU is Springer Nature Customer Service Centre GmbH, Europaplatz 3, 69115 Heidelberg, Germany. If you have any concerns regarding our products, please contact ProductSafety@springernature.com

Printed and bound by CPI Group (UK) Ltd, Croydon, CR0 4YY

26/03/2026

02078977-0001